Progress of Recycling in the Built Environment

RILEM STATE-OF-THE-ART REPORTS

Volume 8

RILEM, The International Union of Laboratories and Experts in Construction Materials, Systems and Structures, founded in 1947, is a non-governmental scientific association whose goal is to contribute to progress in the construction sciences, techniques and industries, essentially by means of the communication it fosters between research and practice. RILEM's focus is on construction materials and their use in building and civil engineering structures, covering all phases of the building process from manufacture to use and recycling of materials. More information on RILEM and its previous publications can be found on www.RILEM.net.

The RILEM State-of-the-Art Reports (STAR) are produced by the Technical Committees. They represent one of the most important outputs that RILEM generates – high level scientific and engineering reports that provide cutting edge knowledge in a given field. The work of the TCs is one of RILEM's key functions.

Members of a TC are experts in their field and give their time freely to share their expertise. As a result, the broader scientific community benefits greatly from RILEM's activities.

RILEM's stated objective is to disseminate this information as widely as possible to the scientific community. RILEM therefore considers the STAR reports of its TCs as of highest importance, and encourages their publication whenever possible.

The information in this and similar reports is mostly pre-normative in the sense that it provides the underlying scientific fundamentals on which standards and codes of practice are based. Without such a solid scientific basis, construction practice will be less than efficient or economical.

It is RILEM's hope that this information will be of wide use to the scientific community.

For further volumes:
http://www.springer.com/series/8780

Enric Vázquez
Editor

Progress of Recycling in the Built Environment

Final Report of the RILEM Technical Committee 217-PRE

Springer

Editor
Enric Vázquez
Materials Section of the Department
of Construction Engineering
Universitat Politècnica de Catalunya
Barcelona, Spain

ISSN 2213-204X ISSN 2213-2031 (electronic)
ISBN 978-94-007-4907-8 ISBN 978-94-007-4908-5 (eBook)
DOI 10.1007/978-94-007-4908-5
Springer Dordrecht Heidelberg New York London

Library of Congress Control Number: 2012951847

Printed on acid-free paper

Springer is part of Springer Science+Business Media (www.springer.com)

RILEM TC 217-PRE: Use of Recycled Materials

Acknowledgements

The Scientific Editor and chairman of the TC thanks the members of the committee and contributors for their valuable contributions and availability. Likewise, thanks to M. Eng. Carolina Sarmiento Mantilla for her work on the adequacy of the definitive text of the final report for publication.

Contents

Authors

Scientific Editors

Prof. Dr. Enric Vázquez Materials Section of the Department of Construction Engineering, Universitat Politecnica de Catalunya, Barcelona, Spain

Members of the RILEM Technical Committee TC 217-PRE

Prof. Dr. Enric Vázquez Chairman of TC 217-PRE, Materials Section of the Department of Construction Engineering, Universitat Politecnica de Catalunya, Barcelona, Spain

Mrs. Ir Gabriëlla M.T. Janssen Secretary of RILEM TC 217-PRE, Enviro Challenge, Opoeteren, Belgium

Prof. Dr. Pilar Alaejos Lab. Central de Estructuras y Materiales, CEDEX, Madrid, Spain

Prof. Mario Bassan Politecnico di Milano, Milano, Italy

Mr. Luc De Bock Belgian Road Research Center, Brussels, Belgium

Dr. John Forth University of Leeds, Leeds, United Kingdom

Prof. Dr. Vanderley John Lab. Microestrutura e Ecoeficiencia de Materiais, Escola Politécnica da USP, Sao Paulo, Brazil

Prof. Dr-Ing. Jörg Kropp Hochshule Bremen, Institut für Baustofftechnologie, Bremen, Germany

Dr. Isabel Martins LNEC, Lisbon, Portugal

Prof. Dr-Ing. Anette Mueller Bauhaus Universität Weimar, Weimar, Germany

Prof. Dr. Carmen Zenia Nava Vera Universidad de Tampico, Tampico, Mexico

Prof. Dr. Jianzhuang Xiao Department of Building Engineering, Tongji University, Shanghai, P.R. China

Collaborators of RILEM TC 217-PRE

Dr. Sergio Angulo Instituto de Pesquisas Tecnologicas do Estado de Sao Paulo, Sao Paulo, Brazil

Prof. Dr. Marilda Barra Materials Section of the Department of Construction Engineering, Universitat Politecnica de Catalunya, Barcelona, Spain

Mr. Jan Desmyter Belgian Building Research Institute, Brussels, Belgium

Dr. Angel di Maio LEMIT, Argentina

Mr. Roberto Drummond Kleemann gmbh, Madrid, Spain

Prof. Yu-hui Fan Department of Building Engineering, Tongji University, Shanghai, P.R. China

Mr. Gholamreza Fathifazl Adjeleian Allen Rubeli Consulting Structural Engineers, Ottawa, ON, Canada

Mrs. Cathleen Hoffmann Empa, Concrete & Construction Chemistry, Dübendorf, Switzerland

Prof. Huang X Department of Building Engineering, Tongji University, Shanghai, P.R. China

Prof. Jiabin Li Department of Building Engineering, Tongji University, Shanghai, P.R. China

Prof. Wengui Li Department of Building Engineering, Tongji University, Shanghai, P.R. China

Dr. Hin Ming Ho Hong Kong Polytechnical University, Hong Kong, P.R. China

Prof. Dr. CS Poon Department of Civil and Structural Engineering, The Hong Kong Polytechnic University, Hong Kong, P.R. China

Ir. Johan A.L. Put Enviro Challenge, The Netherlands

Mr. Marco Quattrone Politecnico di Milano, Milano, Italy

Prof. Dr. Ghani Razaqpur Department of Civil Engineering, McMaster University, Hamilton, ON, Canada

Mr. Jorge Rueda CEDEX, Madrid, Spain

Dr. Marta Sánchez de Juan CEDEX, Madrid, Spain

Dr. Ing. Thomas Schnellert Bauhaus Universität Weimar, Weimar, Germany

Prof. Tawana M.M. Department of Building Engineering, Tongji University, Shanghai, P.R. China

Mr. Ignacio Valero Gestora de Runes, Barcelona, Spain

Mr. Jeroen Vrijders Belgian Building Research Institute, Brussels, Belgium

Prof. Jingwei Ying Department of Building Engineering, Tongji University, Shanghai, P.R. China

List of Figures

List of Tables

Chapter 1
Introduction and Activities

Enric Vázquez

Abstract This report is the result of the work done by the members of the TC and collaborators, discussed in nine meetings and the rich contributions of the Symposium of Sao Paulo 2008, the Sao Paulo International RILEM Conference Progress of Recycling in the Built Environment 2009 and Shanghai RILEM 2nd International Conference on Waste Engineering and Management 2010.This events were promoted by members of the TC 217 and co sponsored by RILEM.

Keywords Construction and demolition waste • Concrete • Durability • Life cycle analysis • Production • Recycling • Aggregates • Meetings • Methodology • Symposia

1.1 Objectives of the Book

For centuries, the construction industry has been a voracious consumer of raw materials. The construction industry has become more responsible for the greatest waste stream. According to data published by EUROSTAT, the European Statistical Office, 48% of the waste produced proceeds from the construction and demolition sectors, and third of the prime resources consumed end up transformed into residues and emissions. In other continents, and even within the EU, the proportions can vary according to construction and development activity. In any case, the situation provoked by the amount of waste, can be extrapolated to other countries and even pressing in the future in emerging countries.

Concrete, masonry rubble and stone, stand out as the dominant components of this type of waste. Evidently, such stream of waste can occupy enormous amount of space and ground surface in landfills. As it has been demonstrated for more

E. Vázquez (✉)
Materials Section of the Department of Construction Engineering, Universitat Politecnica de Catalunya, Barcelona, Spain
e-mail: enric.vazquez@upc.edu

E. Vázquez (ed.), *Progress of Recycling in the Built Environment: Final Report of the RILEM Technical Committee 217-PRE*, RILEM State-of-the-Art Reports 8,
DOI 10.1007/978-94-007-4908-5_1, © RILEM 2013

than half a century, these residues have a great potential for reuse in the same field where they proceed from and taking advantage of this can save raw materials.

From a global point of view, some authors estimate that the replacement percentage of prime resources amounts to 15%. The recycling of this waste was heavily promoted in the 1970s, in several countries, on the premise of saving resources and land by slowly reducing the amount o landfills normally available for disposal. Since then recycling of CDW has developed into an ambitious and creative industry that produces recycled aggregate for construction.

Professor Charles F. Hendricks [1], based material and waste management in the Integral Chain Management concept, which relies on keeping prime resources within a single life cycle for the greatest amount of time possible. This cycle accounts for durability, its derivate products and its return to the cycle with the appropriate treatment once concluded its use, "as long as it is appropriate from an environmental point of view" he added. Thus, the environmental impact of closing the life cycle must then be assessed as well.

This issue has been addressed by the Life Cycle Analysis (LCA) through different methodologies. Although it is not the primary goal of this report to go in depth over the Life Cycle Analysis (LCA), which is subject of increasing publications, we believe we should start by addressing the recycling of construction and demolition waste (CDW) from an environmental perspective.

The last decades have shown beyond doubt that recycled materials can be accepted and have a broad demand. Properly processed construction and demolition waste have enabled the production of quality assured products around which an important market has already begun to develop. Nonetheless, the acceptance of recycled products has clashed in several countries. They still face numerous problems and, even with the recognition of their environmental necessity, they are subject to the imperatives of the free market. Local prices, abundance and demand of natural aggregate, determined by the country's economic situation and development of construction industry, govern the competitiveness of recycled aggregate.

Currently (2011), the economic crisis has hit many countries with distinctive intensity, aggravating in some the challenges of their implementation, which range from the decline in construction activity to lack of CDW due to the near disappearance of demolitions. In other countries, among which are found most of the emerging nations, the waste production and recycling data seems to indicate an opposite tendency. For instance, in 2010, China generated 300 t of CDW in demolitions related to the Shangai World Expo alone, while 200 million t is their annual consolidated figure. The status report included in this publication reveals more than just a takeoff.

In Brazil recycling has started long time ago, but his development has been very slow. The coming international sport events will accelerate the construction activities and the demand of aggregates.

Differences in the proportions of recycled CDW in the EU range from 10 to 90%. The EU Waste Framework Directive, which came into effect in 2010, sets an average quota of a 70% recycling for CDW by 2020. In the Netherlands, Belgium, Switzerland and Austria this figure has already been reached, but there are still many other countries left for which reaching this figure seems a complete utopia.

The objective of this publication is to collect the data of advancements, achievements, problems and solutions most representative of the various situations. Thereby, each country can learn from examples and underpin a pedagogy for the effective implementation of recycling.

Apart from some isolated efforts, systematic research on the use of recycled aggregate in concrete road pavements started in the 1970s and appeared in numerous publications. Many suggested their use in structural concrete, promoting what we now call original level and maximum value up-cycling. It was not intended to eliminate the initial practice of using of recycled aggregates in road sub-base (down-cycling), which remains the most widespread and simple application, but to increase the number of applications with greater environmental value.

In 1978, Nixon, P.J. published *Recycled Concrete as an Aggregate for Concrete – A Review* [2]. Later, *Recycling of Demolished Concrete and Masonry* edited by CT Hansen in 1992 [3], became the standard summary of the investigation up to the time on recycling of demolished concrete and masonry rubble, as many researchers and authors used it as a starting point in their investigations. In 1994 RILEM published *Concrete Specifications for Recycled Aggregates* [4], then followed with RILEM symposiums that illustrated remarkable progress until 2000.

The Building Cycle (2000), and *A New Vision of the Building Cycle* (2004), by Ch. Hendricks et al. [5], are books that have strongly contributed to under-laying the knowledge base and technological evolution of recycling of CDW.

Following this tradition, the second objective of this publication is to collect and report on the **advances in concrete recycling produced since 2005**.

Recycled aggregates that are produced in more modern plants, are cheaper than primary aggregates. Cleaning and separation techniques have enabled the removal contaminants, and a secondary crusher can solve the problem of inadequate forms. The generalized presence of gypsum can be reduced to acceptable levels by jigging or using more sophisticated techniques. Today, there are plenty of new features in recycling plants that enable us to obtain quality aggregates.

New products have been developed that extend the field of application of the recycled aggregate. The field of precast concrete products has proven that not only it can absorb its own waste, but that it can produce new successful products based on recycled aggregates. Possibilities have opened up for the application of recycled aggregate in porous concretes, and modular blocks of recycled concrete are being produced.

A complete new approach to the procedure of proportioning concrete mixtures with recycled aggregate has been recently presented by A.G. Razaqpur et al. [6] which treats the residual mortar as part of the aggregate of the total mortar content. The resulting concrete can satisfy the durability, strength, and other physical and mechanical requirements of structural concrete. This concrete requires less cement, which makes it more economical and environmentally attractive. Other teams like the Materials Section of UPC Barcelona are working in the same direction with some new modifications that extends the field of application.

Recent publications, opinions and discrepancies in relation to the durability of recycled concrete, such as freeze-thaw standards, studies of chloride penetration and diffusion, and sulfate attacks, are examined.

In most countries, the proportion of masonry rubble in CDW is elevated and its use is limited to road base overlays of mixed recycled aggregate. This rather limiting application does not provide a definite solution for future markets.

The recovery of material must always focus on the weakest point of the chain in properties and market. Thus informing on the **advances in masonry recycling** is another objective of this book.

The use of the fine fraction, <4 mm in Europe and <2 mm in Japan, remaining from the production of recycled aggregates appears to pose more problems than solutions as its excessive absorption potential can increase shrinkage and creep in recycled concrete. Since this fraction represents between 20 and 60% of production, significantly influencing the economics of the process, offering a broad vision of possibilities for its use will be our goal. Carefully studying its characteristics, possible applications in the field of prefabricated structures can be found.

Finally the **quality assurance of concrete recycled aggregate** is a general necessity. The physical and mechanical properties are checked in accordance with national and international standards. Aspects related to the incoming wastes, crushing processes, new technologies, and original concrete, among others, are considered.

We hope that this report can give information to recyclers, researchers and consumers about new knowledge. Perhaps it can be a useful tool for countries starting his activities in recycling.

References

1. Hendriks, C.F., Nijkerk, A., Van Koppen, A.: The Building Cycle. Aeneas Technical Publishers, Delft (2000)
2. Nixon, P.J.: Recycled concrete as an aggregate for concrete – a review. RILEM TC-37-DRC. Mater. Struct. **11**(65), 371–378 (1978)
3. Hansen, C.T.: Recycling of Demolished Concrete and Masonry (RILEM report 7 DRC). E & FN Spon, London (1992)
4. RILEM: Concrete specifications for recycled aggregates, RILEM TC 121DRG. Mater. Struct. **27**, 557–559 (1994)
5. Hendriks, C.F.: A New Vision of the Building Cycle. Aeneas Technical Publishers, Delft (2004)
6. Razaqpur, A.G., Fathifazl, G., Isgor, O.B., Abbas, A., Fournier, B., Foo, S.: How to produce high quality concrete mixes with recycled concrete aggregate. In: RILEM PRO 73: Construction Waste Recycling and Civil Engineering Sustainable Development ICWEM2010, Bagneux-France, pp. 11–35 (2010)

Chapter 2
Construction and Demolition Waste Recycling in a Broader Environmental Perspective

Abstract The scope of this chapter is to analyse the prospects for recycling the mineral fraction of construction and demolition wastes as aggregates within the frame of sustainable construction, and to identify research needs. Sustainable construction is defined by life-cycle evaluation of environmental, economic and social aspects, and is dependent not only from production processes but also from local constraints and market alternatives. Environmental performance of recycling construction and demolition waste must be evaluated cradle-to-grave, using multi-criteria life-cycle assessment tools. Economic performance depends on life cycle cost and has to consider externalities. Social assessment can be done using the new social life cycle assessment methodology. The flows of aggregates and construction waste in the near future, as well as the presence of informal activities in the supply chain, have strong influence on the sustainability in every location. Attractiveness of recycling CDW as aggregates will depend of local environmental, social and economic conditions. Examples of life cycle thinking applied to CDW aggregates are given and research needs discussed.

Keywords Sustainability • Aggregates production • Recycling • C&D waste • Life cycle thinking • Materials flow

E. Vázquez (ed.), *Progress of Recycling in the Built Environment: Final Report of the RILEM Technical Committee 217-PRE*, RILEM State-of-the-Art Reports 8, DOI 10.1007/978-94-007-4908-5_2, © RILEM 2013

2.1 Sustainability and Recycling Construction Waste

Vanderley John and Sergio Angulo

2.1.1 *Introduction*

Recycling of building materials is not a novelty. Romans use construction waste like ceramic in their concrete [1]. However, nowadays except from large natural disasters or war [2] and in a few European countries actual recycling of mineral fraction of construction materials is often more an exception than a rule. More than that, most of the recycling implies in downgrading, were aggregates are used in less noble functions than the original ones.

The growing interest in sustainable development and, more specifically in sustainable construction, as a reaction of the society to the growing global environmental and social problems, can be a decisive factor to increase recycling rates in the near future. It has been pointed out that recycling has been frequently presented as sustainable "based on a few strength measurements and a few leaching tests on samples tested in the lab" [3] therefore neglecting the complexities of sustainable use of resources.

This chapter aim to analyse the prospects for the recycling of the mineral fraction of construction and demolition waste as aggregates within the frame of sustainable construction.

2.1.2 *Defining and Measuring Sustainability*

A sustainable solution must seek equilibrium between competing social, economic and environmental demands while fulfilling the technical demands, along its life cycle, cradle to grave. The green movement put emphasis in the environmental aspects of sustainability, which are the more relevant in developing countries. However, in developing countries economic and social aspects are also pressing.

Any production and consumption process has simultaneously multiple environmental impacts. Despite being obvious that it is impossible do judge environmental performance (or even sustainability) of any solution base made based only on a single aspect (e.g. content of recycled materials), many of green building guides or

V. John (✉)
Lab. Microestrutura e Ecoeficiencia de Materiais, Escola Politécnica da USP, Sao Paulo, Brazil

S. Angulo
Instituto de Pesquisas Tecnologicas do Estado de Sao Paulo, Sao Paulo, Brazil

labels adopt this kind of criteria. It has been adopted rather frequently to support the need of recycling construction waste as aggregate. The state-of-the-art tool o judge environmental performance is life cycle assessment (LCA), which is a well-defined, standardized, multi-criteria, cradle-to-gave (or cradle) methodology [4]. Traditional LCA methodology aims to be as complete as possible resulting in a rather complex process, which is very intensive on data [5]. Despite being based on a quantitative inventory of flows of energy and materials for each step of service life, the practical application of the methodology requires subjective decisions which can have enormous influence on the practical results. Examples are the selection of the criteria for allocation of environmental loads, particularly when dealing with generated or recycled waste [6, 7], defining scope and system boundaries [4]. Using LCA in a decision process frequently require weighting the different environmental impacts, which depends on the agenda of the decision-maker. This will be the based for all future environmental product declaration (EPD).

Most of the LCA evaluations and technical decisions are based on data from generic life cycle inventories (LCI) available in the market. In other words, decisions are made based without direct measurements of the actual flows of energy and materials from the particular production process in question. The use of LCI data banks is a necessary simplification given that the number of measurements from a complete LCA is large and many of these measurements require very specialized and expensive equipment and personnel. However, considering that there are significant differences between actual process and raw material and fuel formulation and the changes of any given process overtime, this approach can lead to erroneous decisions even if some adjustment is done. Worst, a significant share of LCI data is based on single measurements and is not even representative of the process that was analysed. An alternative to reduce inference errors is to simplify LCA by concentrating on the more important and/or easy to measure impacts, such as CO_2 emissions [8], energy consumption, use of natural & recycled resources, water consumption, waste generation and toxicity. This approach have de advantage of opening the LCA market to institutions that can't afford a professional LCI database and even allow smaller organizations to produce relevant environmental products declarations (EPD).

Economic performance assessment is rather standard in the economy. However the whole life economic performance, Life cycle cost (LCC) [9] a cradle to grave tool is not so popular. While traditional economic decision making tools are concerned with direct, indirect and contingency costs and revenues from the production process, to perform a LCC coherent with LCA it is necessary to include external costs – those who are borne by parties other than the company, e.g. the cost that a local authority has to remove illegally dumped construction and demolition wastes (CDW), which also has social impacts since affects public authority investment capability – as well we intangible costs such as costumer loyalty, worker morale and wellness [10].

Assessing social dimension of sustainability still is a rather underdeveloped area. Only in 2009 the UNEP/SETAC released the first guidelines to perform a Social Life Cycle Assessment (S-LCA) [11], which is defined as "is a social impact (and potential impact) assessment technique that aims to assess the social and socio-economic aspects of products and their potential positive and negative impacts along their life cycle.". It includes all stakeholders during the material service life (Workers/employees); Local community; Society (national and global); Consumers (covering end-consumers as well in the supply chain) and value chain actors It includes aspects like health and safety, transparency, respect to cultural heritage, local employment, fair salary, discrimination [12]. The methodology is designed to measure quantitative and qualitative variables trying to evaluate the degree of corporate social responsibility. The methodology implicitly considers that the producers and its supply chain operates accordingly with local regulations. However that's not the case in most of the developing countries, particularly regarded to natural aggregate production.

The evaluation of the whole life cycle of a product requires the consideration of future impacts, including construction, use, maintenance, dismantling, eventual recycling and final destination of the materials. Forecasting future impacts implies in developing scenarios of utilization concerning environmental aspects (LCA), economic (LCC) and social. It also requires estimating the service life of the products [13, 14], that might be rather complex especially because it depends on the exact function of the product in the construction (that is not clear for an aggregate.) local environmental variables, integration with other construction parts as well as maintenance profile and local environment. In the case of raw materials, like aggregates and cement, not even the application is clear to the producer. Therefore it is a current practice to perform life-cycle evaluation limited to plant gate (cradle to gate), leaving to the client to complete the life-cycle information and taking a decision.

In consequence, even if the cradle-to-gate life cycle analysis demonstrate that the recycled aggregates present better environmental social and economic performance than natural ones it does not implies that they are the more sustainable choice for all uses in all situations.

2.1.3 Environmental Drivers for Recycling – Closing the Materials Flow Cycle

Modern economy depends on a constant flow of materials [15], which starts with the extraction of natural raw materials and goes to the destination of the post-use waste at the end of service life. The ever-improving industrial technology is making possible the production of a progressively large amount of materials goods. Developed countries demand a direct material input between 17-8 tonnes per year per person [15, 16]. However, the total amount of material extracted from nature is much larger, because it is necessary to consider the hidden flows that do not enter in the economy, which are wasted in the process. In consequence, the total materials requirement[1] in this economics is 45–85 tonnes per year per person [15]. Building

[1] Total material requirement includes hidden flows – like deliberate alterations of landscape and quarry cover removal. Direct material input includes only materials that enter in the economy for further processing, like wooden logs, minerals shipped to the mill, etc.

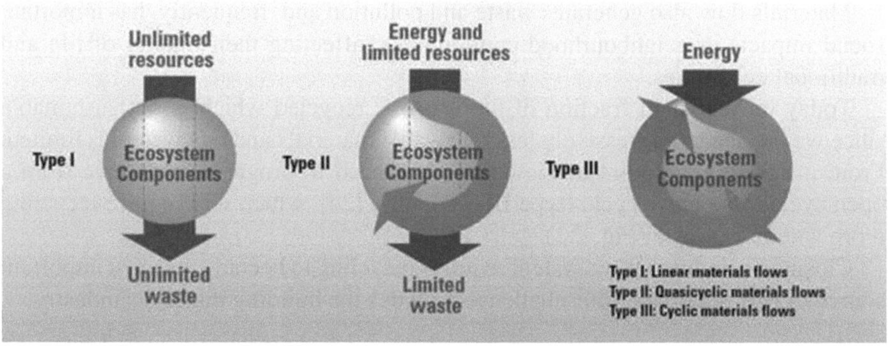

Fig. 2.1 Possible materials flow in industrial economies. The aim is to move to type III close cycle industrial ecosystem [24]

materials represents a substantial share of the total materials consumption. In 1900, 33% of the total non-energy materials (or non-energy carriers) was construction materials and by 1998, it had grown to 70–73% [17].

Materials flow implies in waste generation in every step of its live cycle. A large part of the natural materials extracted become waste: Iron 60%, Copper 99%, Gold 99.99% [18], a significant part of it as a mining tailing. The extraction of commercial wood logs from tropical rain-forest implies selective logging which might destroy 1/3 of the existing biomass [19]. During the processing the wood, logs in the saw mill that 60% of this logs become wood waste [20]. Considering that all used materials become waste eventually, and all production process and even transport activity generates waste, the amount of waste produced in the economy is to up to five [21] times the amount of commercial materials. Waste generation in mining and short-live materials use is important and in consequence ½ to of the direct material input are discharged as waste in less than 1 year time [22]. Construction and demolition waste is therefore a small fraction of total waste generated by flow of the building materials in the economy.

Besides of resource consumption and waste generation the extraction, transportation and processing this large amount of materials have all sorts of impacts on the environment. When measured by weight, materials account for 44% of the United States' resource use, 58% in the case of Japan, and as much as 68% in Germany [18]. The flow of materials is therefore responsible for a significant share of the global ecological footprint being 20% of the ecological footprint in the USA [18].

Extraction of raw materials always has environmental impact on ecosystems, including endangered ones. Materials contribution to global warming is very important. Cement production is responsible nowadays for something around 5% of anthropogenic CO2 emissions and business as usual this share is expected to grow sharply in the near future [23]. In some regions of the Brazilian Amazon rainforest the logging of 20 t of lumber generates in average 85 t of wasted biomass, which will be degraded into 156 t CO_2.

Materials flow also generates waste and pollution and, frequently, has important social impacts in neighbourhood communities, affecting their quality of life and traditional economies.

Today only a small fraction of the waste is recycled which is not sustainable since we are using progressively less renewable materials and our system is limited. From materials flow point of view there is a need to progressively move from a open cycled to a close cycle (type III) economy [24], which will require recycling more and more waste (Fig. 2.1).

Closing the industrial ecosystem requires recycling to become the most important source of row materials. This challenge includes the building materials industry.

2.1.4 The Flow of Mineral Construction Materials and Construction Waste

Materials flow models, which link evolution of social and economic patterns which influences construction demand with service-life modelling can help to forecast construction materials consumption construction and demolition waste generation as well as the potential impact of recycling strategies [25, 26]. To the moment there are no published global studies of future flows of construction and waste materials.

The global demand for construction aggregates was estimated in 2006 to be about 10 tonnes per year per person [27], which is a significant part of the direct materials input. Considering that in 2006 the cement consumption was 397 kg/year per person [28], cementitious materials will require something between 2.1 and 3.5 tonnes per year per person[2] (13 to 23 Gt/y), value which is coherent with the estimative that in UK approximately 40% of aggregates are used for making concrete [29].

Mineral materials wastage rate can be responsible for a significant fraction of this consumption. The reported percentage of waste materials in some developed countries varies from 2 to 15% in UK, 2–22% in Australia [30], 13–15% in Finland and 1–10% in The Netherlands [31]. A comprehensive study conducted in Brazilian multi-storey residential buildings sites found even higher levels of wastage, around 20% of the materials delivered to the site [30]. However, aggregates wastage rate was much higher. Waste rate of coarse aggregate varied between 9 and 294%, with median equal to 38%. Sand waste rate was even higher, between 7 and 311%, with median of 44% [32]. The factors associated to wastage

[2] The average proportion between aggregate and cement varies from country to country and with time – it probably will grow with the tendency of reduction of cement content due to environmental concerns [29, 30]. In the literature this proportion has been estimated from 5.4 t/t for the UK [31] to 7.6–8.9 t/t in the USA [17]. The Sustainable Europe Research Institute data acounts for 4.4 tonnes per year per person [32] for total global aggregate consumption, which is inconsistent with data from cement consumption.

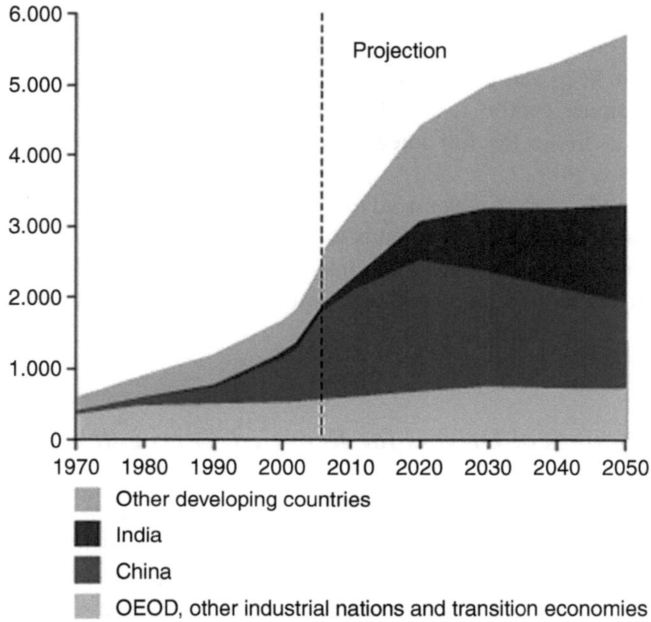

Fig. 2.2 Cement production evolution over time and forecast to year 2050 [17]. The forecast assumes that cement use efficiency will not improve in the future

rates are now all understood, and includes both site management, design decisions, procurement and technical solutions [30, 31, 33]. Preventing aggregate wastage rate is an environmental and economical priority.

The amount of construction materials used in the society has been growing due to the growth of economy and population. Cement production was around 0.7 Gt/year in 1970, 1.17 Gt/year in 1990 and 3.1Gt/year in 2010 in an annual growth rate around in the last 20 years 6% and it is expected to growth to 5Gt by 2030 [23], 1.6 fold grow. This will increase the demand for aggregates to 26–44 Gt/year (3 to 5 tonnes per year per person, considering a population of 9 billion). This estimative is conservative since it neglects the tendency of decreasing binder intensity of cement based materials due to environmental problems in the near future [34] (Fig. 2.2).

Thus, the consumption of construction materials, and particularly, mineral aggregates have a relevant environmental impact, mainly due to the production scale.

2.1.5 *Informal Production of Aggregates*

The informal economy, broadly defined as any economic activity that is not declared to governmental agencies [35–37] represents a significant stake – more than 70% – of the economy in many developing countries [35, 38]. In Brazil about 55% of the total aggregate production was informal in 2006 [39], and even large companies can keep part of their operation informal. After extraction and transportation to consumer centres it is difficult to separate legal aggregates from illegal ones, which facilitate the full integration of illegally produced aggregate in the formal economy [40]

Informal production does not need to respect state regulations, including environmental, social and taxation. This tends to worse environmental impacts and undermines the state capacity to plan and implement policies [37]: reduction of informal aggregate production is a precondition to sustainable policies, including recycling.

2.1.6 Economical Drivers for Recycled Aggregates

Natural aggregates are simple and cheaper to process than recycled aggregates, since natural raw material is more homogeneous and have much lower contamination than recycled one. The low cost of natural aggregates has been certainly a limitation factor for a decisive factor for any material, even for sustainable construction. However in many regions the economic factors are changing.

2.1.6.1 Localized Scarcity and Long Transportation Distances

Even thou natural aggregates are available in large quantities in most of the world regions the combination between more strict natural environmental protection laws and concentrated consumption in large cities can cause sharp increase in costs, especially due to long transportation distances. In almost all developing countries most of the cargo transportation is performed by roads, which is significantly more expensive than rail and river.

In São Paulo metropolitan area, for example, natural sand cost more than USD 39/t (BRL 80) and coarse natural aggregate around USD 25/t, being long distance road transportation a significant fraction of it. In regions were natural rocks are scarce, like the Amazon region in Brazil, prices can be much higher. In Manaus, for example, coarse aggregate cost is USD 66/t. These prices are expected to grow with the increasing demand. These Brazilian prices are much higher than the average price in USA market, respectively USD 7, 70 [41] and 9, 91 [42] or in Europe, were the highest prices are about USD 16/t (~12 Euro), observed in Eastern Europe, particularly in Russia, Hungary, Romania and Bulgaria [43].

This situation can be relevant for a significant fraction of the worldwide aggregate market, since 60% of the urban population of developing countries was leaving in cities with more than one million inhabitants and 24% in cities with more than five million [44]. In the year 2050 it is expected that the number of people leaving in cities will be 6.4 billion, twice the number of 2007 [45].

The reduction of transportation distance by CDW recycling between the production site to construction site and CDW availability evaluated by population density are the main responsible for the economic feasibility of recycling [44].

2.1.6.2 Increase Landfilling Costs for Waste Generator

In cities where illegal dumping is viable cost of landfill is an external cost, since the public (local authority) will pay for the cleaning and landfilling. This can be very costly because it requires moving heavy machinery to illegal dumping sites and long distance transportation to landfill. In city of São Paulo, this represent near 13 Euro/t [46].

When a society succeed to control in illegal dumping cost of construction waste management increases. In large metropolitan areas, space for landfilling is normally high which implies in long transportation distances, which is also expensive. Landfill cost in São Paulo is near 100 Euro/t. A deconstruction plan study conducted at city of Sao Paulo has pointed that CDW recyclable sorting and on site recycling of most CDW were the best economical option, which explains why CDW recycling is fast growing at big cities of under development countries.

2.1.6.3 Landfill and Aggregate Taxes

The implementation of a landfill tax is also an effective way to promote recycling in regions (or countries). This measures are affected only if illegal dumping and natural aggregate extraction market share is small. In Europe, landfilling is taxed in many countries, at values up to 50 Euro/t [43]. In the UK natural aggregate has been subjected a progressive tax [43].

2.1.7 Social Drivers

Reduction of local authorities' costs associated to cleaning illegal dumping is clearly a social benefit from recycling construction demolition waste, especially in developing countries were the rate of illegal dumping is high.

Recycling also generates new green jobs [47] all over the supply chain, from segregation activities, transportation and processing. This can be important especially for countries with high unemployment rates.

2.1.8 *Assessing Sustainability of Recycled Aggregates*

2.1.8.1 LCA of Recycled Aggregates

There are a growing number of LCA assessments of recycled or natural aggregates, and web and Excel data file tools are available.[3,4] However, this tools are cradle-to-gate (does not account for whole life cycle) and probably are not valid to other regions and applications. Even the technology used to demolish the original building and the transportation distance influence the environmental balance [48].

In many regions of the world, alluvial river virgin quartz sand extraction is still available. and resources depletion can be serious involving aggradations and erosions. On the other hand, CDW recycling avoids illegal dumping; one of the responsible for aggradations in urban areas and, it is scarcely quantified to balance environmental advantages of recycling. Waste generated by CDW recycling is not often considered in spite of fine fraction of aggregates ($<$4.8 mm) remains without a noble and defined use. Due to major concerning with climate change, partial environmental evaluations have been done and much attention has been directed to CO_2 emissions.

Despite abundant crushed rocks are also intensely used in world construction and energy and transportation uses are the main environmental concerning due to fossil combustible use. Countries with hydroelectric power domain, environmental impact by energy use is secondary and transportation has a higher relative importance.

Recycled concrete aggregates (RCA) production demands more intense use of machines and combustibles and, as a consequence, higher CO_2 emissions [49, 50]. Other environmental impact categories such as eutrophication, acidification, POC also become worse due to use of combustibles. CDW recycling process generates of large amount of fines [51] that at the present are residues increases the environmental loads of the recycled aggregates, since it reduces the process yield and the effectiveness of the recycling process. Developing practical applications for such fraction is therefore important.

Transportation of aggregates accounts 30–40% of total CO_2 emissions of this specific industry for typical distances and road transportation [52]. The range of CO_2 emissions during transport taking into account consumption of fossil combustibles of different trucks and different emissions factors (GHG, IPCC, etc.) can be 12.8–50.6 g of CO_2/t Km [31].

The environmental advantage of recycled aggregates use appears in terms of CO_2 emission when road transportation of natural aggregates are significantly higher than recycled ones [49]. Recycled aggregates are generated and

---

[3] http://aggregain.wrap.org.uk/sustainability/sustainability_tools_and_approaches/aggregates_lca.html

[4] http://www.ce.berkeley.edu/~horvath/palate.html

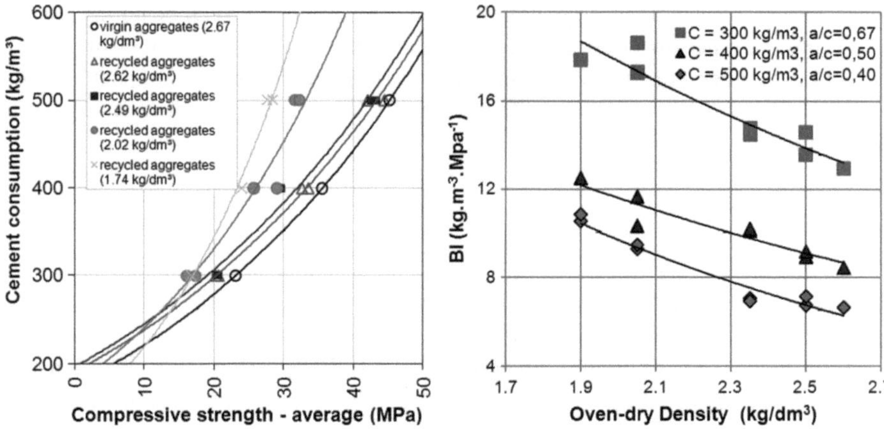

Fig. 2.3 Increase of cement consumption to produce concrete (10–50 MPa) for different oven-dry density aggregates (*left*) and oven-dry density of aggregates × binder index of concrete (*right*)

commercialized inside cities and significantly reduces transportation distance and it is especially interesting of pavement use. Reduction of 1/6 in transportation distance of aggregates (15 Km of recycled aggregates instead of 100 Km of natural aggregates) can be achieved by recycling near building sites and it reduces CO_2 emissions and becomes environmental feasible the use of recycled aggregates.

However, the sustainability of recycled aggregates cannot be judged without considering the entire life-cycle, which includes application. For example, if aggregates are used in concrete, achieving a positive environmental balance is difficult. For almost all practical situations, cement consumption is the main responsible of CO_2 and other key environmentally important emission of concrete life cycle. Depending on the application, the lower mechanical properties of the typical recycled aggregates in comparison with virgin ones will demand an increase of cement consumption to achieve the same mechanical properties.

Depending on several factors, including concrete strength, recycled aggregate porosity (or oven-dry density in Fig. 2.3) [53], as well as degree of natural aggregate replacement, cement content can increase by a factor of two, as shown by Levy and Helene [54]. However, in some situations, low porosity recycled aggregate and low strength concrete, the use of recycled aggregate to produce concrete can reduce overall CO_2 emissions, despite the increase of small increase of cement consumption [49]. Transportation impacts (or distances), environmental impact of cement production, ecological value of the natural aggregate extraction can also change the conclusions.

Consequently, it is not possible to generalize conclusions on environmental sustainability of the use of recycling aggregates and an LCA is needed to every typical situations.

A simple way to evaluate eco-efficiency of concrete in terms of use of scarce binder materials and for CO_2 emissions is by the binder intensity (bi) [34]. From a life cycle analysis methodology point of view, this proposition can be understood as changing the functional unit from a unit of concrete volume or weight (1 m^3 or kg), which is convenient for measuring the overall environmental impact of any concrete construction, to a unit of functional performance, which in many cases is the compressive strength.

$$bi = \frac{b}{p}$$

Where b is the total consumption of binder materials (kg m^{-3}), and b is the performance requirement (MPa^{-1}).

Hence, the CO_2 emissions of concretes with porous (low oven-dry density) recycled aggregates can be very high and easily surpass that of concretes with natural aggregates (Fig. 2.3).

2.1.8.2 Leaching of Recycled Aggregates: In-Use Environmental Impact

The use of recycled aggregates in road exposes a high specific surface area material to pure or contaminated water which cause leaching of chemical species, including hazardous ones. Therefore, from theoretical point of view, leaching has to be included on the LCA results (see [55] for an example).

Hazardous elements may be present in demolition wastes [56], due to inefficient sorting, incorporation of waste in building materials [3] or even exposure to industrial contaminants during use of buildings.

Leaching of metals (Cd, Cu, Mn, Ni, Pb and Zn) of recycled aggregates occurs mainly in acid media [57]. pH media controls the pollutant release due to cement paste solubilisation and it explains most of differences between leaching methods results [58]. Hence, carbonation of cement paste of recycled aggregates will increase leaching [59]. Nowadays, leaching can be predicted for use according to different geochemical scenarios [57] (Fig. 2.4).

Gypsum contaminated recycled aggregates implying in higher sulfate concentrations: 890–1,600 mg/l, especially for fines of CDW [60]. Such concentration surpass limit for secondary drinking water standard of 250 mg/l and may convert to hydrogen sulfide with organic compounds are presented in soils.

Reduction of leaching can be expected for concrete with recycled aggregates (consolidated material) when compared with base pavement with recycled aggregates (unconsolidated material). In this case, leaching is controlled by diffusion inside pores [22, 61].

Leaching of recycled aggregates as well of materials made with this aggregates are therefore an important step for the life cycle assessment.

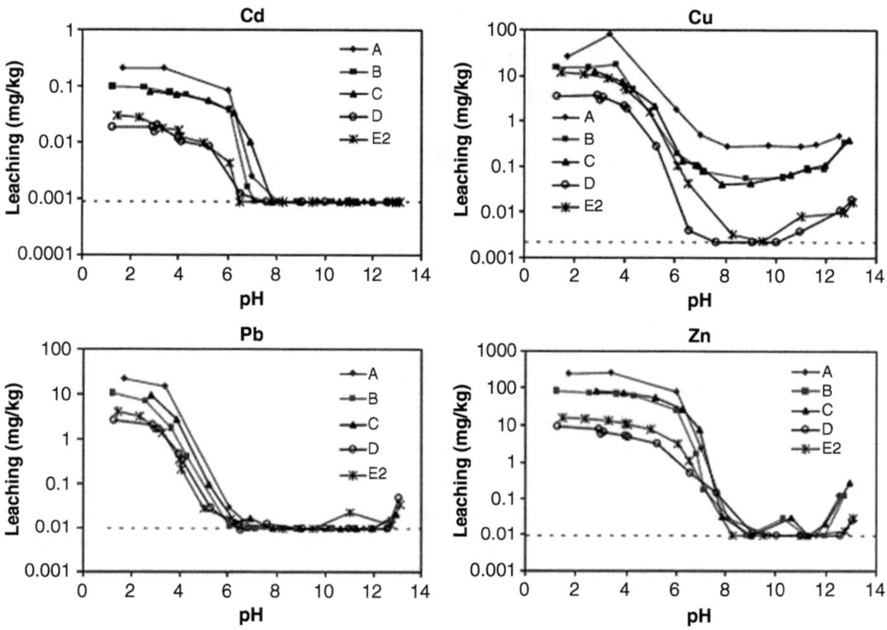

Fig. 2.4 Leaching of metals inside CDW recycled aggregates × pH [54]

2.1.9 *The Need for Further Developments*

There is a need for further development of more comprehensive tools to accesses sustainability of recycled aggregates, including environmental, social and economic dimensions. Materials flow studies can be usefull to plan better policies related to recycling construction and demolition waste.

Traditional recycling technologies are very simple, resulting in an rather heterogeneous products. Even when recycling pure concrete, the resulting aggregate does not have the same performance of the virgin ones. More sophisticated processing technologies aiming to produce better recycled aggregates for every fraction of the mineral construction waste are still needed. In many regions of the world the high cost of natural aggregates and the landfilling costs are increasing the economic viability of such processes.

This implies in a significant investment in research and development, since at the moment there are only limited available knowledge in this field.

2.2 Global Impact Assessment of Urban Renewal Based on Sustainable Recycling Strategies for Construction and Demolition Waste

Ir Gabriëlla M.T. Janssen, Hin Ming Ho, and Ir Johan A.L. Put

Abstract Urban renewal projects merely focus on meeting budgetary constraints. However, nowadays the decision maker cannot afford to ignore the sustainable approach to these projects, whereby he is confronted with the impact of his decisions on sustainability issues, such as global warming and the depletion of natural resources. In this context, strategies for recycling in urban renewal processes have to be considered. In urban renewal, it is obvious that primary raw materials are used. These high specific gravity materials have to be transported over long distances. On the other hand, however, the regeneration of urban areas results in the production of waste materials, as the original buildings have to be demolished first. Such materials can be recycled into secondary raw materials in substitution for primary raw materials. Hence, the application of adequate recycling techniques in urban renewal offers significant savings in CO_2 emissions, as well as considerable reductions in the excavation of natural resources. The paper describes a logistic computation model, which aims at quantifying the possible savings in CO_2 emissions and primary resources, resulting from applying sustainable urban renewal methods. The model, which is part of the Expert System "Sustainable Urban Renewal, Demolition and Recycling", is illustrated by two case studies, based on actual European projects, namely the extension of the residential area at Rotterdam Harbour and the development of the new NATO Headquarters in Brussels. The results of both studies indicate that a saving of up to 80% in CO_2 emissions can be realised by recycling on site, compared to applying only primary raw materials. Moreover, a 70% decrease in the need for natural resources is feasible. Based on the European experience, further case studies will investigate the possibilities of sustainable urban planning and renewal in developing countries, facing similar sustainability issues as referred to in the European case projects.

Keywords Urban renewal • Sustainability • Recycling • Construction and demolition waste • Expert system • Carbon dioxide emission calculation • Primary resources calculation

---

Ir.G.M.T. Janssen (✉)
Secretary of RILEM TC 217-PRE, Enviro Challenge, Opoeteren, Belgium

H.M. Ho
Hong Kong Polytechnical University, Hong Kong, P.R. China

Ir.J.A.L. Put
Enviro Challenge, The Netherlands

2.2.1 Why Sustainable Urban Renewal Is Necessary

Large urban areas worldwide attract more residents, offices and industry. The demand for housing, office blocks, industrial buildings and infrastructure continuously increases. Therefore, the need for building plots rises and so, consequently, do land prices. At the same time, the lifespan of structures without cultural value is only 20–30 years for office blocks and industrial buildings, 50 years for housing and 100 years for infrastructure because they no longer comply with current expectations. When applicable, constructions are renovated.

The above-mentioned trends in urban renewal can be perceived around the world in both developed and developing countries. In urban renewal processes, it is obvious that primary raw materials, such as sand and gravel, are utilised.

These high specific gravity materials have to be transported over long distances. On the other hand, however, the regeneration of urban areas results in the production of waste materials, as the original buildings and infrastructure have to be demolished first, in order to create construction space. These developments result in an increase in the long-distance transportation of raw and waste materials with high specific gravity, e.g. lorries from quarry to building site and back (empty), and from demolition site to landfill and back (empty). This traffic eventuates in a considerable increase of CO_2 emission. For urban development purposes, moreover, substantial amounts of primary resources are required, i.e. crude oil as well as minerals.

However, nowadays the decision maker cannot afford to ignore the sustainable approach to urban renewal projects, whereby he is confronted with the impact of his decisions on sustainability issues, such as global warming and the depletion of natural resources. In this context, therefore, strategies for recycling have to be considered. This implies that the waste materials produced during the demolition phase in urban renewal, are recycled into secondary raw materials in substitution for primary raw materials, and subsequently applied in the same urban renewal project.

In the next paragraph is shown how recycling of construction and demolition waste is currently performed in European countries, such as the Netherlands.

2.2.2 How Recycling Strategies in Urban Renewal Are Applied

According to the "Building cycle" model [62] certain waste materials are released which sometimes can be reused as a component. Usually there is no market available and the materials are recycled. Regarding mass, the most relevant waste material in this cycle is the stony fraction. If selective demolition or dismantling is applied, these stony waste materials are free of unwanted matter such as wood,

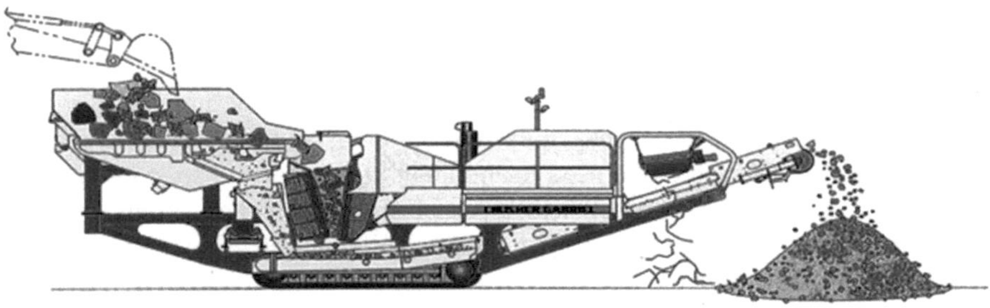

Fig. 2.5 Jaw crusher

insulation, plastics, textiles, unwanted stony materials (e.g. asbestos and aerated concrete), as well as pollutants (e.g. PAHs).

The recycling of the stony waste materials is carried out by two distinct types of crusher, namely the jaw crusher and the impact crusher. The jaw crusher consists of two plates, one of which is moving (see Fig. 2.5). Because of the force of gravity on the stony waste material and the moving plate, the material is crushed. The impact crusher is composed of a sealed chamber and a fast turning shaft fitted with hammers, which crush the stony waste material. Both crushers are supplied with a pre-sieve and a post-sieve, as well as a magnetic separator, in order to obtain end products, i.e. aggregates, of high technical performance.

Figure 2.6 visualises the most important links in the chain of construction and demolition waste, from waste disposal to raw material supply. To obtain an efficient recycling process, prevention of undesirable materials in the construction and demolition waste is always better than cleansing. Therefore, selective and environmentally friendly demolition (dismantlement) is necessary. The recycling can take place "off site", in a plant situated away from the demolition area, or "on site", on the spot employing mobile machines. The freed up waste materials can, providing an adequate process is used, be transformed into recycled raw materials in substitution for primary raw materials. For recycled building materials different high-grade applications are available, such as in road construction (as sub-base), hydraulic engineering and concrete.

2.2.3 How Modelling Supports the Decision-Making Process

In this paragraph a web-based logistic computation model is presented, which quantifies the possible savings in CO_2 emissions and primary resources as a result of applying the above-explained sustainable urban renewal methods. The model is part of the Expert System "Sustainable Urban Renewal, Demolition and Recycling" and is predominantly based on expert values. Table 2.1 shows the relevant steps to follow.

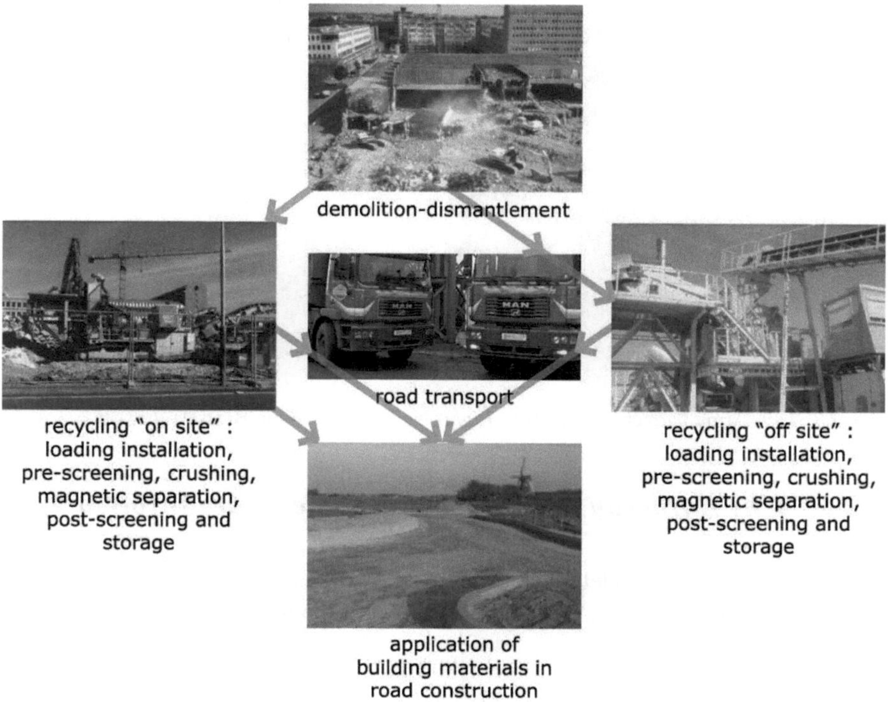

Fig. 2.6 Most important links in the CDW-chain

Table 2.1 Logistic computation model for global impact of urban renewal

1	Demarcation of project area and determination of objects to be demolished
2	Calculation (per demolition object) of environmental risks, divided into: • Non acceptable risk, i.e. waste products that need to be dumped or burned ("red light") • Acceptable risk, i.e. objects needing to be selectively demolished, waste sorting or cleansing processes are necessary ("amber light") • No risk, i.e. recycling without further pre-treatment is feasible ("green light")
3	Determination (per demolition object and per building object) of materials and respective areas, heights, number of storeys and thickness of walls, floors and ceilings, resulting in a volume calculation per type of material
4	Calculation of mass to be demolished and to be built per type of material, taking into account the specific gravity of loosely dumped material
5	Interconnection of environmental risks (red, amber or green light) with mass amounts to be demolished per type of material, resulting in an enumeration of materials and risk categories
6	Determination of distance from project to nearest quarry, waste collector, concrete supplier, asphalt manufacturer, etc.
7	Development of logistic scenarios
8	Choice of rolling stock per scenario (capacity for lorries, type and capacity of crusher); definition of assumptions concerning CO_2 emissions, diesel fuel consumption for lorries and vans (per 100 km) and share of used materials (diesel fuel and primary aggregates) in extraction of raw materials
9	Calculation of amount of trips for lorries to and fro, and total kilometres travelled; calculation of CO_2 emission and equivalent amount of square metres of tropical rain forest necessary for the photosynthesis process; calculation of total diesel fuel consumption and crude oil extraction; calculation of consumption of primary aggregates; calculation of total excavation of raw materials (when in the above-mentioned steps new insights occur or changes are implemented, recalculation is necessary)

Fig. 2.7 *Left*- Army barracks in Brussels. *Right*- New NATO HQ in Brussels [63]

2.2.4 *Why European Urban Renewal Projects Are Involved*

The aforementioned model is illustrated by two case studies, based on actual European renewal projects, which involved one of the authors.

2.2.4.1 NATO Headquarters in Brussels, Belgium (Fig. 2.7)

Obliging NATO's wishes, the government of guest country Belgium, has decided to demolish a Belgian army barracks in close proximity of the old NATO offices, so as to create space for the new NATO Headquarters (further referred to as "NATO Headquarters case"). During demolition, more than 220,000 t of stony waste materials are generated. The Belgian authorities, and moreover the Ministry of Defence, have expressed the explicit wish that traffic movement is monitored so as to minimise the direct and indirect consequences for the surroundings. Therefore, a scenario analysis is executed and recommendations formulated.

2.2.4.2 Rotterdam Harbour, The Netherlands (Fig. 2.8)

The second case involves the extension of the residential area at Rotterdam Harbour [64] whilst the harbour is reconstructed further out to sea (further referred to as "Rotterdam Harbour case"). In the Netherlands, where half the land is below sea level, the awareness of a direct threat to the country, when the sea level is indeed rising, is growing. Because of this social threat, and also thanks to a recent contractual engagement between the City of Rotterdam and the former US-president Bill Clinton to reduce greenhouse gasses, the city developers and politicians wish to put the new city regeneration to tender, taking into account sustainability issues.

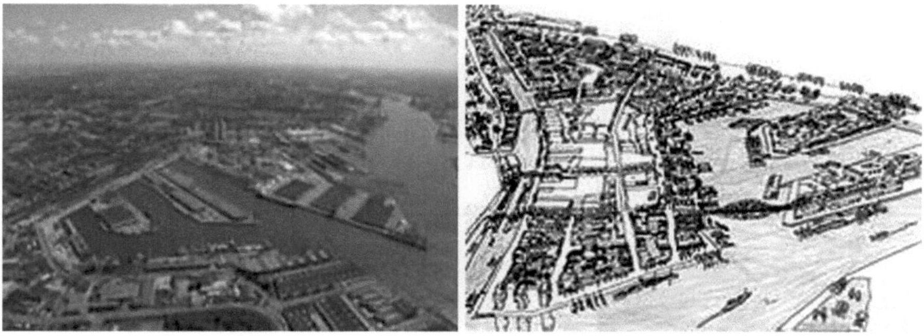

Fig. 2.8 *Left*- Fruitport in Rotterdam. *Right*- New dwellings in Rotterdam

Table 2.2 Relevant absolute CO_2 emission per scenario

Relevant absolute CO_2 emission	Scenario	NATO headquarters (t)	Rotterdam harbour (t)
Calculation is based on:	0	2,420	4,170
• *Step by step method*	1	2,068	2,281
• *CO_2 emission ≈ 880 g/km for lorries*	2	1,725	1,027
≈ 260 g/km for vans			

2.2.5 *How Recycling Reduces CO_2 Emission*

Via the Expert System "Sustainable Urban Renewal, Demolition and Recycling" the CO_2 emissions for different scenarios are calculated. In the expert system, the user can define the scenarios for himself. In the context of this paper, identical scenario analyses are performed so as to be able to compare the results of both case studies. The scenarios applied are:

- scenario 0: "no" recycling and application of primary raw materials only
- scenario 1: "off site" recycling and maximum use of recycled materials on site
- scenario 2: "on site" recycling and maximum use of recycled materials on site

For each different scenario the calculation method, as explained in Table 2.1, is applied via the expert system. In Appendix 1 and 2, the calculations for the two cases are elaborately presented. Only the relevant sources of emission are taken into account. This implies that for each scenario similar parts are abstracted. Subsequently the relevant absolute CO_2 emission per scenario is calculated. For the two cases this eventuates in Table 2.2:

Another approach is to calculate the area of tropical rain forest needed for photosynthesis of the CO_2 emission. This value is called the CO_2 tropical rain forest equivalent and is presented below for the two case studies (Table 2.3):

Interesting for the decision maker is the proportion between the scenarios. The result is CO_2 profitability, which is shown below. With the CO_2 profitability ratio it has to be stated that, when a location with relatively few buildings, like an army barracks or a harbour area, is being replaced by a densely built-up area, the

Table 2.3 CO_2 Tropical rain forest equivalent per scenario

CO_2 tropical rain forest equivalent	Scenario	NATO headquarters (ha year)	Rotterdam harbour (ha year)
$t\ CO_2\ emission\ scenario_x \cdot \frac{1\ ha \cdot 1\ year}{10\ t\ CO_2\ emission}$	0	242	417
	1	207	228
	2	172	103

Table 2.4 CO_2 profitability ratio per scenario

CO_2 profitability ratio	Scenario	NATO headquarters(%)	Rotterdam harbour(%)
$\frac{\begin{bmatrix} ha\ year\ CO_2\ tropical \\ rain\ forest\ equivalent \\ scenario_x \end{bmatrix} - \begin{bmatrix} ha\ year\ CO_2\ tropical \\ rain\ forest\ equivalent \\ scenario_y \end{bmatrix}}{ha\ year\ CO_2\ tropical\ rain\ forest\ equivalent\ scenario_x}$	1 versus 0	14.46	45.32
	2 versus 1	16.91	54.82
	2 versus 0	28.93	75.30

Table 2.5 CO_2 payback period per scenario

CO_2 payback period	Scenario	NATO headquarters (year)	Rotterdam harbour (year)
$\frac{ha\ year\ CO_2\ tropical\ rain forest\ equivalent\ scenario_x}{ha\ used\ by\ project}$	0	3.46	5.96
	1	2.96	3.26
	2	2.46	1.47

profitability ratio is relatively low. This is particularly true in the NATO Headquarters case where a total construction of approximately 220,000 t is replaced by a huge office complex of 850,000 t.

In the Rotterdam Harbour case this phenomenon seems less prevalent, but it needs to be noted that at this stage of the project (strategy building), it is still unclear whether the new residential buildings will consist of two, three or four storeys. Hence the CO_2 profitability ratio is perfectly suitable to compare scenarios within the same project, however not for comparison with other projects (Table 2.4).

Another ratio connects the time and area needed to process the CO_2 emission via photosynthesis with the area of the project. Since this ratio shows how quickly the CO_2 emission could be processed, taking into account the area of the project, the ratio is called the CO_2 payback period, analogous to the financial ratios. For both cases this results in Table 2.5.

2.2.6 *How Recycling Signifies Savings in Primary Resources*

Based on the same scenarios from the previous paragraph, the total excavation of raw materials (crude oil and primary resources for aggregates) is calculated and presented below (Table 2.6):

Table 2.6 Relevant absolute excavation of primary resources per scenario

Relevant absolute excavation of primary resources	Scenario	NATO headquarters (m^3)	Rotterdam harbour (m^3)
Calculation is based on:	0	538,293	199,635
• Step by step method (Table 2.1)	1	412,268	69,137
• Assumption concerning use of primary resources (Table 2.1)	2	411,269	65,489

Table 2.7 Relative savings in primary resources per scenario

Relative savings in primary resources	Scenario	NATO headquarters (%)	Rotterdam harbour (%)
$\frac{\left[\frac{m^3 \, excavation \, primary}{resources \, scenario_x}\right] - \left[\frac{m^3 \, excavation \, primary}{resources \, scenario_y}\right]}{m^3 \, excavation \, primary \, resources \, scenario_x}$	1 versus 0	23.41	65.37
	2 versus 1	0.24	5.28
	2 versus 0	23.60	67.20

Recycling strategies followed by use of the recycled materials on site have a beneficial effect on sustainability. This is confirmed in the following overview showing the savings in primary resources for both case studies (Table 2.7):

2.2.7 How Recycling Contributes to Diminishing the Global Impact of Urban Renewal in Both Developed and Developing Countries

The conclusions on both urban renewal case studies indicate that a saving of up to 80% in CO_2 emissions can be realised by recycling on site, compared to applying only primary raw materials. Moreover, a 70% decrease in the need for natural resources is feasible. One can add to this other advantages of recycling in respect of the reduction of landfill sites and the creation of employment [65].

In this paper, the model is illustrated by two European case studies. The significance of the advantages greatly depends on a number of factors, e.g.:

- the ratio between the mass of the object to be demolished and the mass of the new object (the higher, the better);
- the application of the recycled product on site (the more, the better);
- the possibility to crush on site (if feasible, better);
- the possibility to produce concrete on site (if feasible, better);
- the ratio between the density of the network of recycling plants and the density of the network of quarries (the higher, the better).

A key premise for the above-described calculations of sustainability advantages is that the recycled products are qualitatively equal to primary raw materials, i.e.

that the products are technically, as well as environmentally, equal. This condition can only be fulfilled if the recycling branch in the country concerned utilises an adequate production process such as in Brussels and Rotterdam, with strict acceptance criteria, trained workers, sound machinery, and a quality system to verify the product quality and adjust the process if necessary.

The European experience has originated from necessity, namely a high population density, a lack of space for waste dumping and the absence of primary raw materials in the surrounding areas. Based on the European knowledge and experience, further case studies will investigate the possibilities of sustainable urban planning and renewal in developing countries, facing similar sustainability issues as referred to in the European case projects. It is expected that, even in developing countries, the utilisation of recycled products in urban renewal is likely to offer significant savings in CO_2 emissions, as well as considerable reductions in the excavation of natural resources, especially as they have the European experience to start from.

Appendices

Appendix 1: Calculation NATO Headquarters Case

1. Demarcation of project area and determination of objects to be demolishe:

   carried out by Google Maps and site visit
   ±70 ha and ±70 objects

2. Calculation of environmental risks:

   carried out by site visit and checklists

   - non acceptable risk ("red light"): to incinerator
   - acceptable risk ("amber light"): to sorting/washing plant
   - no risk ("green light"): to crusher

3. Determination of materials and calculation of volumes:
   to be demolished:
   carried out by Forum S.A, total = 111,690 m^3
   to be built:
   carried out by architects of Asar S.A., total = 354,000 m^3
4. Calculation of mass per type of material
5. Interconnection of environmental risks (see Table 2.8):

Table 2.8 Interconnection of environmental risks NATO headquarters case

To be demolished:				
Concrete from road top layers:	12,305 m^3 × 2,400 kg/m^3 =	29,532 t		
Concrete from buildings/roofs, high quality:	19,168 m^3 × 2,400 kg/m^3 =	46,003 t		
Concrete from buildings, normal quality	8,800 m^3 × 2,400 kg/m^3 =	21,120 t		
Mixed masonry and concrete from buildings:	7,179 m^3 × 2,000 kg/m^3 =	14,358 t		
Mixed masonry and concrete from buildings/chimneys:	1,922 m^3 × 2,000 kg/m^3 =		3,844 t	
Mixed masonry and aerated concrete from buildings:	18,000 m^3 × 800 kg/m^3 =		14,400 t	
Asphalt from roads, tar-free:	4,081 m^3 × 2,300 kg/m^3 =	9,386 t		
Asphalt from roads, tar-containing	435 m^3 × 2,300 kg/m^3 =			1,001 t
Original road foundation	39,800 m^3 × 2,000 kg/m^3 =	79,600 t		
TOTAL	**111,690 m^3**	**±200,000 t**	**18,244 t**	**1,001 t**
To be built:				
Different kinds of concrete	354,000 m^3 × 2,400 kg/m^3	**850,000 t**		

6. Determination of distance from project to:

   - nearest quarry: ±30 km
   - waste collector: ±10 km
   - home base of mobile plants: ±50 km

7. Development of logistic scenarios:

   - scenario 0: "no" recycling and application of primary raw materials only

- scenario 1: "off site" recycling and maximum use of recycled materials on site
- scenario 2: "on site" recycling and maximum use of recycled materials on site

8. Choice of rolling stock (type and capacity); definition of assumptions concerning CO_2 emissions, diesel fuel consumption and share of used materials (diesel fuel and primary aggregates) in extraction of raw materials:

• capacity lorries:	20 t per load
• capacity crushers (type: impact crusher):	1,000 t per day per crusher
• CO_2 emissions (estimation):	880 g per km for lorries
	260 g per km for vans
• diesel fuel consumption (estimation):	43 l per 100 km for lorries
	12 l per 100 km for vans
• share diesel fuel in crude oil extraction:	16.79%
• share primary aggregates (sand and gravel) in excavation raw materials:	80%

9. Calculation of total kilometres for lorries and vans; calculation of CO_2 emission; calculation of total diesel fuel consumption and crude oil extraction; calculation of consumption of primary aggregates; calculation of total excavation of raw materials:

scenario 0:

transport of waste = 200,000 t/20 t/load × 2 (to and fro) × 10 km/load = 200,000 km
transport of primary materials = 850,000 t/20 t/load × 2 (to and fro) × 30 km/load = 2,550,000 km
CO_2 emission = 2,750,000 km × 880 g CO_2/km = **2,420 t** = 242 ha year tropical rain forest
crude oil extraction = (2,750,000 km × 43 l/100 km)/16.79% = 7,042,883 l or 7,043 m^3
excavation of primary resources for aggregates = 850,000 t/80% = 1,062,500 t or 531,250 m^3
total excavation of raw materials = 7,043 m^3 + 531,250 m^3 = 538,293 m^3

scenario 1:

transport of waste = 200,000 t/20 t/load × 2 (to and fro) × 10 km/load = 200,000 km
transport of secondary materials = 200,000 t/20 t/load × 2 (to and fro) × 10 km/load = 200,000 km
transport of primary materials = 650,000 t/20 t/load × 2 (to and fro) × 30 km/load = 1,950,000 km
CO_2 emission = 2,350,000 km × 880 g CO_2/km = **2,068 t** = 207 ha year tropical rain forest
crude oil extraction = (2,350,000 km x 43 l/100 km)/16.79% = 6,018,463 l or 6,018 m^3

excavation of primary resources for aggregates = 650,000 t/80% = 812,500 t or 406,250 m^3
total excavation of raw materials = 6,018 m^3 + 406,250 m^3 = 412,268 m^3

scenario 2:

transport of mobile crushers = 2 × 2 × 2 loads × 2 (before and after project) × 50 km/load = 800 km
transport of mobile concrete plant = 2 × 2 loads × 2 (before and after project) 2 × 50 km/load = 400 km
personnel of mobile plants = 3 teams × 100 days × 100 km/day = 30,000 km
transport of primary materials = 650,000 t/20 t/load × 2 (to and fro) × 30 km/load = 1,950,000 km
CO_2 emission = 1,951,200 km × 880 g CO_2/km + 30,000 km × 260 g CO_2/km = 1,725 t = 172 ha year rain forest
crude oil extraction = (1,951,200 km × 43 l/100 km + 30,000 km × 12 l/100 km)/16.79% = 5,018,559 l or 5,019 m^3
excavation of primary resources for aggregates = 650,000 t/80% = 812,500 t or 406,250 m^3
total excavation of raw materials = 5,019 m^3 + 406,250 m^3 = 411,269 m^3

Appendix 2: Calculation City Harbour Rotterdam

1. Demarcation of project area and determination of objects to be demolished:

carried out by Google Maps and site visit
±70 ha and ±60 objects

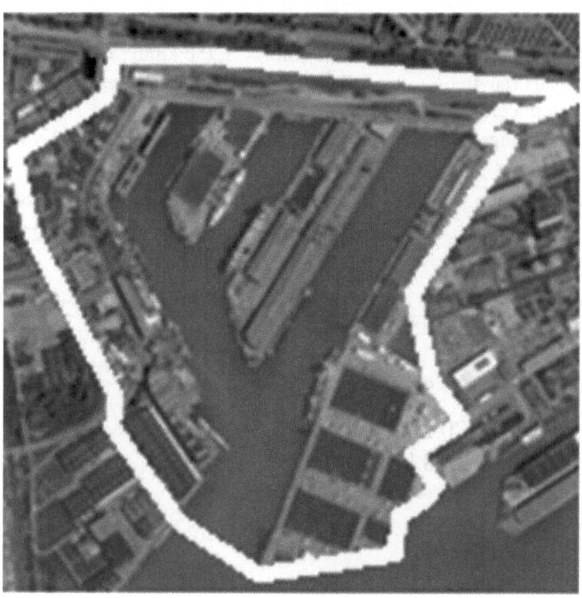

2. Calculation of environmental risks:

   carried out by site visit and checklists (preliminary)

   - non acceptable risk ("red light"): to incinerator
   - acceptable risk ("amber light"): to sorting/washing plant
   - no risk ("green light"): to crusher

3. Determination of materials and calculation of volumes:
   to be demolished:
   carried out by engineers of GWRO, total = 106,960 m^3
   to be built:
   estimated by Enviro Challenge, total = 166,667 m^3
   (assumption: 800 dwellings and 500 t per dwelling)
4. Calculation of mass per type of material
5. Interconnection of environmental risks (see Table 2.9):

Table 2.9 Interconnection of environmental risks City Harbour Rotterdam case

To be demolished:		G	Y	R
Concrete from buildings (walls), high quality	3,929 m^3 × 2,400 kg/m^3 =	9,430 t		
Concrete from buildings (floors), high quality	25,528 m^3 × 2,400 kg/m^3 =	61,267 t		
Concrete from road top layers, normal quality:	4,828 m^3 × 2,400 kg/m^3 =	11,587 t		
Concrete elements from parking lots, normal quality	23,681 m^3 × 2,400 kg/m^3 =	56,834 t		
Concrete from foundation, low quality:	2,427 m^3 × 2,400 kg/m^3=	5,825 t		
Lime stone from buildings (walls)	2,599 m^3 × 1,750 kg/m^3 =	4,548 t		
Asphalt from roads (tar-free/ containing = uncertain)	3,020 m^3 × 2,000 kg/m^3 =		6,040 t	
Masonry from buildings (walls)	17,741 m^3 × 1,800 kg/m^3 =	31,934 t		
Copper slag pavement	14 m^3 × 2,400 kg/m^3 =			34 t
Original road foundation	23,193 m^3 × 2,000 kg/m^3 =	46,386 t		
TOTAL	**106,960 m^3**	**±227,811**	**6,040 t**	**34 t**
To be built:				
Different kinds of concrete	166,667 m^3 × 2,400 kg/m^3	**400,000 t**		

6. Determination of distance from project to:

   - nearest quarry: ±170 km
   - concrete supplier: ±18 km

- waste collector: ±17 km
- home base of mobile plants: ±17 km

7. Development of logistic scenarios:

- scenario 0: "no" recycling and application of primary raw materials only
- scenario 1: "off site" recycling and maximum use of recycled materials on site
- scenario 2: "on site" recycling and maximum use of recycled materials on site

8. Choice of rolling stock (type and capacity); definition of assumptions concerning CO_2 emissions, diesel fuel consumption and share of used materials (diesel fuel and primary aggregates) in extraction of raw materials:

• capacity lorries:	30 t per load
• capacity concrete mixers:	10 t per load
• capacity crusher (type: impact crusher):	1,000 t per day
• CO_2 emissions (estimation):	880 g per km for lorries
	260 g per km for vans
• diesel fuel consumption (estimation):	43 l per 100 km for lorries
	12 l per 100 km for vans
• share diesel fuel in crude oil extraction:	16.79%
• share primary aggregates (sand and gravel) in excavation raw materials:	80%

9. Calculation of total kilometres for lorries and vans; calculation of CO_2 emission; calculation of total diesel fuel consumption and crude oil extraction; calculation of consumption of primary aggregates; calculation of total excavation of raw materials:

scenario 0:

transport of waste = 227,811 t/30 t/load × 2 (to and fro) × 17 km/load = 258,186 km

transport of primary materials = 100,000 t/30 t/load × 2 (to and fro) × 170 km/load = 1,133,333 km

transport of concrete = 300,000 t/10 t/load × 2 (to and fro) × 18 km/load = 1,080,000 km

transport of primary materials (concrete) = 200,000 t/30 t/load × 2 (to and fro) × 170 km/load = 2,266,667 km

CO_2 emission = 4,738,186 km × 880 g CO_2/km = **4,170 t** = 417 ha year tropical rain forest

crude oil extraction = (4,738,186 km × 43 l/100 km)/16.79% = 12,134,723 l or 12,135 m^3

excavation of primary resources for aggregates = 300,000 t/80% = 375,000 t or 187,500 m^3

total excavation of raw materials = 12,135 m^3 + 187,500 m^3 = 199,635 m^3

scenario 1:

transport of waste = 227,811 t/30 t/load × 2 (to and fro) × 17 km/load = 258,186 km
transport of secondary materials = 100,000 t/30 t/load × 2 (to and fro) × 17 km/load = 113,333 km
transport of concrete = 300,000 t/10 t/load × 2 (to and fro) × 18 km/load = 1,080,000 km
transport of secondary materials (concrete) = 100,000 t/30 t/load × 2 (to and fro) × 1 km/load = 6,667 km
transport of primary materials (concrete) = 100,000 t/30 t/load × 2 (to and fro) × 170 km/load = 1,133,333 km
CO_2 emission = 2,591,519 km × 880 g CO_2/km = **2,281 t** = 228 ha year tropical rain forest
crude oil extraction = (2,591,519 km × 43 l/100 km)/16.79% = 6,637,005 l or 6,637 m^3
excavation of primary resources for aggregates = 100,000 t/80% = 125,000 t or 62,500 m^3
total excavation of raw materials = 6,637 m^3 + 62,500 m^3 = 69,137 m^3

scenario 2:

transport of mobile crusher = 2 × 2 loads × 2 (before and after project) × 17 km/load = 136 km
transport of mobile concrete plant = 2 × 2 loads × 2 (before and after project) × 17 km/load = 136 km
personnel of mobile plants = 2 teams × 228 days × 17 km/day = 7,752 km
transport of secondary materials = 27,811 t/30 t/load × 2 (to and fro) × 17 km/load = 31,519 km
transport of primary materials (concrete) = 100,000 t/30 t/load × 2 (to and fro) × 170 km/load = 1,133,333 km
CO_2 emission = 1,165,124 km × 880 g CO_2/km + 7,752 km × 260 g CO_2/km = 1,027 t = 103 ha year tropical rain forest
crude oil extraction = (1,165,124 km × 43 l/100 km + 7,752 km × 12 l/100 km)/16.79% = 2,989,479 l or 2,989 m^3
excavation of primary resources for aggregates = 100,000 t/80% = 125,000 t or 62,500 m^3
total excavation of raw materials = 2,989 m^3 + 62,500 m^3 = 65,489 m^3

References

1. Delatte, N.J.: Lessons from Roman cement and concrete. J. Prof. Issues Eng. Educ. Pract. **127**(3), 109–115 (2001)
2. Lauritzen, E.K.: Emergency construction waste management. Saf. Sci. **30**, 45–53 (1998)

3. Vandecasteele, C., van der Sloot, H.: Sustainable management of waste and recycled materials in construction. Waste Manag. **31**, 199–200 (2011)
4. European Commission – Joint Research Centre – Institute for Environment and Sustainability: International Reference Life Cycle Data System (ILCD), Handbook – General guide for Life Cycle Assessment – Detailed guidance. 1st edn. March 2010 EUR 24708 EN, Publications Office of the European Union, Luxembourg (2010)
5. Guinee, J.B., Gorree, M., Heijungs, R., Huppes, G., Kleijn, R., Koning, A.: Handbook on Life Cycle Assessment. Kluwer Academic Publishers, Dordrecht (2004)
6. Sayagh, S., Ventura, A., Hoang, T., François, D., Jullien, A.: Sensitivity of the LCA allocation procedure for BFS recycled into pavement structures. Resour. Conserv. Recycl. **54**, 348–358 (2010)
7. Ekvall, T., Finnveden, G.: Allocation in ISO 14041–a critical review. J. Clean. Prod. **9**, 197–208 (2001)
8. Bala, A., Raugei, M., Benveniste, G., Gazulla, C., Fullana-I-Palmer, P.: Simplified tools for global warming potential evaluation: when "good enough" is best. Int. J. Life Cycle Assess. **15**, 489–498 (2010)
9. Fuller, S.K., Petersen, S.R.: Life-Cycle Costing Manual, 1995th edn. NIST, Gaithersburg (1996)
10. Norris, G.A.: Integrating economic analysis into LCA. Environ. Qual. Manag. **10**, 59–64 (2001)
11. UNEP, SETAC: Guidelines for Social Life Cycle Assessment of Products: Social and Socio-Economic LCA Guidelines Complementing Environmental LCA and Life Cycle Costing, Contributing to the Full Assessment of Goods. United Nations Environment Programme, Paris (2009)
12. Benoît-Norris, C., Vickery-Niederman, G., Valdivi, S., Franze, J., Travers, M., Ciroth, A., et al.: Introducing the UNEP/SETAC methodological sheets for subcategories of social LCA. Int. J. Life Cycle Assess. **16**, 682–690 (2011)
13. Hans, J., Chevalier, J.: Sustainable tools and methods for estimating building materials and components service life. In: CIB, C. (ed.) 10th International Conference on Durability of Building Materials and Components, Lyon (2005)
14. John, V. M.: Sustainable construction, innovation and durability: trends and research needs. In: Edições, F. (ed.) XII International Conference on Durability of Building Materials and Components, pp. 15–21. Porto (2011)
15. Adriaanse, A., et al.: Resource Flows: The Material Basis of Industrial Economies. World Resources Institute, Washington, DC (1997)
16. Moll, S., Bringezu, S., Schbütz, H.: Resource Use in European Countries. Wuppertal Institute for Climate, Environment and Energy, Wuppertal (2005)
17. Low, M.S.: Material flow analysis of concrete in the United States. Dissertation, Massachusetts Institute of Technology (2005)
18. Gardner, G.: Mind Over Matter: Recasting the Role of Materials in Our Lives. Worldwatch Institute, Washington, DC (1998)
19. Cederberg, C., Persson, U.M., Neovius, K., Molander, S., Clift, R.: Including carbon emissions from deforestation in the carbon footprint of Brazilian beef. Environ. Sci. Technol. **45**, 1773–1779 (2011)
20. Pereira, D., Santos, D., Vedoveto, M., Guimarães, J.: Fatos Florestais da Amazônia 2010. Imazon, Belém (2010)
21. John, V.M.: Reciclagem De Resíduosna Construção Civil: Contribuição à metodologia de pesquisa e desenvolvimento. Dissertation, University of São Paulo (2000)
22. Matthews, E., et al.: The Weight of Nations: Material Outflows from Industrial Economies. World Resources Institute, Washington, DC (2000)
23. Müller, N., Harnisch, J.: A Blueprint for a Climate Friendly Cement Industry. Ecofys Germany GmbH on behalf of WWF International, Nürnberg (2008)
24. Brown III, W. M., Matos, G. R., Sullivan, D. E.: Materials and Energy Flows in the Earth Science Century – A Summary of a Workshop Held by the USGS in November 1998. Circular 1194. U.S. Geological Survey, Denver (2000)

25. Bergsdal, H., Bohne, R.A., Brattebo, H.: Projection of construction and demolition waste in Norway. J. Ind. Ecol. **11**, 27–39 (2008)
26. Hu, M., van der Voet, E., Huppes, G.: Dynamic material flow analysis for strategic construction and demolition waste management in Beijing. J. Ind. Ecol. **14**, 440–456 (2010)
27. Danielsen, S.W.: A Best Available Concept for the Production and Use of Aggregates, in Eco-Serve Seminar. EC Thematic Network European Construction in Service of Society, Warsaw (2006)
28. SNIC: Relatório Anual 2009. Sindicato Nacional da Indústria do Cimento -SNIC, Rio de Janeiro (2010)
29. The concrete centre -TCC, Mineral Products Association-MPA: Concrete Industry Sustainability Performance Report. Based on 2009 Production Sustainable Concrete Forum. The Concrete Centre -TCC, Surrey (2010)
30. Formoso, C.T., Soibelman, L., de Cesare, C., Isa, E.L.: Material waste in building industry: main causes and prevention. J. Constr. Eng. Manag. **128**, 316 (2002)
31. Bossink, B.A.G., Brouwers, H.J.H.: Construction waste: quantification and source evaluation. J. Constr. Eng. Manag. **122**, 55 (1996)
32. Souza, U.E.L., Andrade, A.C., Paliari, J.C., Agopyan, V.: Material waste in the ceramic tiling job. In: 10th Symposium Construction Innovation and Global Competitiveness 10th International Symposium, pp. 536–545. Boca Raton, FL (2003)
33. Jaillon, L., Poon, C.S., Chiang, Y.H.: Quantifying the waste reduction potential of using prefabrication in building construction in Hong Kong. Waste Manag. **29**, 309–320 (2009)
34. Damineli, B.L., Kemeid, F.M., Aguiar, P.S., John, V.M.: Measuring the eco-efficiency of cement use. Cem. Concr. Compos. **32**, 555–562 (2010)
35. Schneider, F., Enste, D.H.: Shadow economies: size, causes, and consequences. J. Econ. Lit. **XXXVIII**, 77–114 (2000)
36. Schneider, F., Enste, D.: Shadow Economies Around the World – Size, Causes, and Consequences. Working paper WP/00/26, International Monetary Fund (2000)
37. Danopoulos, C.P., Znidaric, B.: Informal economy, tax evasion, and poverty in a democratic setting: Greece. Mediterr. Q.: A J. Glob. Issues **18**, 67–84 (2007)
38. Gërxhani, K.: The informal sector in developed and less developed countries: a literature survey. Public Choice **120**, 267–300 (2004)
39. FGV Projetos: Tributação na indústria brasileira de materiais de construção, São Paulo (2006)
40. Dada, J.O.: Harnessing the Potentials of the Informal Sector for Sustainable Development – 850 Lessons from Nigeria. Administrative Staff College of Nigeria (ASCON), Badagry (2007)
41. Bolen, W.: Sand and gravel (construction). U.S. Geological Survey, Mineral Commodity Summaries January 2012, 136–137 (2012)
42. Willett, J.: Stone (crushed). U.S. Geological Survey, Mineral Commodity Summaries January 2012, 152–153 (2012)
43. Neubauer, C., et al.: Aggregates Case Study. Final Report. European Commission's Joint Research Centre/Umweltbundesamt, Vienna (2008)
44. Montgomery, M.R.: The urban transformation of the developing world. Science **319**, 761–764 (2008)
45. Jiang, L., Young, M.H., Hardee, K.: Population, urbanization, and the environment. World Watch **21**(5), 34–39 (2008)
46. Schneider, D.M.: Deposições irregulares de resíduos da construção civil na cidade de São Paulo. Dissertation, Universidade de São Paulo (2002)
47. Bibby, A.: Green jobs in construction: small changes – big effect. World Work. Maga. **70**, 38–41 (2010)
48. Coelho, A., de Brito, J.: Influence of construction and demolition waste management on the environmental impact of buildings. Waste Manag. **32**, 532–541 (2011)
49. Marinković, S., Radonjanin, V., Malešev, M., Ignja, I.: Comparative environmental assessment of natural and recycled aggregate concrete. Waste Manag. **30**, 2255–2264 (2010)

50. Chowdhury, R., Apul, D., Fry, T.: A life cycle based environmental impacts assessment of construction materials used in road construction. Resour. Conserv. Recycl. **54**, 250–255 (2010)
51. Chen, M., Lin, J., Wu, S.: Potential of recycled fine aggregates powder as filler in asphalt mixture. Constr. Build Mater. **25**, 3909–3914 (2011)
52. Fry, C., Wayman, M.: Sustainable Aggregates. Theme 1-Reducing the Environmental Effect. ALSF (2007)
53. Angulo, S.C., Carrijo, P.M., Figueiredo, A.D., Chaves, A.P., John, V.M.: On the classification of mixed construction and demolition waste aggregate by porosity and its impact on the mechanical performance of concrete. Mater. Struct. **43**, 519–528 (2010)
54. Levy, S.M., Helene, P.: Durability of recycled aggregates concrete: a safe way to sustainable 878 development. Cem. Concr. Res. **34**, 1975–1980 (2004)
55. Mroueh, U.M., Eskola, P., Laine-Ylijoki, J.: Life-cycle impacts of the use of industrial by-products in road and earth construction. Waste Manag. **21**, 271–277 (2001)
56. Environmental Protection Agency, and demolition waste landfills.
57. Engelsen, C.J., van der Sloot, H.A., Wibetoe, G., Justnes, H., Lund, W., Stoltenberg, E.: Leaching characterisation and geochemical modelling of minor and trace elements released from recycled concrete aggregates. Cem. Concr. Res. **40**, 1639–1649 (2010)
58. Galvín, A.P., Ayuso, J., Jiménez, J.R., Agrela, F.: Comparison of batch leaching tests and influence of pH on the release of metals from construction and demolition wastes. Waste Manag. **32**, 88–95 (2012)
59. Mulugeta, M., Engelsen, C.J., Wibetoe, G., Lund, W.: Charge-based fractionation of oxyanion-forming metals and metalloids leached from recycled concrete aggregates of different degrees of carbonation: a comparison of laboratory and field leaching tests. Waste Manag. **31**, 253–258 (2011)
60. Jang, Y., Townsend, T.: Sulfate leaching from recovered construction and demolition debris fines. Adv. Environ. Res. **5**, 203–217 (2001)
61. Sani, D., Moriconi, G., Fava, G., Corinaldesi, V.: Leaching and mechanical behaviour of concrete manufactured with recycled aggregates. Waste Manag. **25**, 177–182 (2005)
62. Hendriks, ChF., Janssen, GMT., Xing, WH., Dobbelsteen, AAJF van den.: A new Vision on the Building Cylce. Aeneas, Technical Publishers, Boxtel (2004)
63. In: NATO. Available at http://www.nato.int
64. In: Rotterdam City Council & Port of Rotterdam Authority. Available at http://www.cityports-rotterdam.com
65. In: FOD Economie & BPF : 'De aardoliemarkt', FOD Economie, KMO, Middenstand en Energie. Available at http://economie.fgov.be/energy/balance_sheets/2002/evol2002_nl-08.htm#Raffinage (2007). Accessed 2007
66. Mueller, A.: Closed loop of concrete rubble?. Available at www.uni-weimar.de/Bauing/aufber/Lehre (2007). Accessed 2007

Chapter 3
Overview Regarding Construction and Demolition Waste in Several Countries

Abstract The objective of this chapter is to collect the data of advancements, achievements, problems and solutions most representative of the various countries. Thereby, each country can learn from examples and underpin a pedagogy for the effective implementation of recycling. A short review of the history of recycling of construction and demolition waste is presented. Market situation, applications, standards and characteristics of the production in countries are described.

Keywords Recycling • Aggregates • Concrete • Roads • Research • Processing • Plants • Composition • Problems

3.1 Recycling in Belgium: An Overview of the Present Situation in Belgium from the Recycling Industry and Research About CDW

Jeroen Vrijders and Jan Desmyter

3.1.1 Current Market Situation

Anno 2011, the Belgian recycling sector can be considered 'grown up', with well developed and certified installations. It comprises over 150 fixed recycling locations (crusher plants, sorting plants, etc) and 50 mobile crushers in Flanders and about 40 recycling centers active in the Walloon region. Several of the companies also operate a mixing installation for lean concrete or structural concrete.

J. Vrijders (✉) • J. Desmyter
Belgian Building Research Institute, Brussels, Belgium
e-mail: jan.desmyter@bbri.be

E. Vázquez (ed.), *Progress of Recycling in the Built Environment: Final Report of the RILEM Technical Committee 217-PRE*, RILEM State-of-the-Art Reports 8,
DOI 10.1007/978-94-007-4908-5_3,

On annual basis, about 11 million tons of recycled aggregates are produced in Flanders, and about 3.5 million tons in the Walloon region, resulting in about 14.5 million tons a year (including Brussels).

It can be estimated that 90% of the recycled aggregates are used as sub base and base layers in road construction. The other 10% is used in road-like applications on construction sites and about 100,000–200,000 tons is nowadays used as aggregate for structural concrete in the private market. This means that less than 1% of the recycled aggregate is used in high-grade applications.

Two certification bodies are active in the market, COPRO and CERTIPRO. In order to be recognised as a secondary raw material, specific environmental hygiene legislation has been established in the 1990s by the regional authorities. In the Flemish region, before a recycled aggregate is allowed to the market, it has to be certified on its environmental qualities for specific technical applications.

3.1.2 Standards and Recommendations for Application

3.1.2.1 Aggregates – Standards

The European set of standards for aggregates is used.

NBN EN 13242 – Aggregates for unbound and hydraulically bound materials for use in civil engineering work and road construction

NBN EN 13043 – Aggregates for bituminous mixtures and surface treatments for roads, airfields and other trafficked areas

NBN EN 13285 – Unbound mixtures

NBN EN 14227 series – Hydraulically bound mixtures

PTV 406 – Codification of recycled aggregates, based on the EN standards for aggregates, is also used as national reference. No other Belgian addenda to the EN standards exist.

3.1.2.2 Road Construction

General technical prescriptions exist in Flanders, Brussels and the Walloon region, and can be considered the main reason for the uptake of recycled aggregates in practice: RW99-2004 (version 2009 in preparation) for the Walloon Region and SB250 v2.2 (Standaardbestek 250 voor de wegenbouw) in Flanders (newest version 2010) form the basis for defining specific requirements for road construction or renovation sites.

For illustration, the applications for recycled aggregates in SB250 are mentioned in Tables 3.1 and 3.2.

Table 3.1 Applications for recycled aggregates in SB250-Recycled sand

a. Recycled sand	Technical prescriptions	Sieving sand	Crusher sand	Concrete crusher sand
Backfilling and embankments	Environmental quality (for all applications)	X	X	X
Sand for draining	Grading, MB_F, glauconite		X	X
Sand for subbases	Grading, f_{16}, MB_F,		X	X
Sand for lean concrete for road bases and bases for buildings and structures	SC (calcareous materials), CC (Cl^--ions), Grading, f_{10}, MB_F		X	X
Sand for sand-cement mixes	f_{22}		X	X
Sand for bituminous mixtures	Grading, fines (f and MB), PSV		X	X
Sand for draining foundations in sand-cement-mixtures	Filler <3%, fineness modulus CF		X	X
Sand for concrete for roads	CA (Cl^--ions), Grading, f_3, sand equivalent a, PSV			X
Sand to fill aggregate bases	Grading, methylene blue			X
Sand for pavement layers	Grading, methylene blue			X
Sand for joint filling	Grading and fines			X
Sand for concrete for buildings and structures	Grading, fines, ...			

3.1.2.3 Concrete

The European standards are used, with EN 206-1 as basis. This standard is completed by the Belgian standard NBN B15-001. At the moment (august 2011), there is a pre-standard prNBN B15-001:2011 that contains specifications for the use of recycled aggregates:

On aggregate level (§5.1.3)
The general suitability is demonstrated for recycled aggregates fullfulling EN 12620, with additional requirements:

- $D \geq 10$ mm, $d \geq 4$ mm
- Rc90/Rcu95/Ra1-/ XRg0.5-/FL2- (EN 12620)
- FI20, f1,5, LA40, SS0,2, A10 van (EN 12620)
- $\rho_{rd} \geq 2{,}200$ kg/m;
- water absorption is max. 10%, with a variance of max. 2% on declared value

Remark: for the requirements of (internally produced) aggregates for concrete for prefabricated products, reference is made to NBN EN 13369 and NBN B21-600

Use of the recycled aggregate (§5.2.3.5)
The recycled concrete aggregate can be used in exposure classes X0 and XC1 and (Belgian) environmental classes E0 and EI. The strength class is limited to C25/30. For use in other environments and strength classes, the suitability has to be demonstrated.

Recycled aggregate can be added to a max. of 20%v/v to the total of coarse aggregates.

No other recommendations, certification schemes, technical prescriptions, etc, exist at the moment in Belgium (Table 3.2).

3.1.2.4 Hydraulically Bound Applications

Based upon the NBN EN 14227-x series of standards, recycled aggregates are also used in hydraulically bound applications. A specific certification document, known as TRA21, has been established by COPRO, one of the two certification bodies active in the area of recycled aggregates. Current application of this document is still limited, although the importance may grow in the near future due to the uptake in technical specification documents of the regional authorities.

3.1.3 Principal Problems Related to Standards and Applications

3.1.3.1 Aggregate Level

Lack of experience with EN 933-11 – Classification test for the constituents of coarse recycled aggregate + overlap/interference in requirements set by (environmental) authorities and administrations.

3.1.3.2 Road Construction

Use of recycled aggregates in the road construction is mainly based upon the normative documents and the usual technical specifications of the regional authorities. These specifications are informative, but serve as a basis for defining the specific requirements for most road construction sites.

3.1.3.3 Concrete

Some of the requirements on aggregate level require testing methods that are not familiar to the executing laboratories. This also implies the lack of comparable and reliable threshold values.

- Determination of water-soluble sulphates – EN 1744-1
- Determination of the influence of recycled aggregate extract on the initial setting time of cement – EN 1744-6

Table 3.2 Applications for recycled aggregates in SB250-coarse aggregates

b. Coarse aggregates	Backfill and embankments	Sub base	Not-continuous base	Continuous base	Lean concrete for foundations	Bituminous mixtures	Concrete for road layers and concrete elements	Concrete for buildings and structures
Requirements:		Grading, f_4	f_4, FI_{35}, $C_{50/10}$, LA_{35}	f_4, FI_{35}, $C_{50/10}$, LA_{40} of LA_{50}	f_2, LA_{40}		Natural aggregates, $G_C85/20$, f_4 (2/4 and 4/6,3) and $f_{1,5}$ (other), FI_{20}, LA_{15} or LA_{35} (small roads), PSV_{50}	Natural aggregates
Masonry aggregate	X	X						
Mixed aggregate	X	X		X (bound)	X			
Concrete aggregate	X	X	X	X	X		20% in concrete allowed, on condition that originin (eg. demolition of road) is well known and technical requirements are ok.	
Asphalt-concrete mixed aggregate	X	X						
Asphalt aggregate without tar		X (max. 30%)		X[a]		X[b]		
Asphalt aggregate with tar			[c]					

[a] Treated with cement and addition of 15–20% sand for grading improvement
[b] Max. 50% of total mass of binder (old + new) comes from recycled aggregate
[c] Asphalt aggregate cement foundation

The draft standard prNBN B15-001 is the first standard in Belgium allowing for recycled aggregates.

- The domain of application of recycled aggregate in concrete is limited (first –careful- step). It is possible to go further, when suitability is demonstrated. However, no real manual to demonstrate this suitability exists.
- There is little experience or documentation on quality certification of this type of concrete, while quality certification of concrete is required a lot in the Belgian market.

3.1.4 New Applications and Research

VALRECON20 is a research project, operated by KHBO, Lessius Mechelen and the BBRI to investigate the possibilities to use 100% recycled coarse aggregate in 'normal concrete' (C20/25, C25/30; EE2). Strength evaluation shows good results, durability analysis is underway.

The University of Ghent is working on a new type of concrete, that can be recycled as resource for cement production.

3.2 Recycling in Brazil: An Overview of the Present Situation of CDW in Managing and Recycling

Vanderley John and Sergio Angulo

3.2.1 CDW Generation and Composition

After 1970, Brazil experienced an intense urbanization process, especially concentrated and the surrounding of southern megacities like city of Sao Paulo and Rio de Janeiro. This scenario has changed in the last years with urbanizations of middle-size cities in other Brazilian states (Minas Gerais, Goias, etc). In terms of construction and demolition Waste (CDW) generation, construction sector increases its relative importance for middle and large-size cities Table 3.3.

Construction sector mostly acts like an economic formal sector in small-size and middle-size cities; therefore, official database of constructed area from building

V. John (✉)
Lab. Microestrutura e Ecoeficiencia de Materiais, Escola Politécnica da USP, Sao Paulo, Brazil

S. Angulo
Instituto de Pesquisas Tecnologicas do Estado de Sao Paulo, Sao Paulo, Brazil

Table 3.3 CDW generators and CDW per capita generation for Brazilian cities (MC 2005; IPT 2010)

Cities	Population (10^3 inhab)	Origin of CDW (%) Construction sector	Renovation or demolition sector	CDW per capita generation (kg/hab.ano)
Novo Horizonte (SP)	36	27	73	367
Vitoria da Conquista (Bahia state)	262	18	81	400
Piracicaba (SP)	329	33	67	590
Uberlandia (MG)	501	38	62	680
Santo Andre (SP)	649	47	53	510
Guarulhos (SP)	1.073	44	56	380

Fig. 3.1 Self-construction in city of Sao Paulo http://www.cobrape.com.br

companies has been successfully applied to indirect prediction of CDW generation at building sites (kg/m^2 constructed). Only in large-size cities and megacities, self-construction practices become more common (Fig. 3.1), increasing legalization requests of informal constructed areas in surrounding large-distance urban areas and slums in geological risk areas where environmental accidents are not rare in daily magazines.

A quite different scenario has been found from renovation and demolition sector. **Renovation is mostly done informally by population and especially important for small-size cities.** Small-size cities represent around 90% of municipalities in Brazil. In all cities, **the formal demolition companies are more dedicated for**

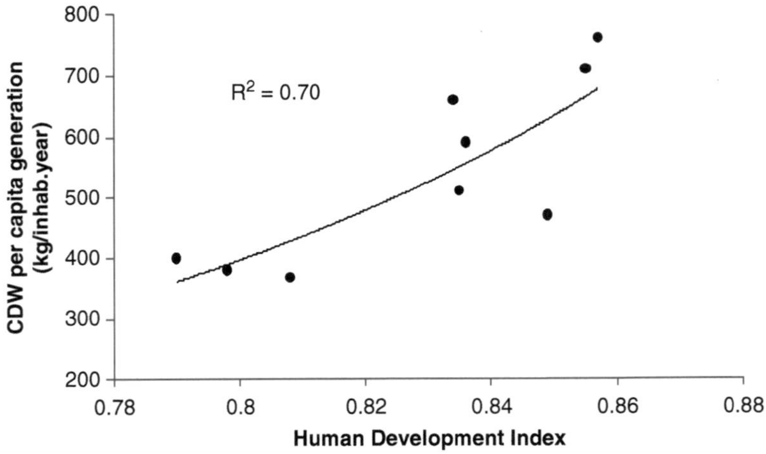

Fig. 3.2 Human Development Index (*HDI*) × CDW per capita generation (kg/inhab.year) (Data from MC (2005) and IPT (2010))

industrial than residential areas and do not exist in small-size cities. Official database of demolished area is very low and all indirect estimations using waste generation indicator (kg/m^2 demolished) are imprecise.

CDW generation predictions for each state or all states are quite difficult. Available data are most concentrated in Sao Paulo state yet. For this state, some predictions of CDW per capita generations have been done by use of Human Development Index (Fig. 3.2), somehow independently of the population density [1].

Analyses of CDW composition of coarse fractions (25.4–4.8 mm) of arriving containers in CDW sorting or recycling sites have shown that **heterogeneity is the major problem for implementing recycling practice**s. First, even in municipalities where CDW managing is implemented, **decontamination is the most important activity**, especially because higher contents of glazed tiles (from renovation sites), of gypsum (from commercial offices), cement asbestos (from self-construction and industrial sites) and intermixed soils (Fig. 3.3).

Twenty-five percent of arriving containers has intermixed soil in CDW, probably due to the fact that this material is classified in the same group of concrete and masonry wastes in accordance with proposed Brazilian management for CDW [2]

The absence of high efficiency sorting practices and technologies will certain affect CDW recycling potential in Brazil. This characterization study and other two studies conducted at IPT demonstrated that, when no sorting practices are implemented, the average contents of gypsum (1.9%) and glazed tiles (2.8%) are higher than the value recommended by Brazilian standard for CDW recycled aggregates [3] becoming impossible recycling practices.

Second, CDW in Brazil is essentially mixed. Arriving containers of pure concrete waste is scarce (18–44% of loads), since concrete pavements are not common and most of reinforced concrete buildings have masonry as wall partitioning, which are not currently sorted by dismantling procedures. Further, the red ceramic

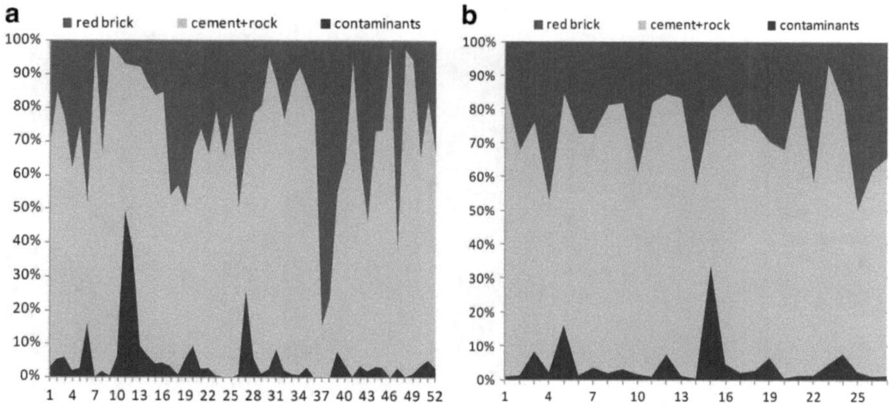

Fig. 3.3 Compositions of CDW in Brazil: (**a**) Macaé-RJ, and (**b**) Maceió-AL

contents are variable for each arriving container in CDW sorting or recycling site increasing porosity becoming recycled aggregates more friable and weaken and, then, downcycling the use (subbase with restricted transportation load, leveling mortar or scrid floor).

3.2.2 CDW Management and Recycling

The first directive to CDW management was approved by National Council for the Environment in 2002 to be totally implemented at 2004 [2]: resolution 307 and 348. The main points are:

Segregation of CDW in four classes: a.1 Class A, composed of concrete and masonry waste; a.2 Class B, mainly steel reinforcement, wood, plastics, glass, and bitumen; a.3 Class C, gypsum and a.4, hazardous wastes (cement lamps and paints with heavy metals and cement asbestos).

Implementation and management of small CDW public collection areas (Fig. 3.4): to receive waste from generators of informal renovations and self-constructions with intention of avoiding illegal dumping at streets.

Obligation of CDW sorting at construction building sites or CDW private recycling network use: to force legal construction sector in reuse and recycling practices at local or municipality level (legal framework of transportation companies, private CDW sorting and/or recycling areas) by use of a public controlled document.

Disposal of reuse or recyclable CDW is prohibited at sanitary landfills: CDW recycled aggregates uses in pavement (Table 3.4: Technical criteria applied to CDW recycled aggregates for pavement [4] Table 3.4) and concrete became technically supported by National standards [3], as well as CDW earthwork fillings (Table 3.5) [4].

Fig. 3.4 Small CDW public collection area in city of Sao Paulo – www.itaimpaulista.com.br – (**a**) and wood materials sorted in city of Guarulhos (**b**). Both are in state of Sao Paulo

Table 3.4 Technical criteria applied to CDW recycled aggregates for pavement [4]

		CDW recycled aggregates		Brazilian standards	
Properties		Coarse	Fine	Coarse	Fine
Particle size distribution		Coefficient of uniformity Cu >10		NBR 7181	
D max		$\leq$63 mm		NBR NM 248	
Shape index		$\leq$3	–	NBR 7809	
Mass passing at sieve 0.42 mm		10–40%		NBR 7181	
Contaminant contents (%)	Material[a]	2		Appendix A	Appendix B
	Sum of materials[a]	3		Appendix A	Appendix B
Application		ISC %	Expansibility %	Compaction energy	
Ground stabilization		$\geq$12	$\leq$1.0	Normal	
Sub-base		$\geq$ 20	$\leq$1.0	Intermediate	
Base[b]		$\geq$60	$\leq$0.5	Intermediate or modified	

[a]Wood, plastics, madeira, refinforcement steel, gypsum, glass, glazed tiles, bitumen
[b]Traffic conditions: N $\leq$106 repetitions with standard axis of 8.2 tf (80 kN)

Recently, a national policy for solid waste management (Brazil 2010) was approved that will probably help in: (a) implementation of legal framework to obligate gypsum and paint waste collection at building and renovation sites by producers; and (b) implementation of regional CD waste management scenarios involving a group of small-size municipalities (Fig. 3.5).

In fact, **only medium-size and large-size cities** (around 50 of 5,565 municipalities) **has implemented national CDW management directives** [5], especially the highest DHI municipalities.

Since renovation or self-construction sites done by informal generators are an important source of CDW, society did not respond in bringing waste for public areas as well as municipalities in the last years could not offer so many public areas to collect the amount of waste generated.

Table 3.5 Some technical and environmental criteria applied to CDW inert landfills [4]

Phase of	Criteria
Project	Area must be located in order to minimize geological and hydrological impacts and has drainage system
	Leaching monitoring in the surroundings are not necessary if volume of CDW landfill is lower than 10,000 m^3
Operation	The filling must be done in layers (never at the top) and landfill stability must be monitored
	Trackability of sorted CD waste, intermixed with soils or not, classified as Class A (only concrete and masonry waste) and dumped at inert CDW landfill by official documents
	CD waste inert landfill leaching monitoring

Fig. 3.5 CDW private recycling plant operating in America, 300 Km far from city of Sao Paulo www.solosreciclagem.com.br

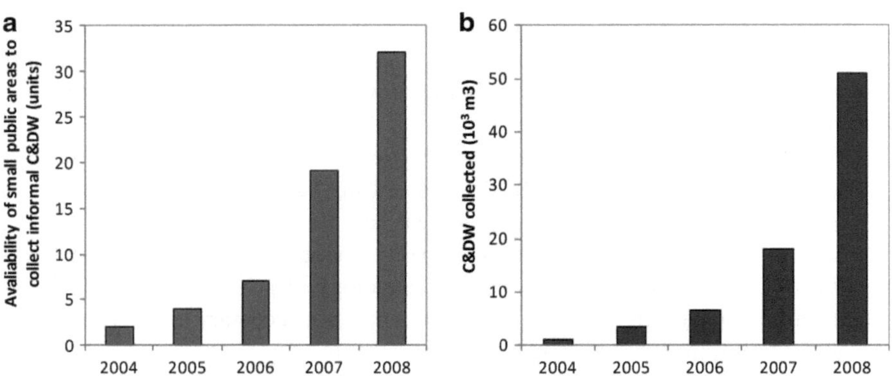

Fig. 3.6 Availability of small public areas to collect CDW (**a**) and CDW volume collect (**b**) in city of Sao Paulo (Data from http://www.prefeitura.sp.gov.br)

Figure 3.6 shows that the public area offer significantly increased in the last 4 years in city of Sao Paulo, but the amount of collected waste had slowly grown. Projections to 2020 indicates that city of Sao Paulo will be able to collect in small

Fig. 3.7 Low cost sorting-sieving CDW recycling unit for low-size municipalities available at IPT

public areas only 30% of CDW generated today. **Dematerization waste practices have to be implemented in future**.

Besides, **CDW sorting at building sites have also achieved partial benefits** [6]. Sorted wood and steel can be recycled, but it represents not more than 10% of CDW generated. Almost 90% of concrete and masonry waste are taking out of building sites without recycling strategies inside, showing clearly the difficult in sustainability comprehension. **Wastage reduction is the most important strategies; however, yet neglected in most cases.**

Marques [7] has pointed that the main barrier to implement CD waste management and recycling in around 5,000 small-size Brazilian municipalities is financial resource and trained human resources to deal with the topic.

An USP, CETEM and Macae municipality cooperation research project was conducted in 2005 about CDW recycling technology. Analyses of Brazilian CDW characteristics have demonstrated that crusher may not necessary in order to recycle for ground reinforcement and subbase of pavements.

A simple mobile sieving system combined with large conveyor belts for hand sorting (Fig. 3.7) implemented by IPT is able to produce 60% (kg/kg) of high purity CDW recycled aggregates (<50 mm) for pavement use and 40% (kg/kg) of geotechnical stabilization materials (>50 mm). A mixture with such aggregates and local soils seem to be a good solution for local soil stabilized roads in the countryside of state of Sao Paulo (IPT 2011), drastically affected by rain during spring.

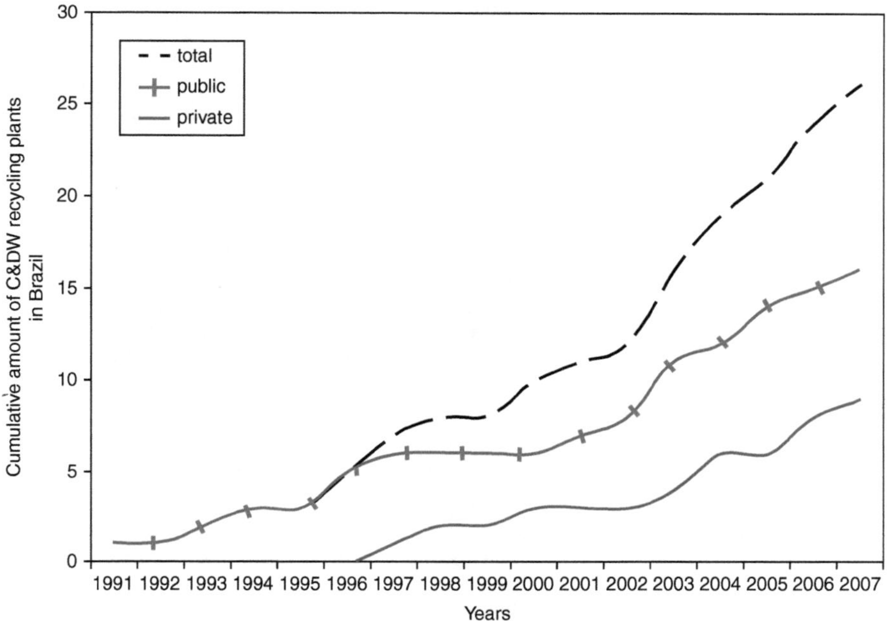

Fig. 3.8 Cumulative amount of CDW recycling plants in Brazil [8]

This low cost technological solution may help to disseminate CDW recycling plants to small-size municipalities in Brazil.

CDW recycling started on 1990s operated by stationary and very simple public owned CDW recycling plants (Fig. 3.8). In the last years CDW recycling is becoming more and more a private business. Evidence is also the **foundation of National Association of CDW recycling companies** (www.abrecon.org.br).

Despite there are public or private efforts to manage and recycle CDW in Brazil, plants until 2007 can recycle lower than 5% of CDW generated. Crusher mobile plants and modern demolition tools (pulverizer and crusher buckets[1]) are being currently offered by international producers and **private demolition sector start to offer recycling integrated solutions for demolition/construction building sites focus on CD recycled aggregates in pavements**.

This dissemination will certainly improve national CDW recycling rates. **The pressure for CDW recycled aggregates use for many geotechnical applications, mortar and concrete will increase**, bringing technology approach to produce high quality CDW recycled aggregates in future (Fig. 3.9).

The large use of CDW recycled aggregates in structural concrete requires technological improvement. It is not sustainable use 100% of low quality recycled aggregates due to high cement demand to achieve a similar mechanical performance

[1] www.mbcrusher.com

Fig. 3.9 Integrated demolition/construction site in Sao Paulo (www.desmotec.com)

of concrete made with natural aggregates [9]. Substitution of natural aggregates by 10–20% low quality recycled aggregates seems to be not interesting for recycling market despite technically feasible.

3.3 Recycling in North America: State-of-the Art of RCA-Concrete in North America

Ghani Razaqpur and Gholamreza Fathifazl

3.3.1 Introduction

In North America recycled concrete aggregate (RCA) is primarily used in non-structural applications, such as backfill material, base or fill for drainage systems, pavement base and sub-base, lean-concrete bases, concrete blocks, highway sound barrier walls, partition walls, embankments, etc. The concrete industry perceives

G. Razaqpur (✉)
Department of Civil Engineering, McMaster University, Hamilton, ON, Canada

G. Fathifazl
Adjeleian Allen Rubeli Consulting Structural Engineers, Ottawa, ON, Canada

a number of fundamental barriers to the use of recycled concrete aggregates in structural grade concrete, including lack of technical data, insufficient specifications with respect to quality control in the processing of RCA and in the production of concrete mixes made with RCA, and the belief that concrete made with RCA is inferior to conventional concrete.

From the regulatory point of view, there is no impediment to the use of RCA in structural concrete. Recycling and reusing demolished concrete as substitute for virgin aggregate has been recognized in North America for nearly 30 years. Since 1982, ASTM C33 [10] has included crushed hydraulic cement concrete in its definition of coarse aggregate while ASTM 125 [11] has permitted crushed hydraulic-cement concrete as manufactured sand; therefore, in principle there is no regulatory barrier to the use of recycled concrete in the United States.

The Federal Highway Administration reported [12] that 41 of the 50 states in the US recycle old concrete into aggregate. In 38 of these states RCA is used primarily for road base/sub-base while in the remaining states it is also used in new concrete. Based on a report of the US Geological Survey, Li et al. [13] reported that 100 million tons of RCA was produced in the US in 2000, 68% of which was used as road base material, 14% as rip rap and other fill, 9% as aggregate in asphalt concrete, 6% in Portland cement concrete, and 7% in other applications.

In Canada, the C-2000 Green Building Standards have set the goal of making recyclable materials up to 75% of the materials used in the construction of new buildings and structures [14], but other than stating that concrete be separated from other demolition waste [15], they do not provide specific guidance about either the processing or the proper use of these materials. According to the Cement Association of Canada, the use of recycled concrete was included in the Ontario Provincial Standard Specification (OPSS) in 1993, which specifies that recycled aggregates may be used as road or parking lot base and sub-base as well as sewer bedding and backfill [16]. According to a report by the Ministry of Natural Resources of Ontario [17], 184 million tons of aggregate was consumed in Ontario in 2007, which included 13 million tons of recycled aggregate, with practically all the recycled aggregate used in non-structural applications. It was also reported that municipalities in Ontario have no specific polices about either recycling concrete or using RCA, but they avoid using recycled materials in general due to either unfavorable past experience, or lack of any experience, or due to the tendency to use high-performance materials, which prohibit use of recycled materials.

In our opinion, the reasons for minimal use of RCA in structural concrete in North America are its already extensive use as base, sub-base and backfill material, lack of evidence to show its economic benefits when used in structural concrete versus the existing applications, the lack of technical specifications and guidelines for quality control and production of RCA and RCA-concrete, the findings of many research works which have consistently reported that RCA-concrete has higher creep and shrinkage, and lower elastic modulus than conventional concrete with similar composition. To overcome the perceived problem of the inferiority of RCA-concrete and to promote its use as structural concrete, more recently more research has been undertaken in various institutions in North America as described in the following section.

3.3.2 Development of a New Mixture Proportioning Method for RCA-Concrete

One of the difficulties with achieving consistent properties for RCA-concrete is lack of a suitable mix design method. Although ACI-555R [18] provides some guidelines for proportioning of concrete mixes containing RCA, neither it nor any other standard/guideline gives a mix design method specifically for RCA-concrete with specified fresh and hardened properties. According to Fathifazl et al. [19], RCA-concrete is traditionally proportioned by applying conventional natural aggregate concrete (NAC) mix proportioning methods, treating RCA akin to natural aggregate (NA). In these methods, RCA is treated as a homogeneous single phase aggregate that can be used to partially or totally replace the NA in a reference mix. In reality, they state, at macro level RCA consists of at least two phases, viz. NA and residual mortar; therefore, a simple replacement of NA by RCA increases the proportion of overall mortar (residual plus fresh mortar) and decreases the proportion of natural aggregate in the RCA-concrete mix compared to the reference mix. They postulated that this difference in the composition of the two types of mixes is generally responsible for the inferior properties of RCA-concrete.

To address the problem, Fathifazl et al. proposed, and experimentally validated, a new mix proportioning method for RCA-concrete mixes and named the method EMV (Equivalent Mortar Volume). In the development of the EMV method, they examined the fundamental principles that govern the mechanical and physical properties of composite materials. In this method RCA is treated as a two-phase material comprising mortar and NA; and the residual mortar in RCA is considered as part of the total mortar (fresh plus residual mortar) in the mix. The method requires less fresh mortar in a RCA-concrete mix compared to the reference mix; therefore, it also requires proportionally less cement. The smaller cement requirement makes RCA-concrete environmentally friendly and possibly more economical.

In order to proportion RCA-concrete by the EMV method, the residual mortar content (RMC) of RCA needs to be quantified. This is crucial to the application of the EMV method. An experimental method was proposed by Abbas et al. [20] to conveniently determine RMC. The quick laboratory method utilizes a combination of mechanical and chemical stresses that disintegrate the residual mortar and breaks up the bond between the mortar and the coarse natural aggregate in RCA. The RMC obtained by this procedure was validated by image analysis techniques [21]. The EMV method allows partial or total replacement of coarse virgin aggregate by RCA, but depending on the RMC content, it may not permit total replacement. In the study by Fathifazl et al. the two types of RCA used had RMC of 41 and 23%, which resulted in mixes with RCA content, as percentage of the total coarse aggregate volume, equal to 63.5 and 74.3%, respectively.

Their investigation into the hardened concrete properties, including, elastic modulus, compressive strength, creep and shrinkage [22], revealed that RCA-concrete mixes proportioned by the EMV method had comparable properties to conventional concrete made with virgin aggregate, but the companion mixes proportioned by the conventional method had inferior properties. Also, in general, the mixes proportioned by the EMV method were found to be more ductile compared to the mixes, with or without RCA, proportioned by the conventional method.

A study by Banthia and Chen [23] involved the use of RCA in shotcrete and involving both the wet and dry processes. The mixes produced by the wet process also contained short steel fibres. The major conclusion of the study was that RCA can be used in shotcrete produced by the wet process, but it will have lower compressive and tensile strength than companion shotcrete made with virgin aggregate. It was also reported that shotcrete made with RCA is tougher and less brittle than that made with natural aggregate, but the addition of steel fibres to the shotcrete made with RCA lowered its toughness and ductility compared to the one made with virgin aggregate.

3.3.3 Structural Behaviour and Strength of RCA-Concrete Members

The flexural performance of reinforced RCA-concrete beams was investigated by Fathifazl et al. [24] and included the study of the following parameters: tensile reinforcement ratios, compression steel ratio, RCA source, and concrete type (with/without RCA). The minimum and maximum reinforcement ratios were calculated according to the Canadian CSA A23.3-04 [25] provisions for conventional concrete, and the reinforcement ratio varied from 0.493 to 3.310%. Based on the results of their investigation, the authors found for beams under flexure no major difference between the failure modes of reinforced RCA-concrete and the control beams made of NAC. For all the test beams, irrespective of the type of concrete, the maximum deflection corresponding to service load level was found to be within the permissible limits specified by ACI 318M-05 [26]. However, reinforced RCA-concrete beams displayed slightly smaller crack spacing but there was no major difference between the cracking patterns of the reinforced RCA-concrete beams and the NAC beams. The cracking moments of the reinforced RCA-concrete beams with/without compression reinforcement were also found to be lower than those of the NAC beams. For the same reinforcement ratio, the ultimate flexural strengths of the reinforced RCA-concrete beams were higher than those of the NAC beams, regardless of the RCA source and the actual tension and compression steel content. The authors concluded based on their test results that provided the RCA-concrete mix is proportioned by the EMV method, existing flexural analysis and design procedures can be applied to RCA-concrete beams without any modification.

To investigate the shear behaviour of reinforced concrete beams made with RCA-concrete, Fathifazl et al. [27, 28] investigated the effect of the following parameters: shear span-to-depth ratio (1.5, 2.0, 2.7 and 4.0), beam size (depth of 250–550 mm), flexural reinforcement ratio (1.0–2.5%) , shear reinforcement ratio (no reinforcement, three times the minimum shear reinforcement and six times the minimum shear reinforcement), RCA source and concrete type (with/without RCA). The minimum shear reinforcement was determined in accordance with CSA-A23.3-04 [25] requirements for NAC. As in the case of flexure, the authors did not observe any major difference between the failure modes, crack spacing, cracking moment, ultimate shear strength, load-deflection response, and load-shear deformation response of reinforced RCA-concrete and NAC beams. Contrary to previous findings reported in the literature, the concrete contribution to shear resistance in RCA-concrete beams was found to be higher than that in the NAC beams and the former beams more ductile than the latter.

The applicability of some major concrete design standards and other pertinent methods for calculating the concrete contribution to the shear resistance of reinforced RCA-concrete beams without stirrups was investigated by Fathifazl et al. [29]. Results of a relatively comprehensive experimental program were used to compare the actual shear strength of test beams with their corresponding predicted values. The concrete mixes for these beams were proportioned by the EMV method. The results showed that the shear capacity of a reinforced RCA-concrete beam is comparable, or sometimes superior, to that of a companion beam made of NAC. Based on these analyses, they concluded that provided the RCA-concrete mixture is proportioned by the EMV method, existing shear design methods, such as those specified by ACI, CSA and Eurocode, can be applied to reinforced RCA-concrete beams without modification.

3.3.4 Bond and Development Length

In accordance with ASTM A944.99 procedure [30], Fathifazl et al. [31] carried out beam end bond tests on 600 × 230 × 500 mm reinforced concrete blocks. The test variables comprised two RCA source, three concrete types (RCA-concrete using EMV and conventional mix design methods and NAC), and two bar sizes with nominal diameters of 15 and 30 mm. Generally, the bond strength of all RCA-concrete mixes proportioned by the conventional method was found to be lower than that of the companion NAC for both bar sizes and RCA sources, while those proportioned by the EMV method showed comparable or higher bond strength than the NAC specimens. Despite the lower bond strength of specimens made of RCA-concrete proportioned by the conventional method and/or specimens containing the larger bar size, the overall bond behaviour was found to be independent of the mix proportioning method and aggregate type.

Butler and Tighe [32] also investigated the influence of RCA on the bond strength of RCA-concrete using the ASTM A944.99 [30] procedures. Two sources of RCA were used and all the mixes were proportioned by the conventional method of ACI. Two types of RCA-concrete mixtures were produced and in both 100% of the coarse aggregate was RCA. The first type of mixtures was designed with constant w/c ratio and the second with constant target compressive strength. On average, NAC specimens were found to have 9–19% higher bond strengths than the equivalent RCA-concrete specimens.

3.3.5 Durability and Long-Term Performance

Abbas et al. [33] studied the freeze-thaw, chloride penetration and carbonation resistances of RCA-concrete mixes proportioned by the EMV method, RCA-concrete mixes proportioned by the conventional method and companion NAC. In compliance with ASTM C666-97 [25] requirements, for each mix three 75 × 100 × 400 mm prisms were prepared for freeze-and-thaw tests. Using the durability factor as a performance indicator, RCA-concrete produced with either conventional mix design method or with the EMV method exhibited strong resistance against freeze-and-thaw action, similar to NAC. The RCA-concrete proportioned by the EMV method was found to have higher resistance against freeze-thaw action than RCA-concrete proportioned by conventional mix design method, likely due to the lower RCA content, and therefore, lower total mortar content of the former mix.

For the chloride penetration, Abbas et al. [33] tested three 100 × 200 mm cylindrical specimens for each of the three types of mixture as per ASTM C 1556-06 [34]. According to their results, the maximum acid soluble chloride concentration (as % of mass) for all specimens was found to be below the limit of 0.2% by mass of cement as specified by ACI 222 [35]. Also, the apparent chloride diffusion coefficients for the three mixtures were reported as being of the same order of magnitude.

Abbas et al. [33] also carried out carbonation tests using a setup recommended by RILEM [36] and using concrete specimens made with the same three mixtures as used for chloride penetration tests. Three 100 × 100 × 406 mm prisms were prepared for each mixture for the accelerated carbonation tests. RCA-concrete specimens showed comparable and lower level of carbonation than NAC specimens. The authors noted that the carbonation coefficient of specimens made of RCA-concrete proportioned by conventional mix design method was lower than that of RCA-concrete prepared based on the EMV method, likely due to the lower cement content, and therefore lower reserve alkalinity, in the concrete proportioned by the EMV method.

Salem et al. [37] studied the freeze-thaw durability of RCA-concrete mixes proportioned by conventional methods. They studied mixes with w/c ratio of 0.47, with or without air entrainment, and another mix with w/c ratio of 0.29 without air entrainment. They concluded that without air entrainment the RCA concrete mixes had inferior durability compared to NAC while air entrainment rendered RCA-concrete as durable as NAC. Salem and Burdette [38] also investigated the freeze-thaw durability of RCA-concrete mixes containing Class F fly ash and 1.5–5% air content. The water/cementitious materials ratio was kept at 0.47 while the cement/fly ash ratio was maintained at 4.8 or 6.0. They concluded that high fly ash content in both RCA-concrete and NAC significantly improved their freeze-thaw resistance and had no negative effect on their physical properties. They observed that air entrainment also increased the freeze-thaw resistance of both types of concrete, but it also negatively affects their physical properties.

Shehata et al. [39, 40] reported the findings of a comprehensive research project that investigated the reactivity of RCA produced from concrete affected by alkali-silica reactivity (ASR). The research was conducted on concrete containing RCA retrieved from a 12-year old concrete block made with reactive aggregate. Reactivity was evaluated using the Concrete Prism Test as the main method. Based on their results, the coarse RCA was found to produce the same level of expansion as that produced by the original virgin aggregate. Fine RCA (materials passing 5 mm sieve) was also found reactive but to a lesser extent. The results demonstrate that not all the reactive silica in the virgin aggregate was consumed during the service life of the old concrete. In addition, processing the demolished concrete to produce RCA exposes fresh surfaces of the virgin aggregate (parts that did not react during the service life of the old structure); these surfaces react with alkalis and form new reaction products that cause expansion in the RCA-concrete. Moreover, fractures formed within the original stones during crushing of RCA serve as channels for transport of alkalis to fresh reaction sites.

Due to the high reactivity of the tested RCA, Shehata et al. [39, 40] found that higher levels of preventive measures were needed to mitigate the expansion in concrete containing reactive RCA, compared to the levels required for concrete containing virgin reactive aggregate. In other words, the level of preventive measures that can suppress expansion in concrete containing virgin aggregate was not sufficient to suppress the expansion in RCA-concrete. One of the reasons for this observation was stated to be the alkalis contributed by the residual mortar in RCA to the pore solution of the new concrete. The elevated alkali content sustained ASR and called for higher levels of supplementary cementitious materials (SCM) to suppress the expansion. The SCM blends that were found effective in suppressing the expansion were ternary blends of 5% silica fume and 20% fly ash with CaO and Na_2O_e contents of less than 20 and 2.0%, respectively. Binary blends containing either 30% fly ash or 50% granulated blast furnace slag, were not effective in suppressing the ASR expansion.

Another practical approach to mitigate expansion in concrete containing reactive coarse RCA was proposed by Shehata et al. [40]. In this method, it was suggested that the reactive RCA be blended with non-reactive natural aggregates. At a blending ratio of 30% non-reactive aggregate to 70% reactive RCA, the level of expansion was found to decrease and the level of SCM required to suppress the expansion was reduced to moderate levels. Indeed, 25% fly ash with CaO <20% or 50% slag were enough to maintain the expansion of the concrete prisms below the specified limit (0.04% at 2 years). Blending reactive RCA with non-reactive natural aggregate is also beneficial for other concrete properties, including creep, shrinkage, strength and workability.

A joint research program was undertaken by Laval University, the University of Wyoming, Ryerson University and Oregon State University to identify the best approach for reliably evaluating the potential for ASR in RCA. RCA was produced by crushing concrete blocks incorporating a number of reactive aggregates. The blocks had been subjected to natural environmental conditions for more than 10 years. Other blocks were retrieved from the different structural elements (massive base, columns and bridge deck) of a large bridge that was demolished in 2010 in Québec City, Canada. The 45 year old bridge was severely damaged by ASR involving local reactive siliceous limestone aggregate. The potential expansion due to ASR for the resulting RCA-concretes was evaluated using the Accelerated Mortar Bar Test (AMBT) [34] and the Concrete Prism Test [34]. A multi-laboratory study was performed to observe the repeatability of the AMBT test between the partner laboratories, to evaluate the effect of different replacement percentages of RCA on the non-reactive aggregate, and to observe the reactivity difference between two types of RCA materials: (1) fine material recovered from the large scale crushing operations used to produce coarse RCA from the concrete blocks (called "crusher's fines") and (2) fine material produced from crushing of coarse RCA (5–20 mm in size) (called "re-crushed RCA"). Specimens with 100% RCA, 50% RCA and 25% RCA content were cast for both the crusher's fines and re-crushed material. Non-reactive natural sand was used for the balance of the fine aggregate in the mixture. The results showed that higher replacement levels of reactive RCA results in higher expansion, but similar to Shehata's [39] observation, lower expansion occurred in the mix containing the crusher's fines than the one containing re-crushed material [41]. Petrographic examination performed on the different fractions used in mortar bars showed that the difference in the behaviour of the above materials is related to the composition of the particles, especially the residual mortar content in the different fractions of aggregate material [42]. The multi-laboratory test results showed that the precision statement in ASTM C 1260 [34] for multi-laboratory testing is applicable to RCA.

3.3.6 Research on RCA in Pavement Concrete

Smith and Tighe [43, 44] investigated the effect of RCA on the coefficient of thermal expansion (CTE) and its impact on pavement performance. CTE is a key property of concrete and relates to the amount of expansion and contraction caused by changes in temperature. CTE testing was conducted on 16 cores containing various amounts of coarse RCA using a simplified methodology. Testing showed that concrete performance improved as the amount of RCA increased.

Pervious concrete with minimal fine aggregate and a high void content is a green alternative to conventional pavements. Pervious concrete allows water to infiltrate through the pavement and thereby reduces the requirement for storm water management systems. This research incorporated RCA into pervious concrete to create a very sustainable concrete product for paving. Cylinders were cast in the laboratory for each percentage of RCA together with a control mix containing only virgin aggregate. The cylinders were tested for compressive strength, permeability, and void content. The results showed that pervious concrete mix containing 15% RCA had strength, permeability, and void content comparable to the control mix [45].

Chan and Tighe [46] utilized some sustainability tools such as the Ministry of Transportation of Ontario GreenPave, and examined the life cycle benefits of using RCA in concrete pavement. In general, GreenPave credits consist of three types: design credits, construction credits, and innovation credits. Design credits are awarded based on pavement design assessment. Each design alternative proposed in a project will be assessed for design credits in GreenPave. Construction credits are awarded at the end of construction. Innovation credits are awarded for sustainable practices that are not identified in GreenPave. RCA provides credits and it appears that it will be incorporated into Ontario practices.

Based on this overview, in North America a reasonable amount of research has been conducted on various aspects of RCA, but despite the many favourable research findings regarding the good short- and long-term performance of RCA-concrete, RCA is still predominantly used as backfill, base and sub-base material rather than substitute for virgin aggregate in structural quality concrete. At this juncture, there does not seem to be any short- or long-term strategy for increasing the use of RCA in structural concrete. However, the finding that the existing design codes analysis and design provisions for the different limit states can be directly applied to RCA-concrete members made of mixes proportioned by the EMV method may remove one of the major hurdles to the use of RCA-concrete in structural applications.

3.4 Recycling in China: An Overview of Study on Recycled Aggregate Concrete

Jianzhuang Xiao, C.S. Poon, Wengui Li, Yu-hui Fan, Jingwei Ying, and M.M. Tawana

3.4.1 Introduction

China is a large resource consumer with shortage of resources and extensive management, its economic growth is increasingly constrained by resources and the environment. According to statistics, China's consumption of cement is 820

J. Xiao (✉) • W. Li • Y.-h. Fan • J. Ying • M.M. Tawana
Department of Building Engineering, Tongji University, Shanghai, P.R. China

C.S. Poon
Department of Civil and Structural Engineering, The Hong Kong Polytechnic University, Hong Kong, P.R. China

million tons and accounts for 55% of the world. It is estimated that approximately 200 million tons of waste concrete are currently produced annually in the mainland of China [47]. Moreover, some natural disasters such as *Wenchuan* earthquake (2008) and *Yushu* earthquake (2010) in China have resulted in a great quantity production of waste concrete. Additionally, in the construction process of *Shanghai Expo 2010*, nearly 300 million tons of construction and demolition wastes were generated by the demolition work.

Consequently, it is becoming a special problem of human environment pollution. The environmental and economic implications of this are no longer considered sustainable. Therefore, a possible solution to these problems is to recycle demolished concrete and produce an alternative aggregate for structural concrete. And Recycled concrete technology is one of the main measures in the development of green concrete and realizing the sustainable development of construction, resources and environment.

To deal with this problem, in recent years, many Chinese researchers have been studied on recycled aggregate concrete and nearly all aspect of mechanical properties and structural performances have been covered. And until now, two national symposiums on recycled concrete and one international Conference on Waste Engineering and Management were held in China. To summarize all of those achievements, this paper is not only written primarily as a state-of-the-art report on the recycled concrete but to provide the theoretical basis of the preliminary for the further study on recycled concrete.

3.4.2 Mechanical Properties of Recycled Concrete: Strength

The strength of concrete is the property most valued by designers and quality control engineers. The water/cement ratio is important in determining the porosity of both the matrix and the transition zone and hence the strength of recycled concrete, factors such as aggregate size and mineralogy as well as replacement rate of recycled aggregate, admixtures, type of stress, and the rate of loading can also have an important effect on strength. In this section the influence of various factors on recycled concrete strength is examined in detail. Since the uniaxial strength in compression is commonly accepted as a general index of concrete strength, the relationship between the uniaxial compressive strength and other strength type such as tensile, flexural, shear, and biaxial strength are discussed.

3.4.2.1 Compressive Strength and Factors Effect on It

Influence of Water/Cement Ratio

Deng [48] studied the influence of the water/cement ratio on the compressive strength of concrete with 100% RCA. It was found that the compressive strength of RAC relies heavily on the water/cement ratio. If the water/cement ratio is higher

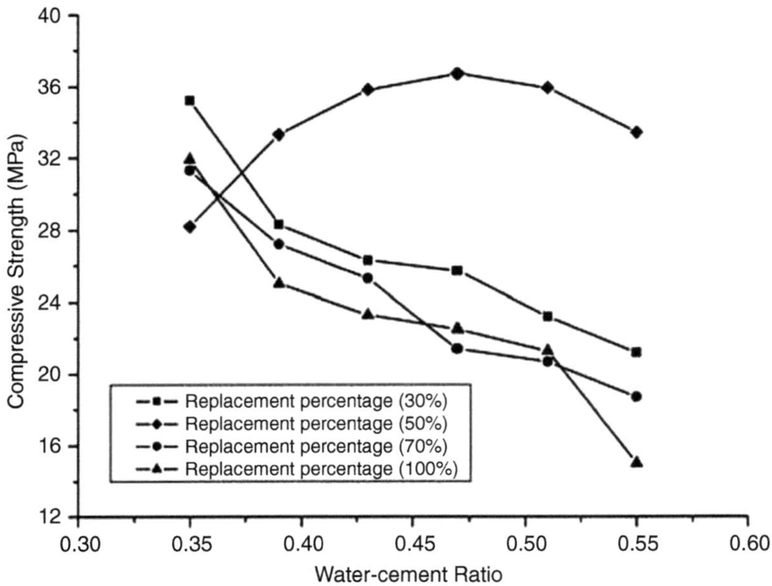

Fig. 3.10 The compressive strength of RAC as a function of water-cement ratio [49]

than 0.57, the compressive strength of RAC decreases with the increase of water/cement ratio. However, if water/cement ratio is below 0.57, the compressive strength of RAC increases with the improvement of water/cement ratio.

A more comprehensive experimental work has been performed by Li [49], who investigated the relationship between the 28 days compressive strength and the water-cement ratio for concrete with different RCA percentages. The test results are illustrated in Fig. 3.10.

It can be seen that the compressive strength is higher with lower water/cement ratio for concretes with RCA percentages of 30, 70 and 100%. However, in case of concrete with 50% RCA, the compressive strength first increases and then decreases as the water/cement ratio increases. It can be seen that there exists a linear relationship between compressive strength and the water/cement ratio for RAC with 30, 70 and 100% RCA; but for concrete with 50% RCA, the relationship is a nonlinear one, which needs further study.

Liu and Chen [50] also reported excellent straight-linear relations for the compressive strength and water/cement ratios for RAC with RCA replacement percentage of 30, 70 and 100%, and a nonlinear relation for RAC with 50% RCA. This phenomenon might be explained from a further investigation on the meso- and micro-level properties of the RAC.

Influence of RAC

Numerous experiments on the influence of RCA on the compressive strength of concrete have been conducted e.g., Li [49], Xiao and Li et al. [51], Tang [52],

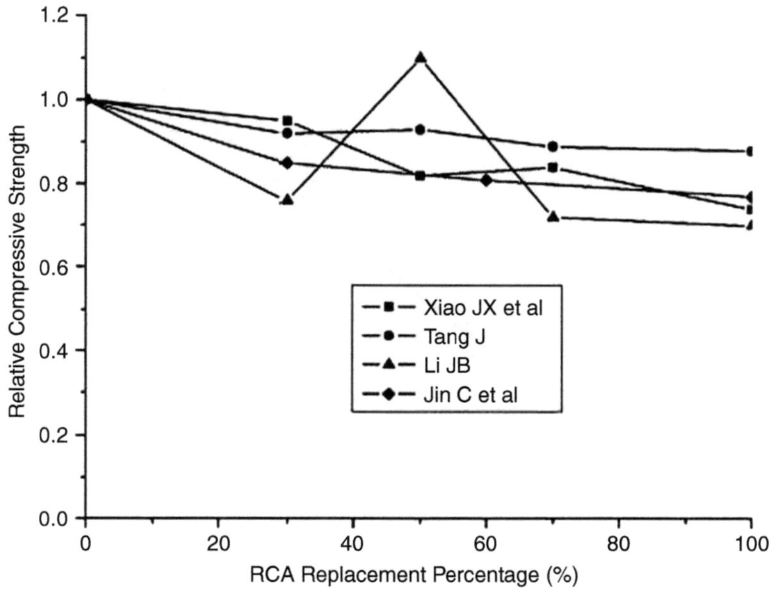

Fig. 3.11 Influence of RCA content on compressive strength [51–53]

Jin and Wang et al. [53]. The results indicate that the amount of RCA has remarkable influences on the compressive strength of concrete. Some typical findings are shown in Fig. 3.11, which illustrates the relationship between the RCA replacement percentage (%, in mass) and the relative compressive strength, defined as the ratio of the compressive strength of RAC to that of the conventional concrete, i.e., with no RCA.

It can be seen from Fig. 3.11 that, in general, the concrete compressive decreases with the increase of the RCA content. With a RCA replacement percentage of 100%, the reduction is between 10 and 30%. However, if the RCA content is less than 30%, the influence on compressive strength is not very remarkable. Several reasons could be responsible for the reduction of the compressive strength for RAC, including an increased concrete porosity and a weak aggregate-matrix interface bond. These observations have been recently explained by Xiao and Liu et al. [54] on the basis of photomicrographs of the fracture patterns of RAC. Furthermore, it was found by Xiao et al. [55] that it is quite possible to achieve a desirable compressive strength by adjusting the waster/cement ratio. This confirmed the earlier results by Buck [56] which also showed that it is possible to make RAC stronger than the parent concretes from which the RCA are derived.

Influence of Density

Xiao et al. [57] found that there exists approximately a linear relationship between the compressive strength (f_{cu}) and the mass density (ρ), as can be seen in Fig. 3.12.

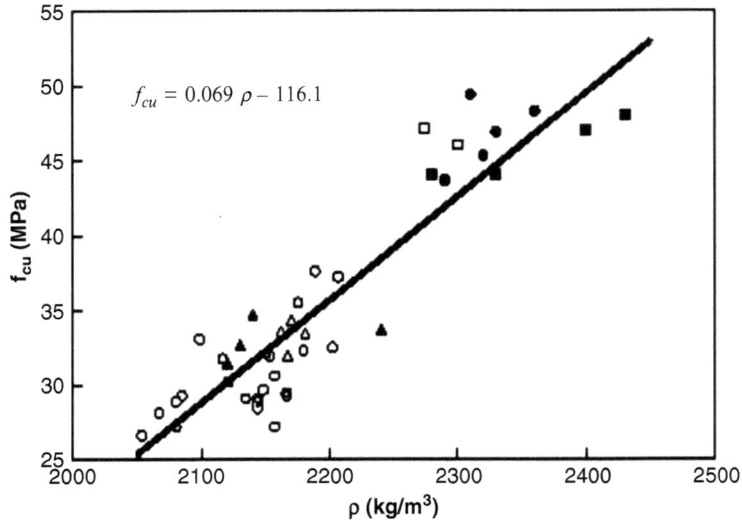

Fig. 3.12 Relationship between the compressive strength and the mass density [57]

Through a statistical regression analysis, the following equation was developed. Similar results were obtained by Li and Zhang et al. [58]:

$$f_{cu} = 0.069\,\rho - 116.1 \tag{3.1}$$

Influence of Mixture

Due to the high absorption characteristic of RCA, some adjustments have to be made for the mix design of RAC. A comprehensive experimental work has been carried out by Xiao et al. [55] and Li et al. [59]. It is found that, in general, the mix design procedure for RAC does not differ much from that for conventional concrete. However, more water is required to attain a similar workability owing to the high absorption of. RCA. It is thus recommended by Xiao et al. [55] and Zhang et al. [60] to divide the water for RAC into two groups: the first part is determined according to the mix procedure for conventional concrete (with similar strength); while the second part is determined according to the water absorption capacity (usually the 10 min one) of the RCA, which is used to compensate the loss of slump of RAC. In practical mixing, the two parts of water are added together. This method has been generally accepted and utilized by many other researchers in China.

Effect of Micro- and Meso-structure on Strength

Poon and Shui et al. [61] investigated the effect of microstructure of ITZ on compressive strength of RAC. SEM observations revealed that the aggregate-cement matrix interfacial zone of RAC consists mainly of loose and porous hydrates

Fig. 3.13 Microstructure of concrete prepared with RCA. (**a**) interfacial zone appeared to consist mainly of loose particles, and the width of the loose zone was approximately 30-60 μm. (**b**) shows at a higher magnification that the interfacial zone was porous with high porosity [61]

whereas the aggregate–cement matrix interfacial zone of conventional concrete consisted mainly of dense hydrates (see Fig. 3.13). It was found that the hydrates in the vicinity of the interface mainly consist of loose granular compounds, and the porosity near the interfacial zone was similar to those observed in the natural aggregate–cement matrix ITZ. A small amount of fine flake-like and whisker-like crystals formed an inter-penetrating mat in the voids.

Du [62] reported that some voids, micro cracks and loose hydrates of $Ca(OH)_2$ crystal exist in the RCA-cement matrix ITZ, and the micro hardness values in the ITZ are lower than those of the cement matrix, but the disadvantageous feature in ITZ is gradually weakened and the ITZ becomes much stronger from the RCA to the cement matrix.

Shui and Pan et al. [63] found that the micro-structural properties of the ITZ is related to the density of the RCA, i.e., excessively dense or loose RCA surface leads to a porous interfacial zone or well-crystallized hydrates, whereas a moderately dense RCA surface results in a relatively dense ITZ.

Xiao and Liu et al. [64] studied the interactions between the old and new interfaces in RAC. The results implied that the qualitative and quantitative changes of the old and new interfaces during the loading process induce a notable fluctuation of the mechanical and the physical properties of RAC.

Xiao, Liu and Li et al. [54] founded that the RCA can be regarded as a composite system consisting of four components: new mortar, old mortar attached to the RCA, mortar-aggregate bond and natural aggregate of RCA (see Fig. 3.14). It was shown that the mortar attached to RCA has influence on the mechanical properties of RAC, and the mortar-aggregate bond strength plays a vital role in the failure mec.

3.4.3 Behaviour of Concrete Under Various Stress Stages

3.4.3.1 Uniaxial Tensile Strength

Xiao and Lan [65] investigated the behavior of RAC with different amounts of RCA under uniaxial tension. The test results, given in Fig. 3.15, indicate that an increase

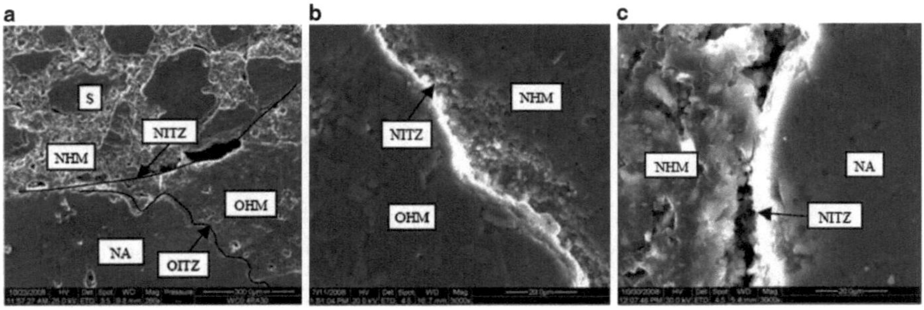

Fig. 3.14 The micro-and meso-structure of RAC [54]. (**a**) Old ITZ and new ITZ (**b**) Old mortar and new mortar (**c**) Natural aggregate and new mortar. Note: *NA* natural aggregate, *S* sand, *NHM* new hardened mortar, *OHM* old hardened mortar, *NITZ* new interface transition zone, *OITZ* old interface transition zone

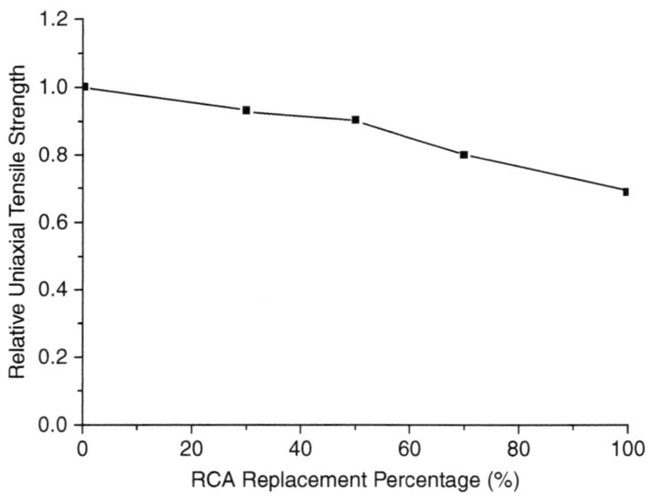

Fig. 3.15 Uniaxial tensile strength as a function of RCA replacement percentage for RAC [65]

of the RCA content leads to a reduction in uniaxial tensile strength. When the RCA replacement percentage is 100%, the strength reduced by 31% compared to that of conventional concrete with similar composition. Based on a regression analysis, the following equation was proposed for predicting the uniaxial tensile strength from the compressive strength:

$$f_t = (0.24 - ar)f_{cu}^{2/3} \tag{3.2}$$

Where f_t and f_{cu} are the uniaxial tensile strength and cube compressive strength of RAC, respectively. Parameter r is the RCA replacement percentage, and parameter $a = 0.06$.

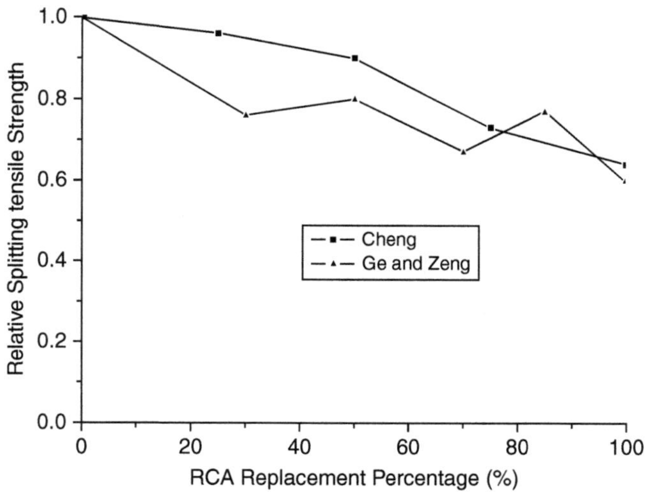

Fig. 3.16 Splitting tensile strength as a function of RCA replacement percentage for RAC [67, 68]

3.4.3.2 Splitting Tensile Strength

Owing to the difficulties in carrying out uniaxial tensile test, the splitting tensile strength is usually used to evaluate the behavior of concrete in tension. Yang and Wu et al. [66], Cheng [67] as well as Ge and Zeng [68] investigated the influence of RCA on the splitting tensile strength of concrete. In their tests, cube specimens of 150 × 150 × 150 mm were used. Their test results are presented in Fig. 3.16. From the figure, it is evident that the splitting tensile strength reduces as the RCA content increases. For RAC with 100% RCA, the reduction in the splitting tensile strength is approximately 30–40%.

Xiao et al. [57] proposed the following equation to describe the relationship between splitting tensile strength (f_{sp}) and cube compressive strength (f_{cu}). This equation was adopted in the Technical Code for "*Application of Recycled Concrete" (DG/TJ08-07)*:

$$f_{sp} = 0.24 f_{cu}^{0.65} \tag{3.3}$$

Xiao and Lei et al. [69] investigated the splitting failure and splitting tensile strength distribution and the relationship between splitting tensile strength and compressive strength of RAC with different RCA. It was found that the splitting failure of RAC initiates not only from the interfaces between RCA and new cement paste, but also from some of the RCA itself. With the same water/cement ratio, the splitting tensile strength of RAC with different kinds of RCAs was smaller than that of concrete with only one uniform RCA, while the difference of the standard deviations as well as the coefficients of variation was not distinct. It was confirmed that normal model can be used to describe the distributions for the splitting tensile strength of concrete with different RCAs (see Fig. 3.17).

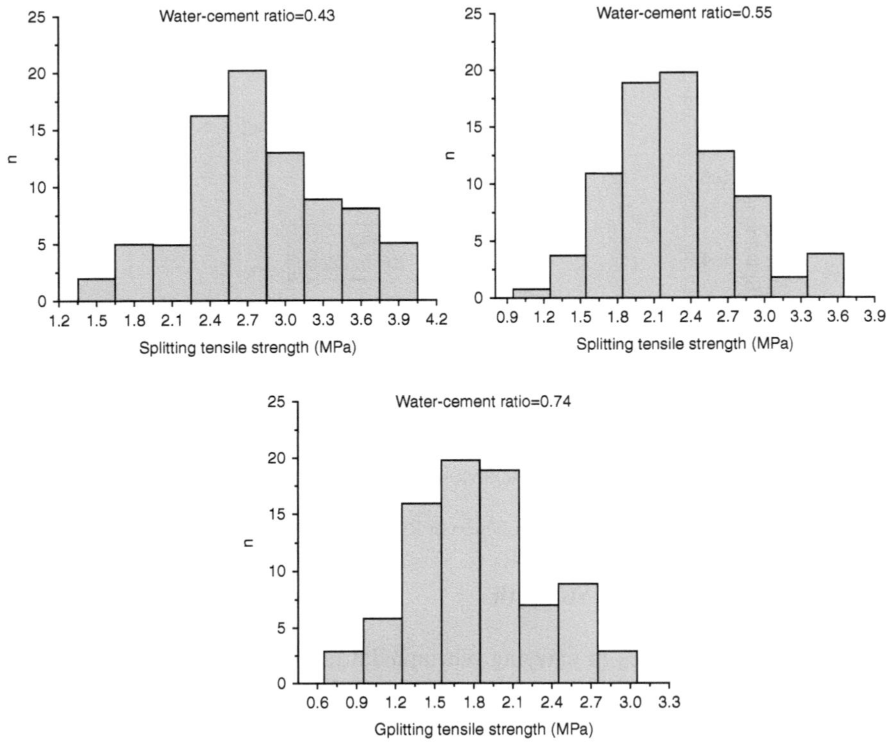

Fig. 3.17 Histograms for the distribution of splitting tensile strength of RAC [69]

3.4.3.3 Flexural Tensile Strength

Flexural tensile test is another approach for evaluating the behavior of concrete subjected to tension. Xiao and Li [70], Zhou and Liu et al. [71] found that the RCA replacement percentage has only marginal influence on the flexural strength of the RAC, see Fig. 3.18. Similar results were also obtained by Hu [72]

Based on the regression analysis of test data, the equation below is suggested for describing the relationship between the flexural tensile strength (f_f) and the compressive strength (f_{cu}) of the RCA. This equation was firstly developed by Li [49] and later adopted in the Technical Code for "*Application of Recycled Concrete" (DG/TJ08-07).*

$$f_f = 0.75\sqrt{f_{cu}} \tag{3.4}$$

3.4.3.4 Shear Strength

Huang, Deng and Luo et al. [73] investigated the shear behavior of RAC with different amounts of RCA. The shear test was performed on four-point-load beam

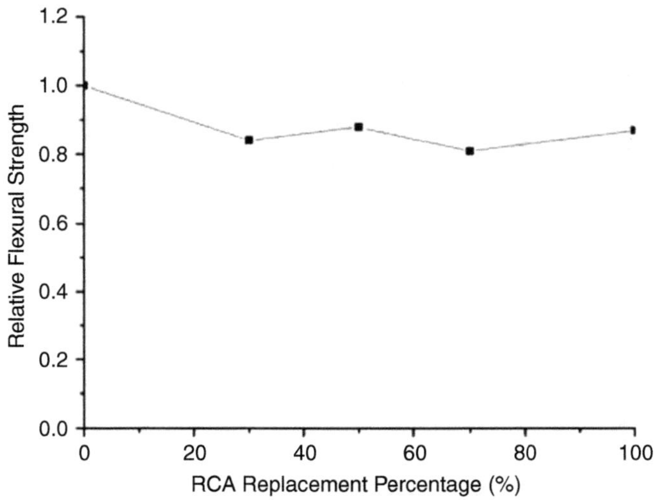

Fig. 3.18 Flexural tensile strength as a function of RCA replacement percentage for RAC [70]

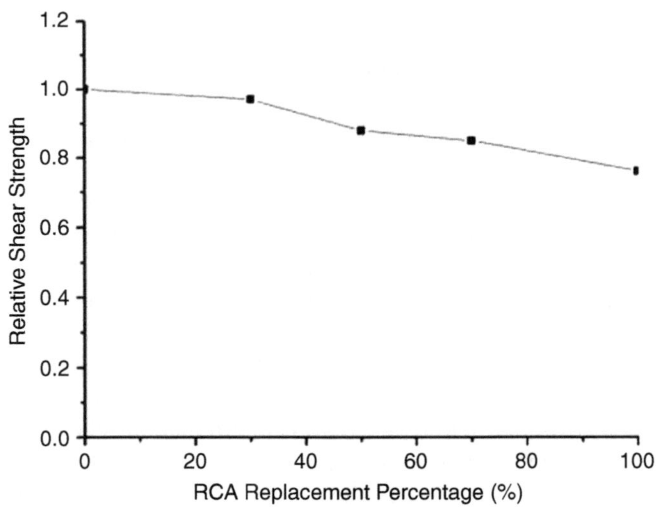

Fig. 3.19 Shear strength as a function of RCA replacement percentage for RAC [73]

with the same height and different width, e.g. 650 × 120 × 150 mm [74]. It is found that the shear failure mode of RAC is similar to that of conventional concrete. However, the shear strength decreases with the increase of the RCA content (see Fig. 3.19). For concrete with 100% RCA, the shear strength is approximately 30% lower than that of conventional concrete.

Based on a regression analysis of the test data, the relation below was set up for the shear strength f_v and the cube compressive strength f_{cu} of RAC:

$$f_v = 0.18 f_{cu}^{0.79} \tag{3.5}$$

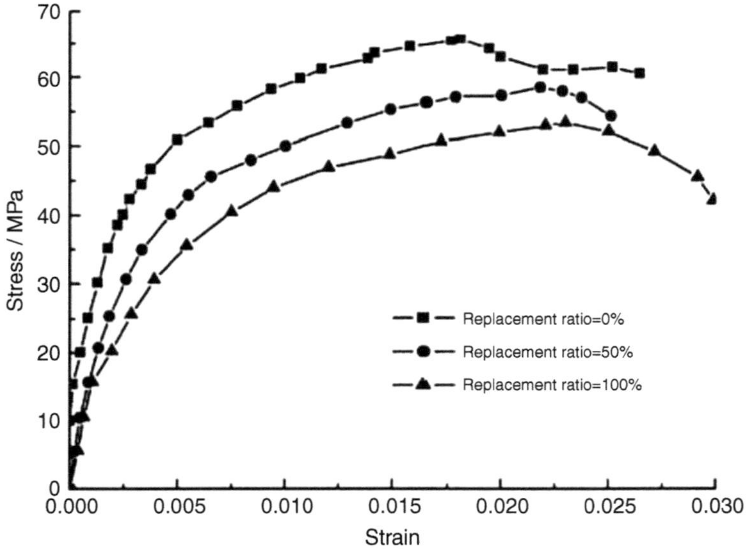

Fig. 3.20 The stress-strain relationships as a function of RCA replacement percentage [77]

3.4.3.5 Multiaxial Compressive Strength

So far, no sufficient experimental data are available in China for evaluating the strength of RAC in biaxial and triaxial compression directly at the material level, e.g. on plate, cube or cylinder specimens. However, the behavior of RAC in triaxial compression has been investigated through RAC filled steel tube and GFRP (glass fiber reinforced plastics) confined RAC specimens.

Based on the related axial tests on conventional concrete and RAC-filled steel tube columns, Wu and Yang [75] as well as Yang [76] reported that the loading capacity of the RAC column is lower than that of conventional concrete column. When the RCA replacement percentage increases, a distinct reduction of the loading capacity is found. However, the load-deformation relationship for RAC columns is similar to that of conventional concrete columns.

Xiao and Yang [77] investigated the property of RAC confined GFRP (glass fiber reinforced plastics) under axial compression. It was found that the confinement offered by the GFRP greatly improves the ultimate and deformation capacity as well as the post-peak ductility. The ultimate strength of (f_{cc}) GFRP confined RAC is approximately 30% higher compared to that of unconfined RAC. The test data also indicates that with the increase of the RCA content, the loading capacity of the columns decreases (see Fig. 3.20).

3.4.3.6 Bond Strength Between Recycled Concrete and Steel Rebar

Xiao and Li et al. [78] studied the bond behavior between RAC and steel rebar through pull-out tests according to RILEM standard. In their tests, three RCA

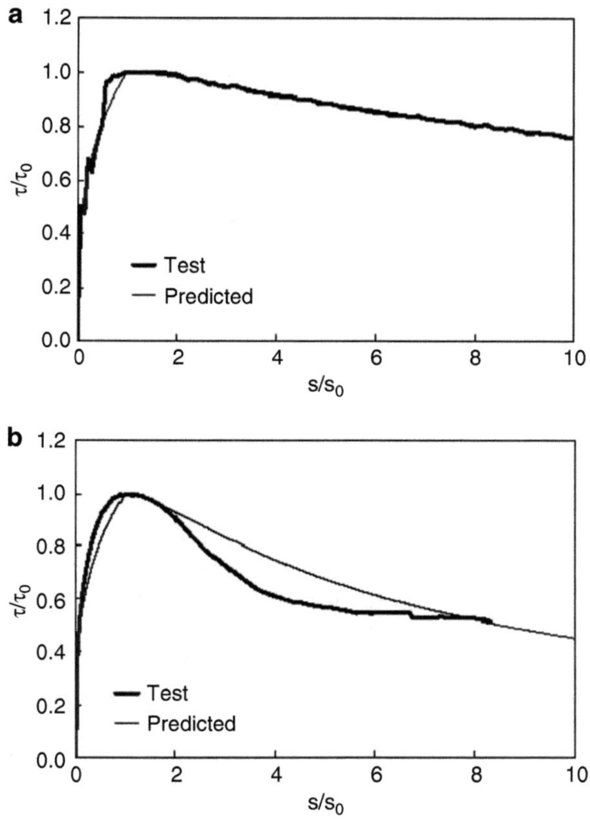

Fig. 3.21 Comparison for the normalized bond-slip curves of RAC with 100% replacement percentage [78]. (**a**) Plain rebar (**b**) Deformed rebar

replacement percentages (0, 50 and 100%) and two kinds of steel rebar, i.e., plain and deformed with a diameter of 10 mm, were used. It was found that, the bond strength decreases 12 and 6% with RCA replacement percentage of 50 and 100% when plain rebar is used. Whereas the amount of RCA has only a minor influence on the bond strength when deformed rebar is used. This is because the bond strength between RAC and deformed rebar depends mainly on the mechanical anchorage and friction resistance, while the bond between RAC and plain rebar depends mainly on the adhesion between rebar and concrete. The latter is strongly influence by the RCA content. It is also observed that the bond strength between deformed rebar and RAC is approximately 100% higher than that between plain rebar and RAC, see Fig. 3.21.

The following mathematical model was proposed by Xiao and Li et al. [78] for the bond-slip relationship of RAC:

$$\frac{\tau}{\tau_u} = \begin{cases} (s/s_u)^a & (s/s_u) \leq 1 \\ \dfrac{s/s_u}{b(s/s_u - 1)^2 + s/s_u} & (s/s_u) > 1 \end{cases} \tag{3.6}$$

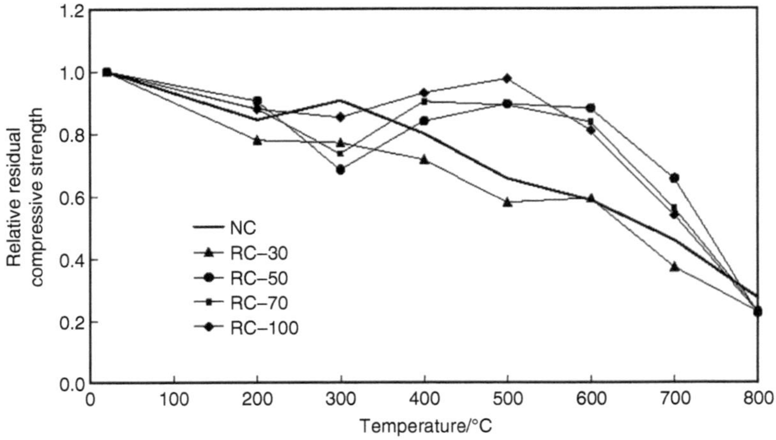

Fig. 3.22 Relative of residual compressive strength of RAC with different RCA replacement percentage [80]

Where parameters τ, s are the bond stress (MPa) and slip (mm), respectively; τ_u, s_u are the bond strength (MPa) and corresponding slip (mm); parameter a is 0.3 for all tests. The parameter b depends on the type of the steel rebar and the RCA amount. It is 0.038 for plain rebar with all RCA replacement percentages; while for deformed rebar, b is 0.10 for RAC with 0 and 50% RCA, while for RAC with 100% RCA, b is 0.18.

3.4.3.7 Strength After High Temperatures

Residual Compressive Strength

Xiao and Huang [79] studied the behavior of RAC exposed to high temperatures. In their tests, five RCA replacement percentages, i.e., 0, 30, 50, 70 and 100%, were used and the concretes were heated from 20 to 800°C. A total of 160 cube specimens with 150 × 150 × 150 mm sizes were tested.

Figure 3.22 presents the influence of RCA replacement percentage on the relative residual compressive strength of RAC with different RAC replacement percentages. It can be seen that, when the temperature is lower than 300°C, the residual compressive strength for all the concrete follow the similar trend; however, after 300°C, the influence of RCA begins to be remarkable. For RAC with 30% RCA replacement, the residual compressive strength decreases sharply, which is similar to conventional concrete; while for RACs with 50, 70 and 100% RCA, the residual compressive strengths tend to increase until 500°C. After that the residual compressive strength begins to decrease.

The following equations were proposed to describe the relationship between the residual compressive strength (f_{cu}^T) of RAC and elevated temperature.

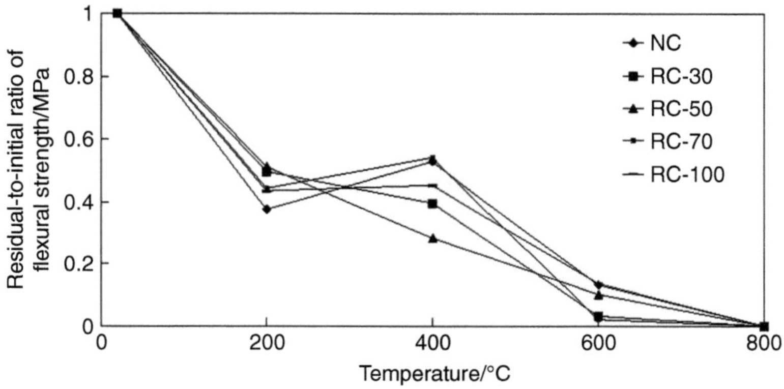

Fig. 3.23 Relative of residual flexure strength of RAC with different replacement percentage [79]

For RAC with RCA replacement percentage of 30%:

$$\frac{f_{cu}^{T}}{f_{cu}^{20}} = \begin{cases} 1.018 - 0.088(T/100) & (T \leq 300°C) \\ 0.93 - 0.059(T/100) & (300°C < T < 600°C) \\ 1.62 - 0.174(T/100) & (600°C < T \leq 800°C) \end{cases} \quad (3.7)$$

For RAC with RCA replacement percentage more than 50%:

$$\frac{f_{cu}^{T}}{f_{cu}^{20}} = \begin{cases} 1.015 - 0.075(T/100) & (T \leq 300°C) \\ 0.489 - 0.096(T/100) & (300°C < T < 600°C) \\ 2.086 - 0.224(T/100) & (600°C < T \leq 800°C) \end{cases} \quad (3.8)$$

Where is the residual cube compressive strength of RAC after elevated temperature T (MPa), is cube compressive strength of RAC at room temperature (MPa), and T is the temperature $\left(f_{cu}^{T}\, f_{cu}^{20\circ}\mathrm{C}\right)$.

Residual Flexural Strength

Xiao and Huang et al. [81] further investigated the residual flexural strength of RAC with different replacement percentages of RCA (i.e. 0, 30, 50, 70 and 100%) after fire. The specimens were prism with 100 × 100 × 400 mm and 150 × 150 × 550 mm and they were exposed to temperatures from 20 to 800°C. The test data are illustrated in Fig. 3.23. The results reveal that the residual flexural strength of RAC decreases with the exposed temperatures; however, the effect of the RCA replacement percentage on the residual flexural strength is less remarkable in comparison with the residual compressive strength.

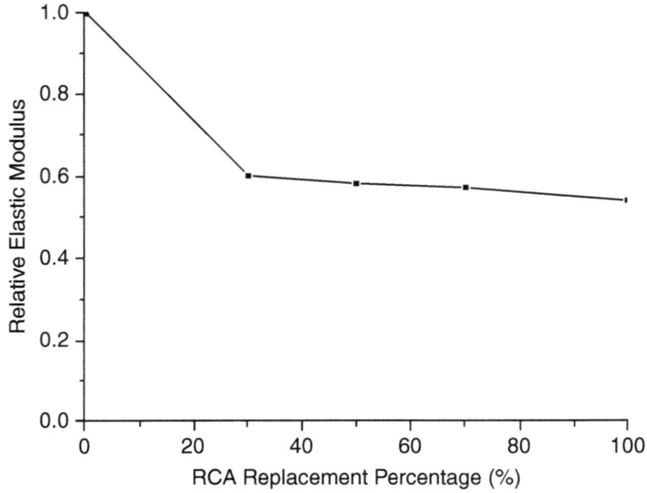

Fig. 3.24 Elastic modulus as a function of RCA replacement percentage [82]

The following equations were suggested to describe the variation of the residual flexural strength (f_f^T) for RAC with the elevated temperatures:

$$f_f^T \Big/ f_f^{20} = \begin{cases} 1.058 - 0.0029T & 20\ ^\circ C \le T < 200\ ^\circ C \\ 0.53 - 0.00027T & 200\ ^\circ C \le T \le 400\ ^\circ C \\ 1.117 - 0.00174T & 400\ ^\circ C < T \le 600\ ^\circ C \end{cases} \tag{3.9}$$

Where is the residual flexural strength of RAC after elevated temperature (MPa), is flexural strength of RAC at room temperature (MPa) and T is the temperature $\left(f_f^T f_f^{2C}\right)$°C.

3.4.4 *Deformation Characteristics*

3.4.4.1 Modulus of Elasticity

The modulus of elasticity for RAC is always lower than that of corresponding conventional concrete. This is mainly due to the fact that a large amount of old mortar is often attached to original aggregate particles in RCA and these mortars usually has relatively smaller elastic modulus. Many tests on the modulus of elasticity for RAC have been carried out.

Xiao [82] found up to 45% lower modulus of elasticity for RAC made with the 100% RCA in comparison with that of conventional concrete. Figure 3.24 illustrates the relationship between the modulus of elasticity and the RCA content. Hu et al. [83] reported the elastic modulus of RAC is 15–25% lower than that of

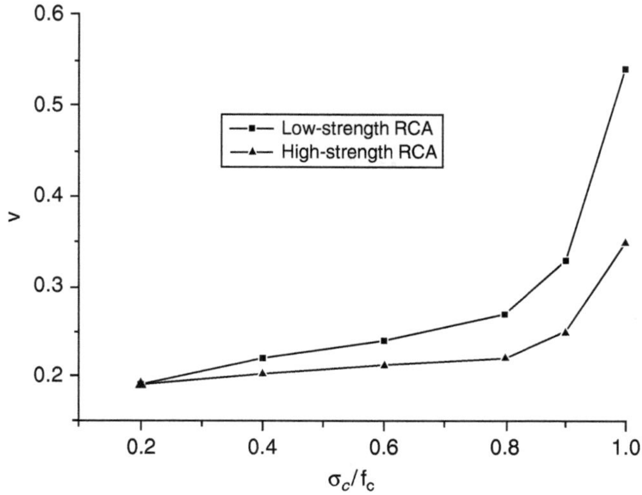

Fig. 3.25 Poisson's ratio of RAC with different types of RCA [83]

conventional concrete, depending on the qualities of the parent concrete and the properties the RCA as well. Moreover, Xiao and Du [84] stated that the elastic modulus of RAC is mainly controlled by the RCA with lower parent concrete strength grade.

Different equations have been suggested for describing the relationship between the elastic modulus (E_c) and the compressive strength (f_{cu}) of RAC. To accurately predict the modulus of elasticity for RAC, it is believed that the influence of RCA content should be taken into account. For instance, Xiao and Li et al. [57] proposed an equation as given in Eq. (3.10). A former equation was firstly recommended for determining the modulus of elasticity for RAC by Ravindarajah and Tam [85], see Eq. (3.11). Figure 3.25 illustrates that the elastic modulus decrease with the increase of the replacement percentage of RCA. It should be noted that Eq. (3.10) is very close to the results given in Eq. (3.11).

$$E_c = \frac{10^5}{2.8 + (40.1/f_{cu})} \tag{3.10}$$

$$E_c = 7770 f_{cu}^{0.33} \tag{3.11}$$

3.4.4.2 Poisson's Ratio

The Poisson's ratio of RAC reflects the ratio of the lateral to axial deformation. Hu et al. [83] found that the initial Poisson's ratio of RAC varies between 0.15 and 0.23, which is in fact similar as that of conventional concrete. The change of the Poisson's ratio with the stress level was also investigated by Hu et al. [83]

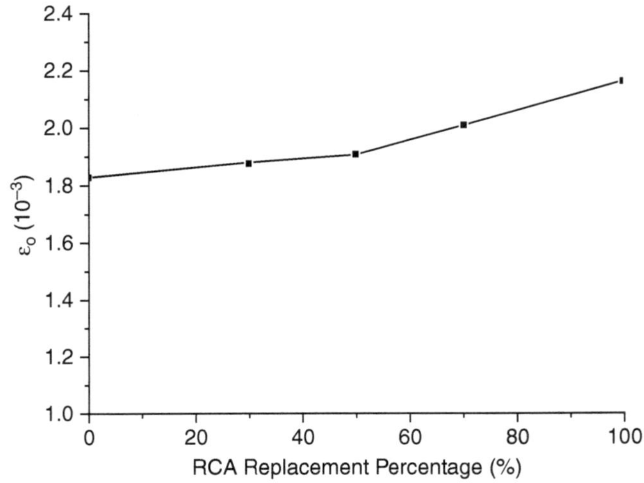

Fig. 3.26 Peak strain as a function of RCA replacement percentage [87]

and Song [86]. The influence of types of RCA on the Poisson's ratio is shown in Fig. 3.25.

In the Technical Code for "*Application of Recycled Aggregate Concrete" (DG/TJ08-07)*, a value of 0.2 is recommended for the initial Poisson's ratio of RAC, irrespective of the RCA replacement percentage.

3.4.4.3 Strain at Peak Stress

Xiao and Li et al. [87] studied the peak strain ε_0 of RAC under uniaxial compression with different RCA contents. The peak strains of RAC with different RCA contents are shown in Fig. 3.26. It can be seen that the value of the peak strain increases as the RCA content increases. For the RCA replacement percentage of 100%, the peak strain is increased by 20%. The main reason for the increase of the peak strain of RAC is due to the reduced elasticity modulus of RCA, which leads to a larger deformation. Xiao [82] also found the peak strain is approximately 20% higher for the RAC with 100% replacement percentage than for the corresponding conventional concrete.

3.4.4.4 Ultimate Strain

The ultimate strainε_u is taken as the longitudinal strain in the stress-strain curve at a stress level equals to 85% of the peak stress. Xiao [82] and Li [49] reported the ultimate strain on different RCA replacement percentages, i.e., 0, 30, 50, 70 and 100%. It can be seen from Fig. 3.27 that the ultimate strain may decrease or increase with the increasing RCA replacement percentage, depending on the value of

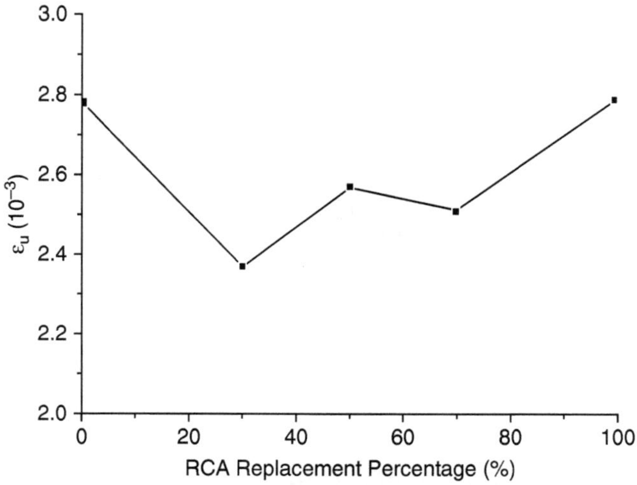

Fig. 3.27 Ultimate strain as a function of RCA content of RAC [82]

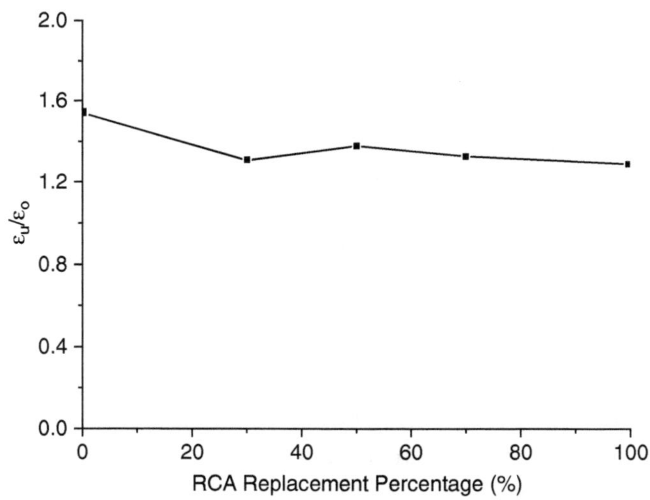

Fig. 3.28 $\varepsilon_u/\varepsilon$-value as a function of RCA content of RAC [82]

replacement percentage. For a low value, the ultimate strain decreases with increasing of replacement percentage, while the opposite may be the case for a high value of replacement percentage.

Xiao [82] also studied the influence of the RCA replacement percentage on the $\varepsilon_u/\varepsilon_0$ -value of RAC. It can be concluded that the $\varepsilon_u/\varepsilon_0$ -value of RAC is smaller than that of the conventional concrete, as shown in Fig. 3.28. This means that under the same loading and deformation conditions, the energy absorption capacity of RAC is reduced. Compared to the conventional concrete, the RAC is more brittle with a poor ductility.

Fig. 3.29 Complete stress-strain curves of RAC under uniaxial compression [87]

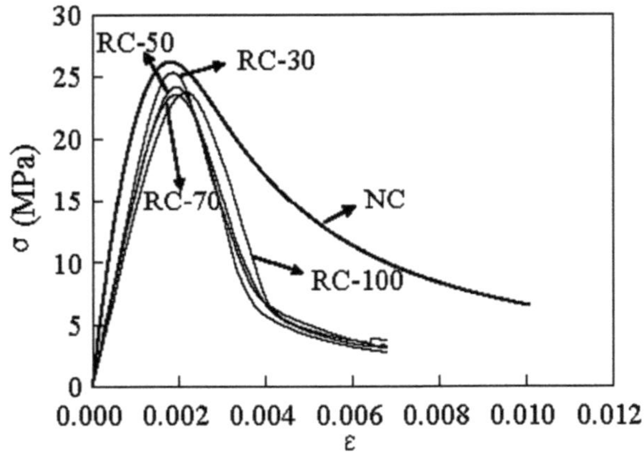

3.4.5 *Stress – Strain Relationship*

3.4.5.1 Uniaxial Compression

The stress-strain curve of RAC in uniaxial compression is very important for nonlinear analysis of this kind of concrete structure. Several studies have been carried out to investigate the influence of RCA content on the complete stress-strain curves, e.g. Xiao and Li et al. [82, 87], Xu [88], Hu and Song et al. [83] and Song [86]. Prism specimens of 100 × 100 × 300 mm were usually used in these tests. The typical test results obtained by Xiao and Li et al. [82], [87] for concrete with various replacement percentages of RCA, namely 0, 30, 50, 70 and 100%, are illustrated in Fig. 3.29.

The following important characteristics for the complete stress-strain curves of RAC under uniaxial compression were found:

1. The RCA content has a remarkable influence on the stress-strain curves of concrete. Nevertheless, the shape of the stress-strain curves for all the RAC was similar to that of the control concrete, irrespective of the RCA contents;
2. The complete stress-strain curve of RAC with various RCA content can be divided into three characteristic parts. The first part represents the linear portion, the second represents the nonlinear portion of the ascending branch, and the third part is the descending branch;
3. The curvature of each ascending branch of the stress-strain curve increase with the increasing of the RCA content, while the slope of the descending branch of the complete stress-strain curves decrease as the RCA content increases.

The above findings are based on tests on concrete made of RCA with relatively uniform quality. If the RCA has a non-uniform quality, the diversity of stress-strain curves might become greater. This has been confirmed by Xiao and Du [84]. In their tests, the RCA used was a mixture of three RCAs produced from six different kinds of parent concrete (see Fig. 3.30).

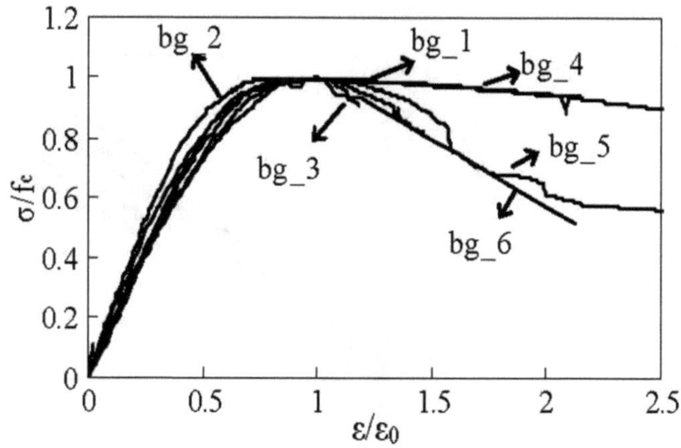

Fig. 3.30 Dimensionless stress-strain relationship curve of RAC with different parent concrete [84]

Various mathematic models for the stress-strain relationships of RAC under uniaxial compression have been proposed, e.g. Xiao and Li [82, 87], Hu and Song et al. [83], Deng and Yang et al. [89]. In the Technical Code for *"Application of Recycled Concrete" (DG/TJ08-07)*, the following analytical model was recommended:

$$y = \begin{cases} ax + (3 - 2a)x^2 + (a - 2)x^3 & 0 \leq x < 1 \\ \dfrac{x}{b(x-1)^2 + x} & x \geq 1 \end{cases} \tag{3.12}$$

Where $x = \varepsilon/\varepsilon_0$, $y = \sigma/f_c$, as ε_0 is the peak strain and is the prism compressive strength. The parameters a and b are determined, depending on the RCA replacement percentage ($f_c r$).

$$\begin{aligned} a &= 2.2\left(0.748r^2 - 1.231r + 0.975\right) \\ b &= 0.8(7.6483r + 1.142) \end{aligned} \tag{3.13}$$

Equation (3.12) is in fact from Chinese Code for *"Design of concrete structures"* (GB5001-2002) for conventional concrete under uniaxial compression [80]. The above model has been used to by some researches in their nonlinear finite element analysis of RAC structures, e.g. Yang [76]. Fairy good results have been achieved.

3.4.5.2 Uniaxial Tension

Xiao and Lan [65] evaluated the stress–strain curve of RAC through an intensive test by five group of prism specimens with dimension of 100 × 100 × 400 mm.

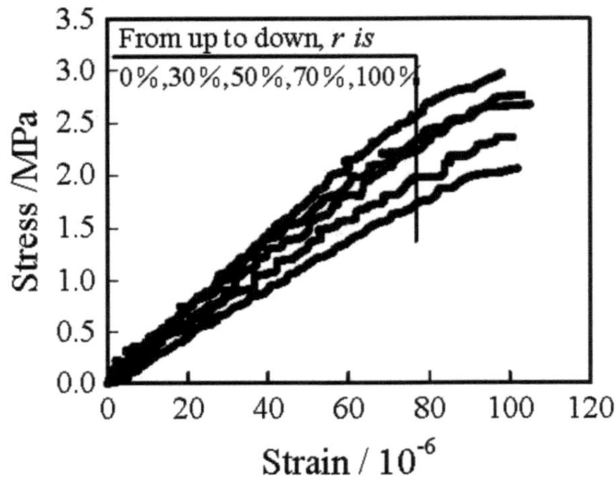

Fig. 3.31 Uniaxial tensile stress-strain curves of RAC with different replacement percentage [65]

The experimental investigation on the uniaxial tensile behavior of RAC were carried out with the same effective water-cement ratio but different RCA contents.

The uniaxial tensile stress-strain relationship curves of RAC with different replacement percentage are shown in Fig. 3.31. With an increase of the RCA replacement percentage the peak tensile strain of the RAC is slightly increased whereas the tensile strength and the tangent modulus of the RAC are decreased.

The analytical model which was firstly proposed by Guo et al. [74] for the uniaxial tension stress-strain relationship of conventional concrete was extended by Xiao and Lan [65] for RAC under uniaxial tension, as shown below:

$$y = ax - (a - 1)x^6 \tag{3.14}$$

Where $x = \varepsilon/\varepsilon_0$, $y = \sigma/f_i$, the parameter a is related to the RCA replacement percentage (r) and can be determined by statistical regression analysis, given as:

$$a = 0.07r + 1.19 \tag{3.15}$$

3.4.5.3 Pure Shear

The behavior of RAC subjected to pure shear was investigated based on the shear test of four-point loading beam with the same height and different width by Huang, Deng and Luo et al. [73]. The result shows that the failure mode of the sheared RAC specimens is similar to that of conventional concrete, but shear strength and deformability of the girders are lower than those of conventional concrete.

The following analytical formulation was used to describe the stress–strain curve of RAC under pure shear:

$$y = ax^4 - 2ax^3 + (a + 1)x \tag{3.16}$$

Table 3.6 Relationship between RCA replacement percentage and parameter a [73]

Replacement percentage (%)	0	30	50	70	100
The parameter *a*	0.8	0.4	0.3	0.3	0.3

Where $x = \gamma/\gamma_p$, $y = \tau/f_v$, γ is shear strain, is peak shear strain, γ_p τ is shear stress (MPa), f_v is shear strength (MPa). The parameter describes the influence of the RCA replacement percentage and can be determined by regression analysis, as given in Table 3.6.

3.4.6 Durability

The durability of recycled concrete is a very important area in the research of recycled concrete. It is defined as its ability to resist weather action, chemical attack, abrasion, or any other process of deterioration. Chinese researchers had carried out a lot of investigations with various effects of environment on durability of recycled concrete. However, the results show that the low density, high absorption, high crushing index and poor durability of recycled aggregate have led to the recycled concrete have poor carbonation resistance, freezing and thawing resistance, permeation resistance, abrasion resistance, and so on. The promotion and application of recycled aggregate concrete is greatly limited by the defects of recycled aggregates. Based on the Analysis of literature, this paper introduce the development status of durability of recycled concrete in China from carbonation resistance, Chloride penetration, freezing and thawing resistance, drying shrinkage, creep and abrasion resistance, and give some suggestions for the further research on the durability of recycled concrete.

3.4.7 Carbonization

The carbonization of concrete is a phenomenon that CO_2 in the air and water interact with hydration product (mainly $Ca(OH)_2$) of set cement in concrete, producing carbonate or some others, thus lowering the concrete alkalinity (neutralization, in other words). The density of recycled aggregate concrete is lower than the natural concrete for the porosity of recycled aggregate is higher than natural aggregate, which will undoubtedly reduce the carbonation resistance. On the other hand, the old mortar on the surface of recycled aggregate increase the total amount of may be carbonized material, which is a beneficial effect to the carbonation resistance of recycled concrete.

The experiment by Yuan (2010), Zhuang (2009), Sun (2006) and Lei (2008) show that the depth of recycled concrete carbonation Increase with replacement rate of recycled aggregate. Yuan (2010) has carried out experimental investigations on the durability of recycled concrete. The results showed that the anti-carbonizing capability of recycled aggregate is worse than that of normal concrete. Fly ash have adverse effects on the carbonation resistance of recycled concrete, the influence of fly ash should be considered when its content more than 10%. The Experiment of Zhang (2009) concluded that the carbonation resistance of recycled concrete decrease with the increase of replacement rate of recycled aggregate, but the carbonation depth of recycled concrete will reduce when the replacement rate of recycled aggregate is more than 70%. Increasing water-cement ratio will reduce the carbonation resistance of recycled concrete, but the carbonation resistance of recycled concrete can be improved when the water-cement ratio is constant. What's more, the recycled concrete will have better carbonation resistance by some technical measures, such as reduce the maximum size of recycled aggregate, the second mixing process, using recycled aggregate with state of semi-saturated surface dry, Adding mineral admixture(Fly ash or slag), etc. Through experiment, Lei (2008) concluded that carbonation depth of recycled concrete is proportional to the square root of carbonization time; the carbonation rate slow with the increase of age; the carbonation depth of recycled concrete is inversely proportional to the strength of original concrete; and the carbonation depth of recycled concrete will increase when the recycled coarse aggregates mixed with brick aggregate. Liu (2009) studied the influence of cement content, the type and replacement rate of recycled fine aggregate to the carbonation of recycled concrete. The result show that the same replacement rate of high-quality recycled fine aggregate concrete is superior to simple crushed recycled fine aggregate concrete in carbonation resistance, but inferior to the natural fine aggregate concrete carbonation.

3.4.8 Chloride Penetration

Chloride ingress is one of important reasons causing steal rust, therefore the durability of concrete structure is effected and even the structure will be destroy. As recycled concrete, chloride resistance capability is also an important factor for the durability of structure. As the existence of old cement mortar in the recycled aggregate and cracks of recycled aggregate produced in the broken process, which makes the chloride penetration resistance decreased some extent. What's more, the old interface in the recycled aggregate have an impact on the chloride resistance of recycled concrete also.

Zhang (2009) has studied the effect of some factors (such as the replacement rate of recycled coarse and fine aggregate, mineral admixture and so on) on chloride permeability of RC. The results showed that the permeability of concrete with recycled fine aggregate of high quality surpassed that of natural aggregate concrete, whereas there is little difference between the permeability of concrete with recycled coarse aggregate of high quality and that of natural aggregate concrete. It is also concluded that blending in superfine mineral powder and silica fume would remarkably improved the permeability of RC, nevertheless, the fly ash had adverse impact on the permeability of RC. After experimental study, Wu (2010) found that fly ash can improve the chloride resistance of recycled concrete, and the best quantity of fly ash is from 10 to 15%. Ye (2009) studied the influence of high-quality recycled aggregate (using the production technology of Wang (2007)) to the chloride resistance of recycled concrete. The results show that the chloride resistance of high-quality recycled concrete is better than normal recycled aggregate concrete, but worse than normal concrete. The experiment operated by Sun (2009) showed that the chloride resistance of recycled concrete reduced with the increase of replacement rate of recycled aggregate. This is because the porosity and interface of recycled concrete will increase with the replacement rate of recycled aggregate for the porosity of recycled aggregate is higher than normal aggregate and the old interface existence in the recycled aggregate. The experiment performed by Du (2006) showed that recycled concrete has good resistance to chloride penetration, but its resistance to chloride penetration is somewhat less than normal concrete. The chloride penetration resistance of recycled concrete will be enhanced with the increase of strength of recycled concrete; it can be improved when the Mineral admixture (Slag, fly ash, silica fume, etc) was mixed also, the composite mixing of mineral admixture is better than separately and the effect of silica fume best. Through experiment, Wu (2009) concluded that the chloride penetration resistance of recycled concrete will be improved with the increase of strength of recycled concrete. The chloride penetration resistance of recycled concrete and normal concrete only has little difference under same strength grade.

3.4.9 Freeze and Thaw

The frost resistance of concrete is the property of concrete (with saturation condition) which could withstand freeze-thaw cycle several times with no failure occurred, and the decrease in concrete strength is limited. Recycled aggregate in recycled concrete always Freeze-thaw damage before the new cement matrix for the old mortar have a higher absorption and it absorb water very fast, which makes the frost resistance of recycled concrete very poor. In the other hand, the larger porosity of recycled aggregate can also play a role in micro-conservation to improve the frost resistance of recycled concrete.

At present, the research of Chinese scholars on the frost resistance of recycled concrete is still in the initial stage. The study result on the frost resistance of recycled concrete is quite different, there are not a unified conclusion until now. Wang (2009) has researched on the frost resistance of RC with 100% recycled aggregate. The results showed that, along with the increasing of freeze-thaw cycle times, the declining rate of quality and relative dynamic elastic modulus of RC is less than that of NC, while the declining rate of strength of RC is greater than that of NC. Cheng (2008) have studied the effect of RA, which is strengthened by water glass solution with different concentration, on the frost resistance of RC. The results showed that higher concentration of water glass was of benefit to improving the frost resistance of RC if the concentration of water glass is less than 7% (by mass). Dai (2007) taking advantage of waste concrete recycled aggregate studied the freezing-thawing resistance of recycled concrete by using recycled aggregate from 50 age period of waste concrete, and the replacement ratio of recycled aggregate respectively by 25, 50, 75 and 100%. The result show that the recycled concrete incorporated different percentages of recycled aggregate and normal concrete have little difference in mass loss after some freezing and thawing cycles. The reason is due to the large water absorption of recycled aggregate and freeze-thaw cycle experiment was carried out mainly in the water, so recycled concrete absorb more water than normal concrete. The large water absorption of recycled aggregate has a harmful influence for itself Frost heave, which lead to the strength loss of recycled concrete higher than normal concrete. The experiments of Cui (2006) and Zou (2010) concluded that the loss rate of relative dynamic elastic modulus of recycled concrete is higher than normal concrete, and loss rate of relative dynamic elastic modulus increase with the replacement rate of recycled aggregate. Through using ordinary method, pre-wet slurry method and increasing cement method three mix proportion design method, Zhang (2005) studied the frost durability of recycled concrete. The result show that recycled concrete can achieve even surpass normal concrete in frost durability after air-entraining agent Incorporated, and the effect is best when adopt increasing cement method. In addition, reducing water-cement ratio can improve the frost resistance of recycled concrete. Li (2003) studied the influence of fly ash to the frost resistance of recycled concrete, and found the effect is not obvious when 12% cement replaced by fly ash. Qin (2005) studied the performance of recycled concrete after one-time freeze in the early hardening. The result show that the pre-curing time (curing period before cold) and freezing day has a greater influence to the later stage strength of recycled concrete; difference freezing temperature has a smaller influence to later stage strength of recycled concrete but a large influence to the strength when one-time freeze was thawed; recycled is not suitable for winter construction for it need a longer time to make the late strength was not damaged.

3.4.10 Abrasion

Abrasion resistance is an important indicator in judging the performance of concrete pavement. The abrasion resistance of concrete in pavement is determined by its strength and hardness, especially in the surface layer. Coarse aggregate play a important role in the abrasion resistance of concrete. Recycled aggregate has an adverse effect to the abrasion resistance of recycled concrete for its strength is lower than nature aggregate.

As shown in the work by the authors Li (2010) and Peng (2009) stated that the abrasion resistance of recycled concrete is lower than normal concrete when their s same. Peng (2009) studied the influence of blending fly ash and slag into cement and its content to the abrasion resistance of recycled concrete. The result show that the influence of blending fly ash and slag into cement to the abrasion resistance of recycled concrete is not obvious in early stage, but it has significant influence to abrasion resistance in the late stage; the best ratio of fly ash and slag is 4 to 1; the abrasion resistance of recycled concrete decrease with the increase of replacement rate of fly ash and slag in the early stage, and the best replacement rate in improving the abrasion resistance of recycled concrete in late stage is 20%. After the study the influence of recycled coarse aggregate, polypropylene fibers and air entraining and water reducing admixture to the abrasion resistance of recycled concrete. Chen (2010) conclude that the replacement rate of recycled aggregate has the greatest impact on the abrasion resistance of recycled concrete and the replacement rate of recycled aggregate is not suitable more than 40%.

3.4.11 Structural Performance of Recycled Concrete

To make RAC as a widely accepted structural material, a comprehensive experimental study has been carried out in recent years in China. The study covers not only the monotonic behaviour of the common reinforced RAC members (including beams, columns and slabs) but also the seismic performance of reinforced RAC elements and structures, such as beam-column joints, shear walls and frame structures. These investigations provide valuable information on the structural performance of RAC.

Based on the test results, some recommendations for the design of reinforced RAC members are given in the first specifications and standard for RAC in China—Technical Code for Application of Recycled Aggregate Concrete (DG/TJ08-2018-2007) [90]. This paper gives a summary and discussion on the related findings and also a brief introduction on the design provisions in the technical code (DG/TJ08-2018-2007).

3.4.12 Reinforced RAC Elements Under Monotonic Loadings

3.4.12.1 Flexural Behaviour of RAC Beams

The flexural behaviour of reinforced concrete beams is the basic and very important property. A large number of experiments on the cracking behaviour, the failure mode, the flexural capacity as well as the deflection of RAC beams, have been undertaken in China. Simply-supported RAC beams were incorporated in most of the testes. Figure 3.32 shows the test setup of Xiao and Lan [91].

The most important mechanical parameters and the corresponding test conditions collected are briefly described and given in Table 3.7.

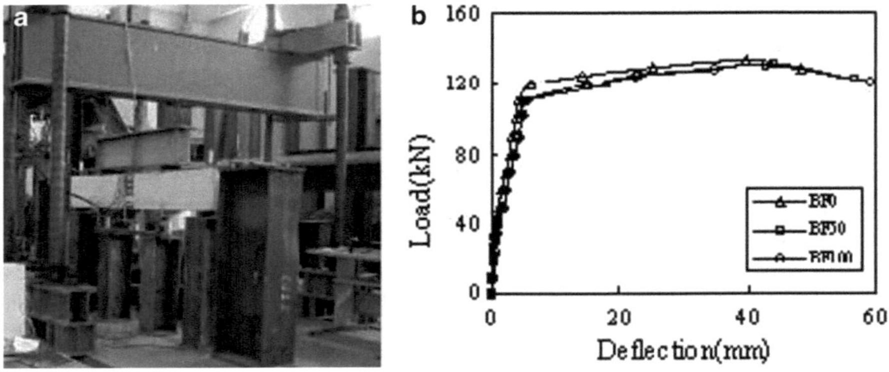

Fig. 3.32 Bend test and results of RAC beams [91]. (**a**) Test setup (**b**) Curves of load versus the mid-span deflection

Table 3.7 Collected test results on the flexural behaviour of RAC beams

Investigator	RCA content (%)	28 day compressive strength (cube) (MPa)	Size(mm) L × B × H	Reinforcement ratio (%)	Ultimate load(kN)/ moment (kN.m)
Xiao and Lan [92]	0 (BF0)	31	2,300 × 150 × 300	0.77	133/33.3
	50(BF50)	35			131/32.8
	100(BF100)	34			128/32.0
Li and Song et al. [93]	0	41.9	2,000 × 150 × 300	0.82	155/46.5
	30	45.6		0.82	153/45.9
	70	43.7		0.82	159/47.6
	100	39.1–50.5		0.55–2.43	114–317/ 47.1–95.1
Hu et al. [94]		38.4	2,300 × 150 × 300	0	18/–
		36.4		0.56	62.5–70/–
		36.8		1.00	100–125/–
		37.0		1.58	150–170/–
		38.4		2.49	320–340/–

The following variables were involved:

- The influence of recycled coarse aggregates (RCA) replacement ratio, which is defined as the weight percentage of RCA within the gross coarse aggregates. (Xiao and Lan [91], Ding et al. [95] and Li et al. [96]).
- The influence of longitudinal reinforcement ratio (Li et al. [96], Hu et al. [97])

Based on the experiments noted above, the following conclusions can be drawn:

- The reinforced RAC beams show very similar crack patterns and failure feature, irrespective of the recycled aggregate percentage (Fig. 3.33).
- The assumption of "plane sections remain plane" in the classic flexure theory, that is, sections perpendicular to the axis of bending which are plane before

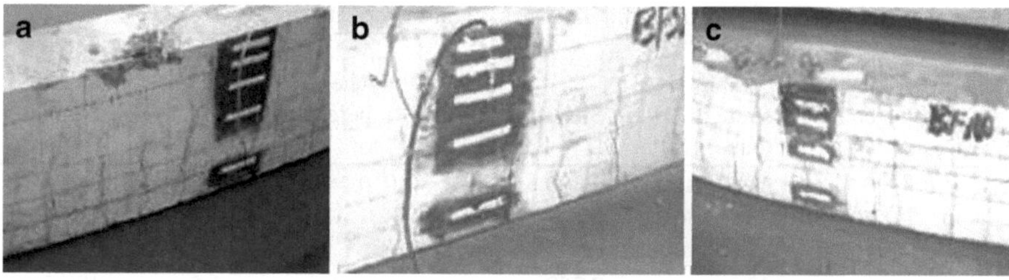

Fig. 3.33 Crack patterns and failure mode of RAC beams [92]. (**a**) BF0 (**b**) BF50 (**c**) BF100

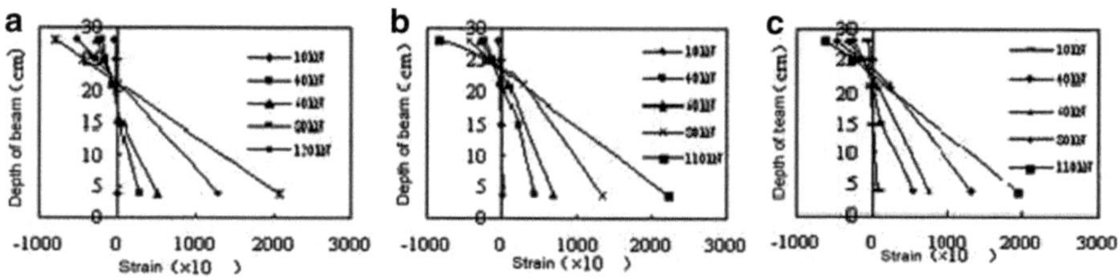

Fig. 3.34 Beam depth-stain relationship [92]. (**a**) BF0 (**b**) BF50 (**c**) BF100

bending remain plane after bending, is also valid for reinforced RAC beams with various recycled aggregate percentages (Fig. 3.34). That means the procedure developed for reinforced conventional concrete beams could be suitable for the analysis of RAC beams subjected to flexure.

- Under same conditions, the cracking and ultimate moment of the RAC beam are similar to the natural aggregate concrete (NAC) beam, nevertheless, the stiffness of the RAC beam reduces to a certain extent.
- Comparing the RAC beam to NAC ones, the research results show that under the action of the same load, the maximum width of cracks of the former, are slightly larger.

To estimate the flexural capacity of reinforced RAC beams, the predictions from the design equations recommended in EC 2 [98] and Chinese Code for Design of Concrete Structures (GB 50010-2002) [80], and the test data are compared.

In the Technical Code for Application of RAC (DG/TJ08-2018-2007), the following formula is recommended to calculate the flexural bearing capacity of RAC beams.

$$M_u = \alpha_m f_c b\left(h_0 - \frac{x}{2}\right) \tag{3.17}$$

$$\alpha_m f_c bx = f_y A_s \tag{3.18}$$

Where is the ultimate moment (kNm); b, M_u h_0 are the width and the effective height of the section (mm); are the yield strength (MPa) and the area (mmf_y A_s^{2}) of

Table 3.8 Collected test results on the shear behaviour of RAC beams

Investigator	RCA content (%)	28 day compressive strength (cube) (MPa)	Size(mm) L × B × H	The percentages of stirrups	The shear-span ratio	Ultimate load (kN)
Xiao and Lan [60]	0	31 (BS0)	2,300 × 150 × 300	0.25%	1.5	330
	50	35 (BS50)				298
	100	34 (BS100)				274
Zhou et al. [51]	0	42.2	2,250 × 150 × 300	0.16%	1.5	190.5
	10	40				188.0
	20	37.8				185.4
	30	34.9				181.4
	40	34.2				174.0
Zhang et al. [52]	0	28.7–38.7		0	1.0–3.0	114–530
	30	32.4	2,400 (2,000) × 150 × 300		1.5	336
	50	33.3			1.5	350
	70	30.4			1.5	300
	100	29.6–34.1			1.0–3.0	119–509

the longitudinal reinforcement; are the cylinder compressive yield strength (MPa) of the RAC; f_c α_m, the modified parameter for, can be taken as $0.95.f_c$

In addition, Xiao et al. [55] carried out the reliability analysis of the recycled concrete beams bending capacity by the Montcalo method. The main findings of the reliability analysis can be summarized as the following:

- The reliability index of recycled concrete beams is proved to meet with the requirement of the Unified Standard for Reliability Design of Building Structures (GB 50068-2001) [99].
- The effect of uncertainty coefficient associated with resistance is obvious, while the partial safety factor of recycled concrete compressive strength on the reliability index of ideally reinforced recycled concrete beams bending capacity is neglectable.

3.4.12.2 Shear Behaviour of Reinforced RAC Beams

Unlike in reinforced concrete beams under flexure, where the longitudinal reinforcement governs the structural response, concrete plays a more significant role in beams subjected to shear loadings. Taking into account the difference between RAC and conventional concrete, the study on the shear performance of reinforced RAC beams is of more importance.

To investigate the influence of the recycled aggregates on the shear behaviour of the reinforced concrete beams, experimental work has been carried out, both on beams with and without web reinforcements. The details of the experiments are presented in Table 3.8.

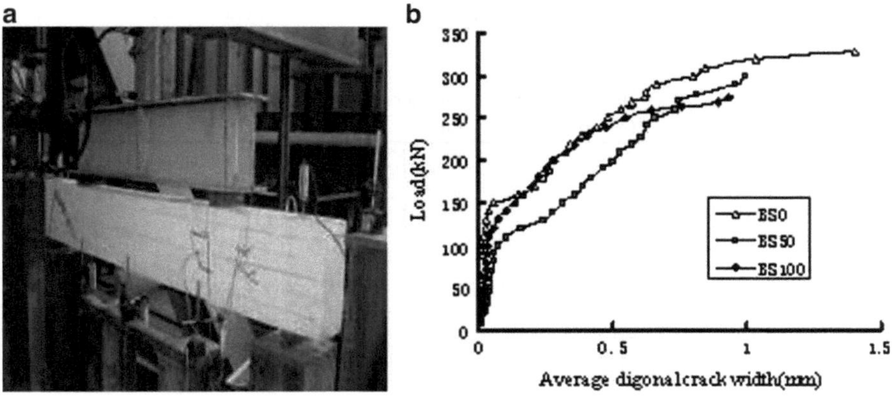

Fig. 3.35 Shear test and results of RAC beams [100]. (**a**) Test setup (**b**) Curves of load versus the diagonal crack width

As shown in Fig. 3.35, Xiao and Lan [100], Zhou and Jiang [101] tested the shear behaviour of RAC beams with web stirrups. On the basis of the test results, the characteristics of shear behaviour of RAC beams were discussed. By using a general finite element method, the influences of recycled aggregates' replacement percentage, the shear span to depth ratio and the percentage of stirrups on the shear behaviour of RAC beams were theoretically analyzed [100]. These studies show that the shear capacity of recycled concrete beam reduces with the increase of recycled coarse aggregates content, whereas, the deflection and diagonal crack width is similar to that of the conventional concrete beam. The shear failure pattern of RAC beam is typical shear-compression failure when the beam is properly designed. The diagonal section cracking load of RAC beams is slightly lower than that of NAC beams, and average diagonal crack width of RAC beams is slightly larger than that of NAC beams.

It is evident that the influence of RAC can be better reflected using reinforced beams without web reinforcements. To investigate the shearing capacity of RAC beams without stirrups, Zhang et al. [102] conducted an experiment using 13 RAC beams and a natural aggregate concrete (NAC) beam. The effects of replacement percentages of recycled coarse aggregates and shear-span ratio on the characteristics of shear capacity and deformation of RAC beams were studied. The test result shows the deflection and diagonal crack of RAC concrete beams are bigger than NAC beams, the shear capacity of RAC beams reduces with the increase of recycled coarse aggregates' content, the shear capacity reduces with the increase of shear-span ratio.

In the Technical Code for Application of Recycled Aggregate Concrete (DG/TJ08-2018-2007) [103], the following formula is used to calculate the shear capacity of RAC beams,

$$V_u = a_v\left(0.7f_tbh_0 + 1.25f_{yv}\frac{A_{sv}}{s}h_0\right) \tag{3.19}$$

Table 3.9 Collected test results on the compressive performance of unconfined RAC column

Investigator	RCA content (%)	28 day compressive strength (cube) (MPa)	Slenderness ratio	Eccentricity (mm)	Section size(mm)	Ultimate load(kN)
Xiao et al. [53]	0 (C-e-1)	31–34	7	0 (C-0)	180 × 120	393–501
	50 (C-e-2)			30 (C-1)		501–523
	100 (C-e-3)			82 (C-2)		315–340
				100 (C-3)		280–310
Zhou et al. [71]	0	42	3	0	200 × 200	908–920
	5	40.76				890–903
	10	40.2				872–884
	15	38.3				825–863

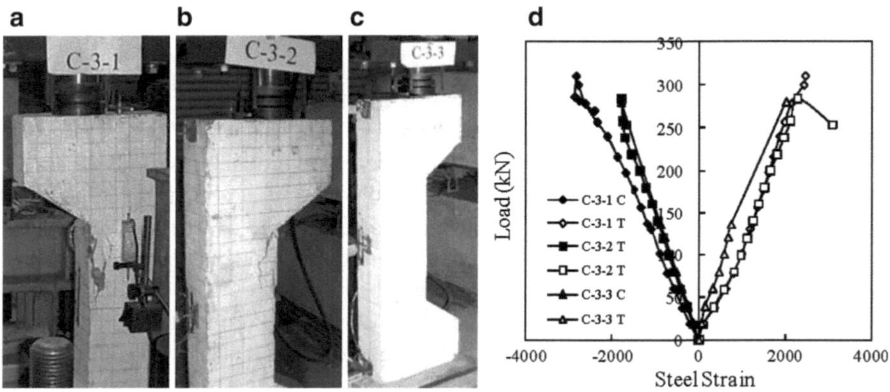

Fig. 3.36 Performance of RAC columns when eccentricity is 100 mm [53]. (**a**) C-3-1 (**b**) C-3-2 (**c**) C-3-3 (**d**) Load-steel strain relationship. Note: *T* tensile steel bar, *C* compressive steel bar

In which, a_v is the reduction coefficient considering the RCA content influence; f_t is the tensile strength of RAC (MPa); f_{yv}, A_{sv}, s are the yield strength (MPa), the area (mm^2) and the spacing of the stirrups (mm). The other signs share the same meanings as in Eqs. (3.17) and (3.18).

3.4.12.3 Reinforced RAC Columns Under Axial and Eccentric Compression

The compressive behaviour of RAC columns under axial loading was studied by Zhou et al. [104]. The compressive behaviour of reinforced RAC columns under axial and eccentric compression was studied by Xiao et al. [105]. A summary of the specimens tested is presented in Table 3.9.

Based on the experimental results from Table 3.9 as well as from Fig. 3.36, it was proved that, in comparison with the NAC columns, the failure modes of RAC columns still have compression failure, tensile failure and a balance failure under compressive loading with different eccentricities. The assumption “plane sections

Table 3.10 Collected test results on the compressive performance of confined RAC columns

Investigator	28 day compressive strength (cube) (MPa)	Dimension of section D × t (mm)	L (mm)	e (mm)	e/r_0	N_{ue} (kN)
Yang and Han [54]	42.7(0% RCA)	165 × 2.57	1,650	0	0	1,090–1,158
	41.8(25% RCA)			20	0.24	795–855
	36.6(50% RCA)			40	0.48	600–607
		150 × 2.94	1,732	0	0	1245–1273
				20	0.27	825–875
				40	0.53	625–686

remain plane" is also suitable for RAC columns. The load capacity calculation of RAC columns can be estimated by the equation for NAC columns used in the current Chinese Code GB50010 (2002) [80]. In the technical code for RAC [101], the following equation is used to predict the compression bearing capacity of RAC column under axial loading.

$$N \leq 0.9\alpha_m \varphi \left(f_c A + f'_y A'_s \right) \tag{3.20}$$

Where α_m is adjustable coefficient of RAC column; φ is the stability coefficient. The other signs share the same meanings as in Eqs. (3.17) and (3.18).

3.4.12.4 Compressive Performance of RAC Filled Steel Tube Columns

For concrete filled steel tube columns, numerous experiments have verified the benefits due to the confinement of the steel tube on the core concrete. To investigate the possible confinement on RAC, Yang and Han [106] tested the compressive performance of RAC filled steel tube(RACFST) columns. In their tests, 56 RACFST members under short-term loadings, including 24 stub columns, 8 beams and 24 eccentrically loaded columns, were tested. The corresponding normal CFST members, including 6 stub columns, 2 beams and 6 eccentrically loaded columns, were also tested to evaluate the differences between them. The specimens were designed to investigate the effects of the following parameters: (1) cross-section type, circular and square; (2) RCA replacement ratio, from 0 to 50%; (3) load eccentricity ratio, from 0 to 0.53 [107].

The test results (Table 3.10) show that all the columns failed due to overall buckling of steel tube. Comparisons were made with predicted ultimate bearing capacity of RACFST columns using the existing codes, such as ACI Committee 318 [108], AIJ-1997 [109]. Figure 3.37 shows the effects of RCA replacement ratio (r) on bearing capacity of stub columns (Nue,s) and columns under eccentric loading (Nue,e), where D and B are the outside diameter and width of circular and square members respectively, e is the load eccentricity, and r_o is D/2 or B/2 for circular or square specimens respectively. It can be seen that, with increase of RCA replacement ratio, the bearing capacity of the tested specimens gradually decreases.

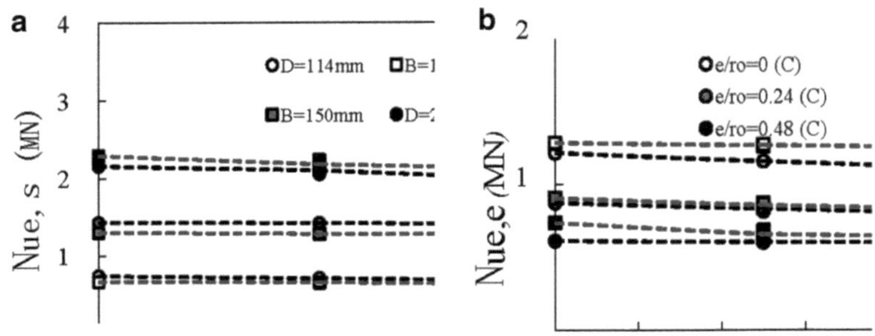

Fig. 3.37 Effects of RCA replacement ratio on bearing capacity of the tested specimens [109]. (**a**) Stub columns (**b**) Eccentrically loaded columns

Table 3.11 Summary of specimens for flexural performance of RAC slabs

Investigator	RCA content (%)	Cracking load (kN)	Ultimate load (kN)
Zhou et al. [87]	0 (RC0)	17.4–18	69.8–72.1
	5 (RC5)	15.9–16.9	63.5–67.5
	10 (RC10)	14.5–15.6	58.0–62.4
	15 (RC15)	13.2–14.2	13.2–14.2

A theoretical model for normal concrete filled steel tube (CFST) columns is adopted for RACFST columns. Similar to normal CFST, confinement factor (ξ_r) is suggested to model the interaction between outer steel tube and core RAC of RACFST columns, and it is defined as:

$$\xi_r = \frac{A_g \cdot f_y}{A_c \cdot f_c} \tag{3.21}$$

Where A_s and A_c are the cross-sectional area of the steel tube and core concrete (mm^2); f_y and f_c are the yield strength of steel and the characteristic compressive strength of RAC, respectively (MPa).

3.4.12.5 Flexural Performance of RAC Slabs

Zhou et al. [110] studied the performance of RAC slab which are simply supported on four sides. In their tests, 12 slabs divided into four groups were tested. All the slabs shared the same cross-section and reinforcement ratio but with different recycled aggregates' replacement percentage, i.e. 0, 5, 10, 15%, respectively. A summary of the specimens tested is listed in Table 3.11.

The specimens were designed to be subjected to concentrated load. The test results indicate that the recycled aggregates replacement percentage influences the load mid-span deflection behaviour, as shown in Fig. 3.38.

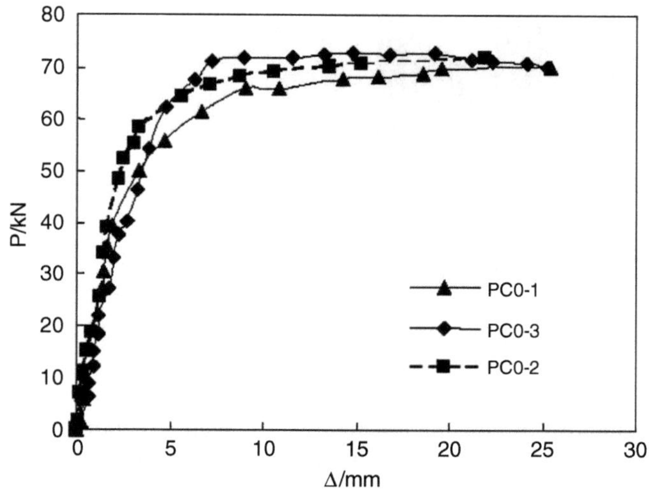

Fig. 3.38 Load mid-span deflection curve of RAC slabs [110]

It is also found that with the increase of recycled aggregates' replacement percentage, the cracking load and the ultimate load of the slab decrease while the deformation of the slab increase; however, the decreasing tendency of the ultimate load is greater than that of the cracking load.

By studying the deflection of the RAC slabs, the general method (Eq. 3.22) used for determining the deflection of conventional concrete slab is not suitable for estimating the deflection of the RAC slabs.

$$\Delta = \alpha \frac{Ml_0^2}{B_+} \tag{3.22}$$

Where Δ is the deflection value of the slab (mm); α is deflection coefficient; B_s is the stiffness of the slab (N/m); l_0 is the effective length of the section (mm); M is the ultimate moment (kNm). Nevertheless, the test results show that the deflection of the RAC slabs can be also obtained by multiplying the value of Δ with an amplification factor which can be taken as 1.4.

3.4.12.6 Steel Deck RAC Composite Slabs

In order to investigate the promotion of RAC in the steel deck concrete composite slabs, 6 cold formed steel deck-RAC composite slabs without stud were tested under static loading by Xiao et al. [111]. The test specimens and various stages load of the composite slabs are shown in Table 3.11. Where p_a is the critical value of the slab; p_b is the ultimate load; p_{20C} is the load value when the deflection value of the slab is $L/200$.

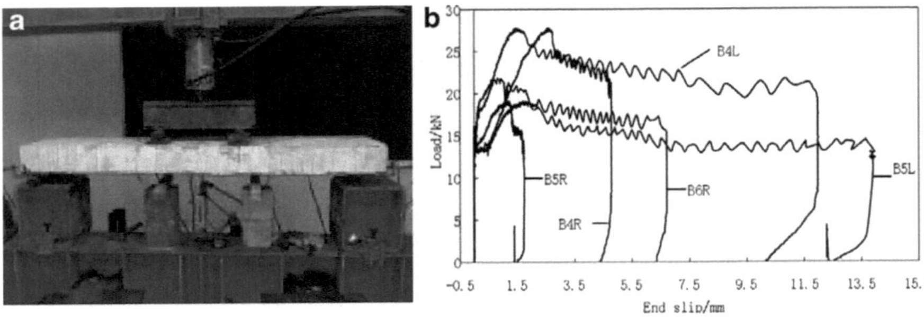

Fig. 3.39 Test setup and main load-relative slip curve of composite slabs [57]. (**a**) Test setup (**b**) Load – relative slip curve of the some composite slabs

Based on the test results, the effects of shear-span to depth ratio to the failure model as well as the relationship between the longitudinal slide and load were investigated. The test setup and the main test result are given in Fig. 3.39.

It was found that the slip between RAC and steel deck was relatively large. The failure mode was a longitudinal splitting failure. By analyzing the measured load-slip curves, the influence of recycled aggregate content on the bearing capacity of steel deck RAC composite slabs was investigated. With increasing RCA content, the bearing capacity of the slabs first increased and then decreased. Nevertheless, the RCA content had nearly no influence on the ultimate shear bearing capacity of the slabs. Longitudinal shear capacity became lower as the shear span ratio increased. A formula related to the shear capacity of the slabs was proposed and recommended as expressed in Eq. (3.23).

$$V_u = \alpha_{lv} \cdot \left(175.406\, \rho h_0 / L_v + 0.0391\sqrt{f_c}\right) \tag{3.23}$$

Where α_{lv} is the shear capacity coefficient; ρ is the steel reinforcement ratio; is the shear span (mm); L_v h_0 is the effective height of the section (mm); f_c is the cylinder compressive yield strength of the RAC (MPa).

3.4.13 *Reinforced RAC Structures Under Cyclic Loadings*

3.4.13.1 RAC Beam-Column Joints

Xiao and Zhu [112] investigated the behaviour of reinforced beam-column joints under low frequency reversed lateral loads. In their study, 3 half-scaled specimens were tested. The objective of the test was to investigate the influence of the recycled aggregate on the seismic performance of the reinforced concrete frame joints. The used recycled aggregates' replacement percentages were 0, 50 and 100% respectively. Figure 3.40a shows the typical test specimens. Based on the

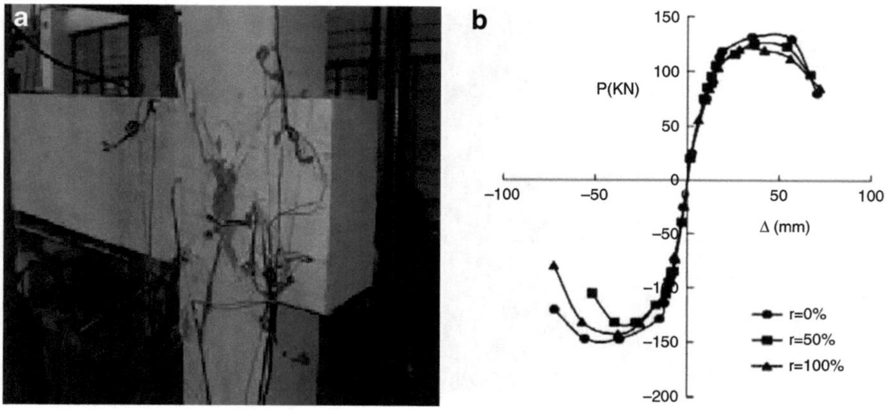

Fig. 3.40 Failure mode and results of RAC column-beam joints [58]. (**a**) Failure mode (**b**) Skeleton curve of specimens

Table 3.12 Summary of specimens for seismic behaviour of RAC column-beam joints

Investigator	RCA content (%)	28 day compressive strength (cube) (MPa)	Ductility coefficient	Ring stiffness
Xiao and Zhu [58]	0	42.0	4.85	3.95
	50	42.5	4.25	3.94
	100	40.0	4.34	4.11

test observations, the failure features, the hysteresis loops, the ductility and the degeneration of strength and stiffness under seismic loads were comparatively investigated.

The test results in Table 3.12 indicate that the seismic behaviour of the reinforced beam-column joints made of RAC is slightly weaker than that with conventional concrete, as can be seen from the skeleton curves of the specimens illustrated in Fig. 3.40b. However, it is found that the ductility coefficient and the stiffness of the RAC joints are able to meet the related requirements for the current earthquake-resistance design provisions.

On the basis of the test data, Xiao and Zhu [112] concluded the following equation in Chinese Code for Design of Concrete Structures (GB50010-2002) [80] can be used to calculating the shear capacity of joints made of RAC.

$$V_u = 0.1\left(1 + \frac{N}{f_c b_c h_c}\right) f_c b_j h_j + f_{yv} \frac{A_{sv}}{s} (h_o - a_s) \tag{3.24}$$

Where f_c is the cylinder compressive yield strength (MPa); b_c,h_c is the width and the effective height of the column section (mm); b_j,h_j is the effective width and the effective height of the joint core section (mm); N is the axial compression of the column (kN); a_s is the cover thickness (mm). The other signs share the same meanings as in Eq. (3.19).

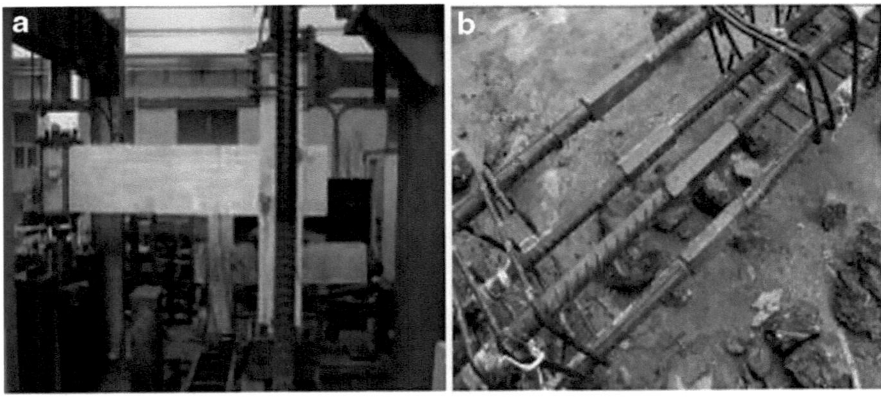

Fig. 3.41 Test setup and steel reinforcement connection of pre-cast joint [114]. (**a**) Test setup (**b**) Steel reinforcement connection

Table 3.13 Collected test results on the seismic behaviour of RAC pre-joints

Investigator	Name of specimens	Types of steel reinforcement connections	28 day compressive strength (cube) (MPa)	Ductility coefficient	Stiffness deterioration coefficient
Xiao et al. [114]	PJ-0	Welding	42.0	7.54	0.77
	PJ-1	Welding	36.8	9.34	0.65
	PJ-3	Beam connector	38.1	2.13	1.89
	PJ-4	column connector	37.3	8.59	0.88

3.4.13.2 Pre-cast RAC Beam-Column Joint

To explore the potential of RAC in precast concrete structures, Xiao et al. [113] investigated the behaviour of RAC frame joints with precast beams and columns. The testing variables were the types of steel reinforcement connections, i.e., welded reinforcement, beam reinforcement connector and column reinforcement connector. Four specimens were tested under lateral reverse loadings with simulated practical boundary conditions. The test setup and steel reinforcement connection are displayed in Fig. 3.41.

The test results (Table 3.13) indicate that the behaviour of the pre-cast RAC joints is similar to that of joints with conventional concrete. The failure process can be divided into five stages: crack primarily, crack entirely, beam yielding, and limit state and ultimate failure. The type of the steel reinforcement connections has a significant effect on the bearing capacity, ductility and energy consumption of the joints. However, the ductility, energy consumption of the RAC joints is found to be able to meet the design requirements for precast concrete structures. This study gives a positive result towards the use of RAC in precast concrete structures.

Table 3.14 The properties of the shear wall specimens [114]

Specimens		Percentage of recycled aggregate replacement (%)		Reinforcement percentage in the walls (%)	Axial-force ratio
		Fine	Coarse		
Low-rise shear wall	RCSW1.0-1	0	0	0.25	0.2
	RCSW1.0-2	100	33	0.25	0.2
	RCSW1.0-3	100	67	0.25	0.2
	RCSW1.0-4	100	100	0.25	0.2
	RCSW1.0-5	100	100	0.25	0.4
	RCSW1.0-6	0	100	0.25	0.2
	RCSW1.0-7	100	100	0.25	0.2
High-rise shear wall	RCSW2.0-1	0	0	0.25(0.25)	0.2
	RCSW2.0-2	0	0	0.25(0.15)	0.2
	RCSW2.0-3	50	50	0.25(0.15)	0.2
	RCSW2.0-4	100	100	0.25(0.15)	0.2
	RCSW2.0-5	100	100	0.25(0.15)	0.4
	RCSW2.0-6	100	100	0.25(0.15)	0.2

3.4.13.3 RAC High Rise Shear Walls

CAO et al. [115, 116] evaluated cyclic behaviour of high rise shear walls made of RAC (Table 3.14). The aims of their study were to investigate the influence of the recycled aggregates on the structural performance of the concrete shear walls and the role of concealed bracing in enhancing the cyclic behaviour of the RAC shear walls. To this objective, 6 high-rise shear wall specimens with shear-span ratio 2.0 were designed. The percent of coarse and fine recycled aggregates' replacement of the specimens were changed from 0 to 100%. The reinforcement percentage in the lower half part of the walls was 0.25 while that in the upper half part was 0.25 or 0.15. The axial-force ratio was 0.2 or 0.4. Specimen RCSW2.0-6 was designed the same as RCSW2.0-4 except that the extra inclined steel bracings were embedded in the walls. The size of the specimen, as can be seen in Fig. 3.42, was 160 × 1,000 × 2,000 mm. The properties of all the specimens are listed in Table 3.14.

Based on the test findings, the load-carrying capacity, the stiffness and its degradation, ductility, hysteretic behaviour, energy dissipation and failure phenomena of each shear wall were analyzed. The results indicate that compared to the high-rise NAC wall, RAC shear wall show poorer performance, and the more of the coarse aggregates replacement, the poorer the performance of the recycled aggregate concrete shear wall, but it is still able to satisfy seismic requirements through rational design.

3.4.13.4 RAC Low Rise Shear Walls

Cao et al. [65, 66] further examined the cyclic behaviour of low rise shear walls made of RAC with a size of 160 × 1,000 × 1,000 mm. Seven specimens (Table 3.14)

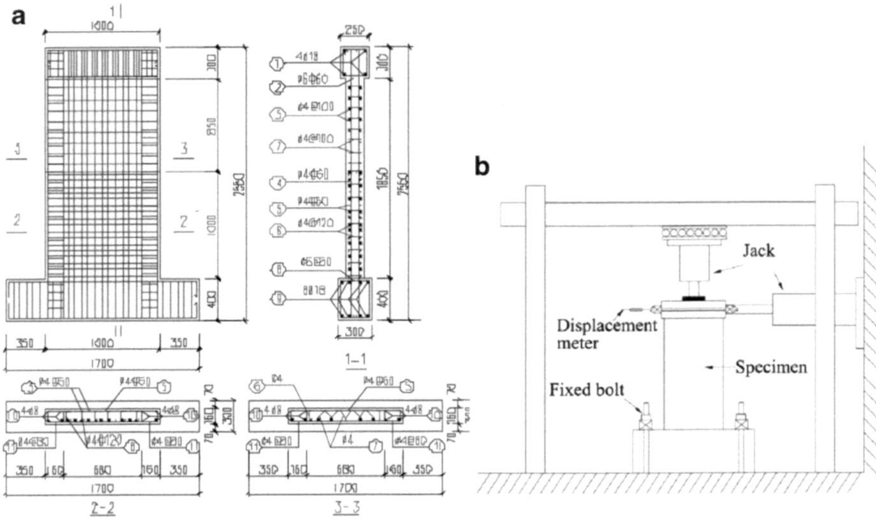

Fig. 3.42 Steel details and test setup of RAC high rise shear wall [65]. (**a**) Steel details of RCSW2.0-2 (**b**) Schematic view of test arrangements

Fig. 3.43 Failure mode of some RAC low rise shear walls [65]. (**a**) RCSW1.0-1 (**b**) RCSW1.0-4

were tested. The influence of the recycled aggregates' replacement percentage, the axial load ratio, as well as the presence of the concealed bracing was investigated. The failure mode of some specimens is shown in Fig. 3.43.

The main findings of the tests (Table 3.15) are:

- Compared to the low-rise NAC wall, the RAC shear walls with recycled coarse aggregates show almost the same performance on load-carrying capacity, stiffness, ductility and energy dissipation (Fig. 3.44). It is feasible to use recycled coarse aggregates as a replacement to natural coarse aggregates in the shear wall.
- Compared to the low-rise NAC wall, the seismic performance of RAC shear walls with recycled fine aggregates shows a little decline. Compared to the

Table 3.15 Corresponding conditions for shear wall test

Investigator	Specimens	Research parameters	Ductility coefficient	Stiffness deterioration coefficient
Cao et al. [115]	high-rise RAC shear walls	RA content, replacement ratio, concealed bracing	7.359–8.789	0.188–0.197
Cao et al. [65]	low-rise RAC shear walls	The axial compression ratio, RA content, replacement ratio, concealed bracing	5.108–7.817	0.088–0.153

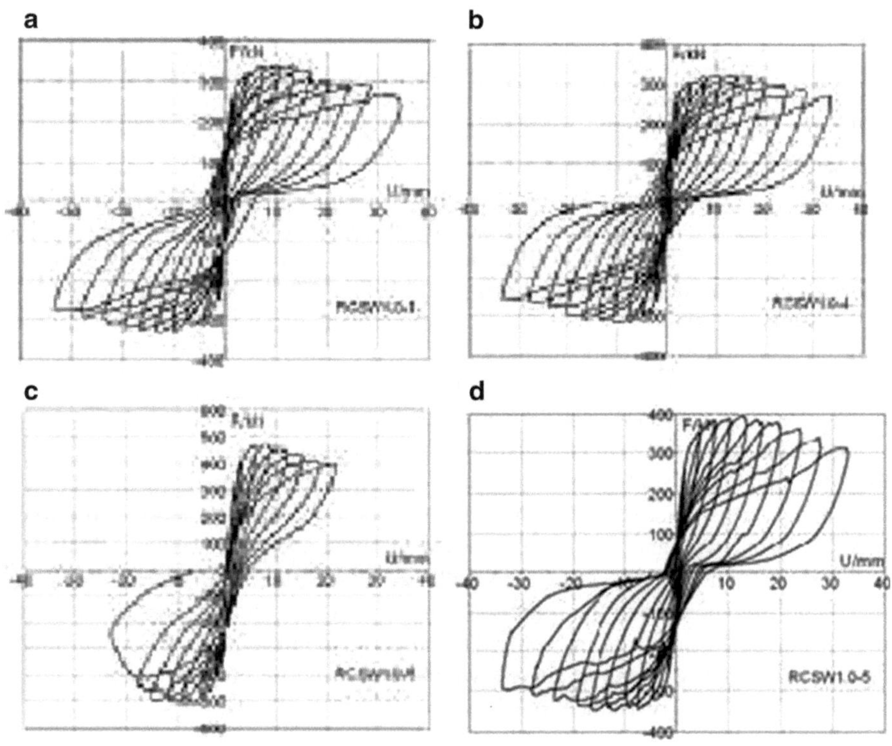

Fig. 3.44 The measured load-displacement hysteretic loops of some specimens [65]. (**a**) RCSW1.0-1 (**b**) RCSW1.0-4 (**c**) RCSW1.0-5 (**d**) RCSW1.0-7

recycled coarse aggregates, the recycled fine aggregates have slightly larger influence on the seismic performance of the shear wall. The replacement ratio of recycled fine aggregates should be strictly controlled.

- The recycled concrete shear wall with concealed bracing shows good performance on load-carrying capacity, stiffness, ductility, energy dissipation. With setting of concealed bracing in the low-rise shear wall with recycled concrete, seismic performance was improved obviously.

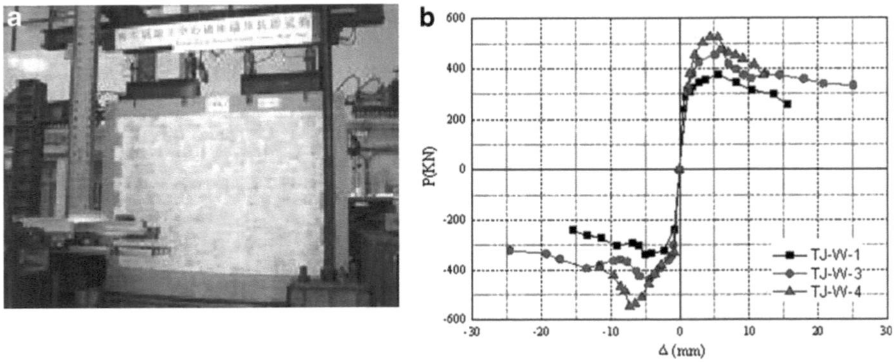

Fig. 3.45 Seismic test and results of recycled concrete block walls [67]. (**a**) Test setup (**b**) Skeleton curve of specimens

3.4.13.5 RAC Block Walls

Xiao et al. [117] reported the seismic behaviour of small-sized RAC hollow block walls. The test, including four specimens confined by tie column-beam systems, was carried out under low cyclic horizontal loadings, as shown in Fig. 3.45.

The testing variables were axial-load ratio and the different ratios of reinforcement at tie column. Corresponding test results are listed in Table 3.16, where P_{cr}, $P_{\max}$, P_u are the critical value, maximum value, and ultimate value of the load, respectively; Δ_{cr}, $\Delta_{\max}$, Δ_u are the corresponding displacements.

Based on the experiment observations, the damage process, the failure mode, the ultimate load and the deformation capacity as well as the energy dissipation of each specimen were studied. In addition, the interaction mechanism between the RAC block masonry wall and the tie column-beam confined system was investigated.

It can be seen that, generally, the small-sized RAC block wall has good seismic behaviour; the increase of the reinforcement ratio in the columns does not improve the seismic performance of the walls. The inflection point was found to occur in the descending stage of skeleton curves, which means the tie column-beam confined system affects the resistance mechanism of the small-sized RAC block wall specimens.

3.4.13.6 RAC Plane Frame

Xiao, Sun and Falkner [118] investigated the seismic behaviour of plane frames. The important parameters and objective of the test are presented in Table 3.16. Based on the tests of four large scaled frame specimens under low frequency cyclic lateral load with constant vertical actions, as shown in

Figure 3.45, the failure pattern, the hysteresis curved, the skeleton curves, the energy dissipation capacity, and the stiffness degradation laws of the frame structures made with recycled coarse aggregates were investigated. The effects of

Table 3.16 Measured bearing capacity and displacement of specimens

Investigator	Specimen size (mm)	The vertical stress (MPa)	P_{cr} (kN)	P_{max} (kN)	P_u (kN)	Δ_{cr} (mm)	Δ_{max} (mm)	Δ_u (mm)	Ductility coefficient
Xiao et al. [67]	3,200 × 2,200 × 220	0.3	239.29	357.09	303.16	0.666	5.263	10.113	$\mu_1 = 7.9$
		0.3	305.44	456.22	387.64	1.056	5.266	15.064	$\mu_1 = 4.99$
		0.6	328.64	536.23	455.68	1.056	5.874	8.593	$\mu_1 = 5.56$

different recycled aggregates' replacement percentages (i.e., 0, 30, 50 and 100%) on the performance of the reinforced RAC frames were examined. It is concluded that the general seismic performance of the frame structure made of RAC declines with an increase of the recycled aggregates' replacement percentage. Nevertheless, a frame structure with a higher content of recycled aggregates still behaves well enough to resist an earthquake attack.

The interesting test results indicated that:

- All the investigated frames behaved similarly in the aspects of the failure pattern under low frequency lateral loading regardless of the recycled aggregates' replacement percentage.
- The presence of recycled aggregates reduced the yield, the maximum and the ultimate loads of the frames; however this reduction was less than that of the mechanical properties of the concrete material.
- The characteristic displacements among the test specimens proved that there are no obvious differences between the frames with RAC and conventional concrete, particularly in the aspects of the ductility coefficients and lateral rotations.
- From the hysteresis loops, the energy dissipation and the rigidity degradation points of view, the seismic performance of the frames made with RAC is comparable to that with conventional concrete. And the frames with a properly designed RAC according to Chinese Code GB50011 (2001) [119] is able to resist an earthquake.

In addition, Sun and Xiao et al. [120] further examined the influence of the filling of RAC blocks on the performance of the RAC frames subjected to cyclic lateral loading (Fig. 3.46). The main finding can be summarized as the following: The infilled wall can work together well with the frame, stiffness and capacity of resistive horizontal loads of frame are improved obviously, but deformation and energy dissipation performance are weaker than frames without the filling of RAC blocks. The influences of infilled wall of frame on main frame structure should arouse attention of designers.

3.4.14 Shaking Table Tests on RAC Structures

To more realistically check the seismic behaviour of RAC, the much expensive and complex shaking table tests were also carried out in China on RAC structures. This work was performed towards the reconstruction of the damaged areas during the Wenchuan Earthquake.

3.4.14.1 Small-Sized RAC Hollow Block Structure

Xiao et al. [71] investigated the behaviour of a RAC masonry structure with the shaking table test. In the study, the Wenchuan Wolong seismic wave parameters

Fig. 3.46 Test specimen filled with recycled concrete blocks [70]

were used to study the behaviour of RAC masonry structure of 'tie column + ring beam + cast-in-place slab' under seismic conditions. The test, as shown in Fig. 3.47, included not only the dynamic characteristics of structure, but the damping ratio, the frequency changes and the top floor displacement of the structure under different levels of seismic activity. Figure 3.48 shows the displacement time history of the top floor under three occurrences

On the basis of the test results, it is found that maximum drift of the model structure is less than the allowable value of earthquake resistant structure in the Code for Seismic Design of Buildings [119]. Although tiny cracks were found on the walls, the test results prove that RAC masonry structure confined by a 'tie column + ring beam' system exhibits fairly good seismic behaviour.

3.4.14.2 RAC Frame Structure

Xiao et al. [122] also investigated the seismic performance of RAC frame structures using the shaking table test, see Fig. 3.49.

The structure has six stories, and non-symmetrical lay out. The seismic fortification intensity is 8°, the design group is the second group of earthquake, according to the related Chinese Code [69]. After the shaking table test, the failure mechanism of the model structure was analyzed. The natural frequency and damping ratio of the model structure were calculated. In addition, the acceleration response and displacement response of the model for each loading case were also obtained.

Fig. 3.47 RAC block masonry model [121]

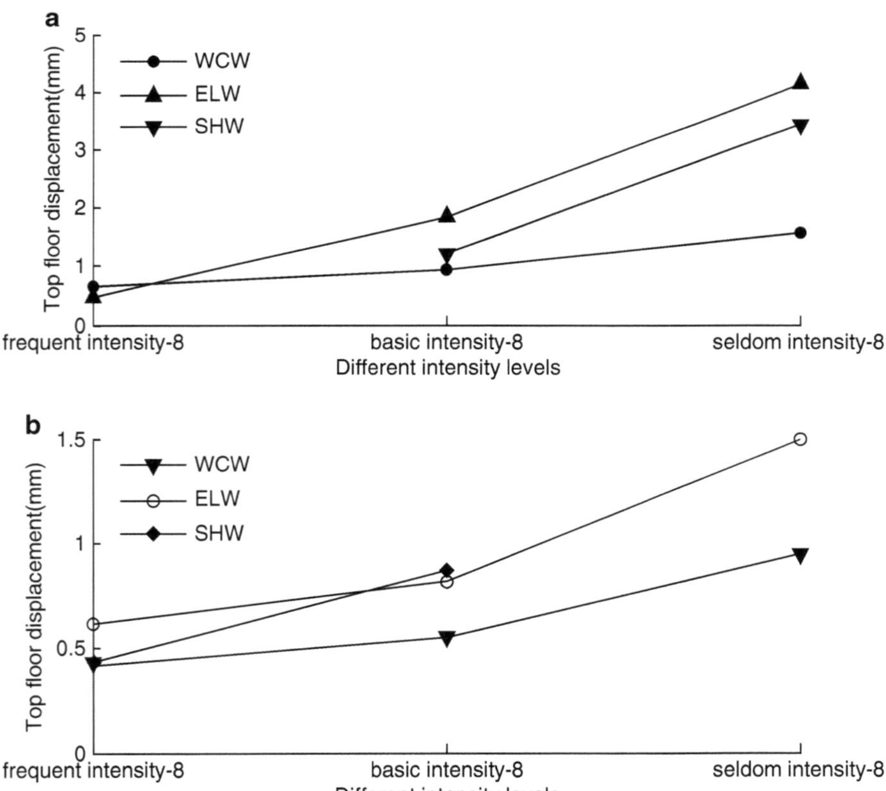

Fig. 3.48 Displacement time history of the top floor under three occurrences [121]. (**a**) Maximum storey displacement in the X-direction under different test phases (**b**) Maximum storey displacement in the Y-direction under different test phases

Fig. 3.49 General view of the RAC frame model [122]

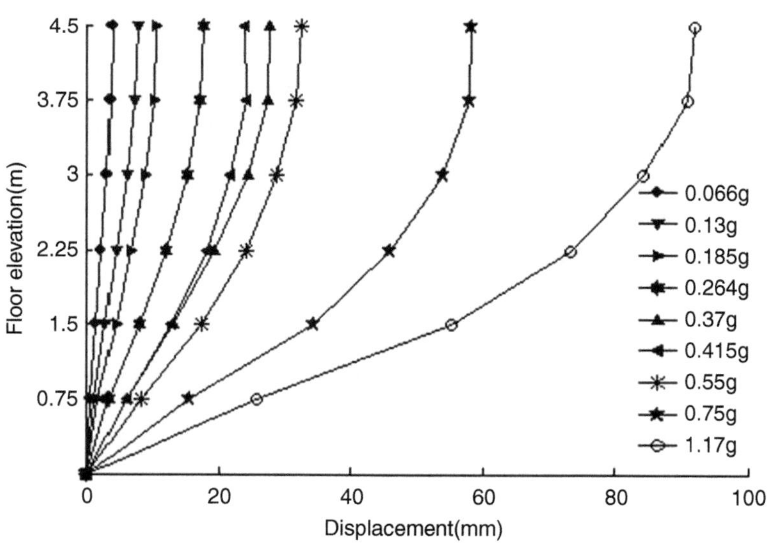

Fig. 3.50 Maximum storey displacement under WCW [72]

The maximum storey displacements obtained from WCW during earthquake simulation tests with PGA of 0.130, 0.185, 0.370, 0.415, 0.550 and 0.750 g were compared and shown in Fig. 3.50. The maximum inter-storey displacements obtained from WCW were demonstrated in Fig. 3.48. The capacity curve of the

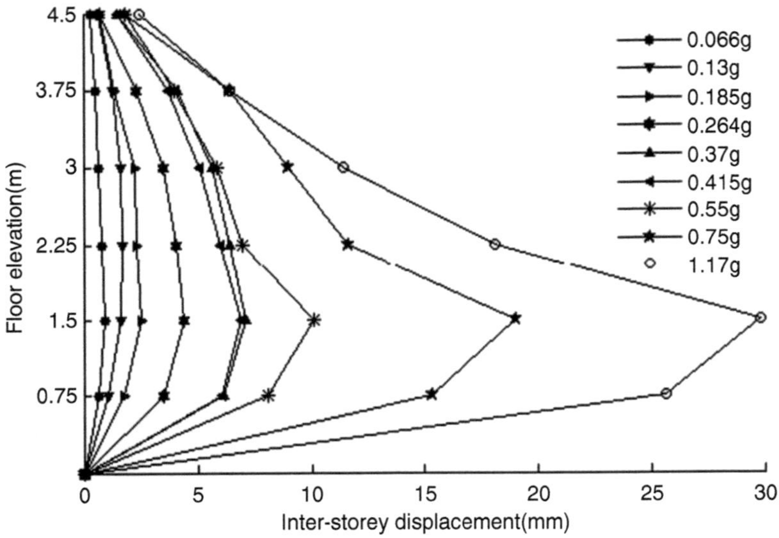

Fig. 3.51 Maximum inter-storey drift under WCW [72]

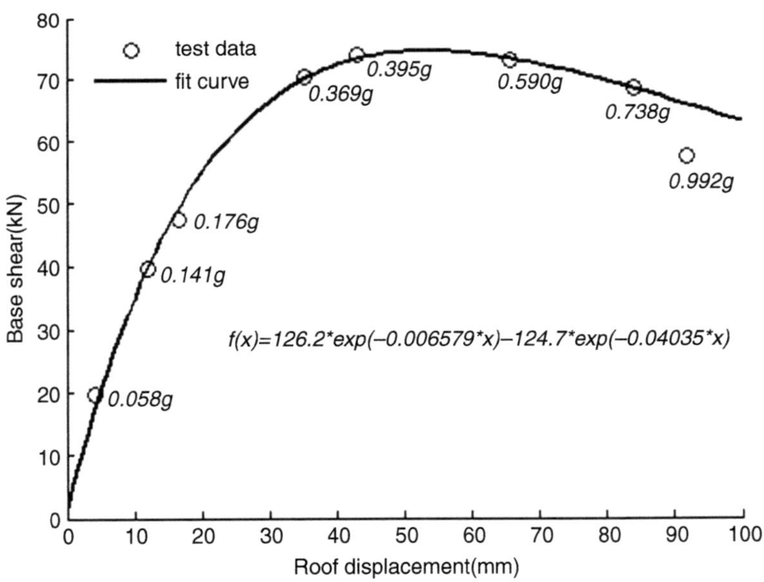

Fig. 3.52 Capacity curves of the RAC frame model structure under different test phases [122]

RAC frame model structure was obtained through fitting of the form of exponential function for the maximum reaction values as shown in Fig. 3.51. Each date point as demonstrated in Fig. 3.52 represents an earthquake simulation test case, and the label beside the data point is the actual input peak acceleration.

Based on the intensive analysis of test results, the following conclusions and corresponding suggestions are derived:

- In the shaking table model tests, the RAC frame structure with 100% RCA replacement, failure first occurred at the end of the beams, and then at the bottom of columns, which is characterized in a manner of "strongest joints, stronger columns and weaker beams".
- The structural displacement response is considerably influenced by the lower order vibration modes, and the distribution of the maximum storey displacement assume inverted triangular form basically along the height of the RAC frame model structure throughout the shaking table model tests.
- The structural system in the RAC frame model demonstrates good performance in resisting earthquakes. The test result indicates that the RAC frame model structure has good ultimate deformation capacity and enough ductility.
- The RAC frame structure with proper design and construction has good load-bearing capacity, deformation capacity and other seismic performances. It is feasible to apply and popularize the recycled concrete frame buildings in the Sichuan Wenchuan post-earthquake reconstruction area.

3.4.15 Conclusions and Recommendations

This paper gives a state-of-the-art report on the relevant researches and findings on the mechanical properties and structural performance of RAC in the mainland China. The main conclusions can be summarized as the follows:

- The compressive strength of RAC relies heavily on the water/cement ratio.
- The quality of RCA has a considerable effect on the probability distribution for the strength of RAC.
- The compressive, tensile and shear strength of RAC are generally lower than that of conventional concrete.
- The modulus of elasticity for RAC generally reduces as RCA increases; however the strain at peak stress is larger than that of conventional concrete.
- The RCA replacement percentage has nearly no influence on the bond strength between RAC and deformed rebar.
- The RCA has a remarkable influence on the residual strength of RAC after high temperatures.
- The carbonation resistance of recycled concrete decrease with the increase of replacement rate of recycled aggregate and that the anti-carbonizing capability of recycled aggregate is worse than that of normal concrete.
- The chloride resistance of recycled concrete reduced with the increase of replacement rate of recycled aggregate and fly ash can improve the chloride resistance of recycled concrete, and the best quantity of fly ash is from 10 to 15%. Recycled concrete has good resistance to chloride penetration, but its resistance to chloride penetration is somewhat less than normal concrete.

- Recycled concrete absorbs more water than normal concrete and harmfully influence frost resistance.
- The replacement rate of recycled aggregate has the greatest impact on the abrasion resistance of recycled concrete and the abrasion resistance of recycled concrete is lower than normal concrete.
- The structural behaviour of RAC elements/members is generally weaker in comparison to that of structures made of NAC;
- Through proper design, RAC can be used as a structural material from the view point of the loading capacity behaviour.
- More tests on the serviceability properties of RAC elements and structures need to be carried in the future.
- The reliability index of recycled concrete beams is proved to meet the requirements of the Unified Standard for Reliability Design of Building Structures (GB 50068-2001).

Acknowledgements The present work was sponsored by the New Century Excellent Talents in Ministry of Education Project (NCET-06-0383), the Key Projects in the National Science & Technology Pillar Program (2008BAK48B03), as well as the support by the National Science Foundation (50738005) and the Shanghai Science and Technique Committee (No. 10231202000).

3.5 Recycling in Germany: Overview Regarding CDW in Germany

Anette Mueller

3.5.1 Amounts of Construction and Demolition Waste and Fields of Utilization

The volume of Construction and Demolition Waste can be taken together with informations on raw material consumption of the construction industry from the regular monitoring reports submitted by the Kreislaufwirtschaftsträgers Bau, a working group of the German trade and industry associations involved in the construction sector [123, 124]. The 2007 report [124] shows that in Germany in 2004 72.4 million tons of building waste were generated divided in the different groups mineral debris from buildings (50.5 million tons), concrete and asphalt debris from road construction (19.7 million tons) and construction site waste, including gypsum-based construction waste (2.2 million tons). The recycling rate as the ratio of the recycled quantity and the generated quantity was over 70%.

A. Mueller (✉)
Bauhaus Universität Weimar, Weimar, Germany

Table 3.17 Quantities of CDW in Germany, fields of application and rates of recycling [123, 124]

	1996	1998	2000	2002	2004	2006
Amount (million tons)						
Construction rubble (mostly from buildings)	58.1	58.5	54.5	52.1	50.5	57.1
Rubble from road construction	17.6	14.6	22.3	16.6	19.7	14.3
Mixed rubble (from 2004 inclusive rubble of gypsum)	7.5	4.0	11.8	4.3	2.2	11.3
Total	83.2	77.1	88.6	73.0	72.4	82.7
Fields of application (million tons)						
Road construction	38.2	40.4	42.5	35.5	32.9	
Earth works	13.4	11.8	11.9	9.9	12.3	
Concrete	1.6	0	1.9	0.8	2.4	
Other	5.3	3	5.1	4.9	2	
Total	58.5	55.2	61.4	51.1	49.6	56.2
Rate of recycling (%)						
	70.3	71.6	69.3	70.0	68.6	68.0

Fig. 3.53 Rubble from demolished masonry material and concrete before processing. (**a**) Masonry (**b**) Concrete

A total of 548.5 million tons of raw materials were consumed in the construction sector, which were taken at about 85% of natural resource deposits. So Construction and Demolition Waste represents the largest waste stream in Germany, at the same time the construction industry has a considerable demand of resources.

Looking at the data available since 1996 for the amount of Construction and Demolition Waste and its use (Table 3.17) it is clear that at the figures for the volume of construction waste some fluctuations occur, which can be cyclical. The recycling rates are relatively constant, however. The reasons for this may be that large amounts are generated under intense construction activity. At the same time an intensive building activity attracts a high demand for raw materials and building materials which is sometimes covered with recycled materials.

The treatment of CDW to recycling building materials can be made in mobile or in stationary plants. The task of the processing is to generate a usable product with defined characteristics from the rubble delivered. Already during the delivery distinction is made between masonry material and concrete Fig. 3.53.

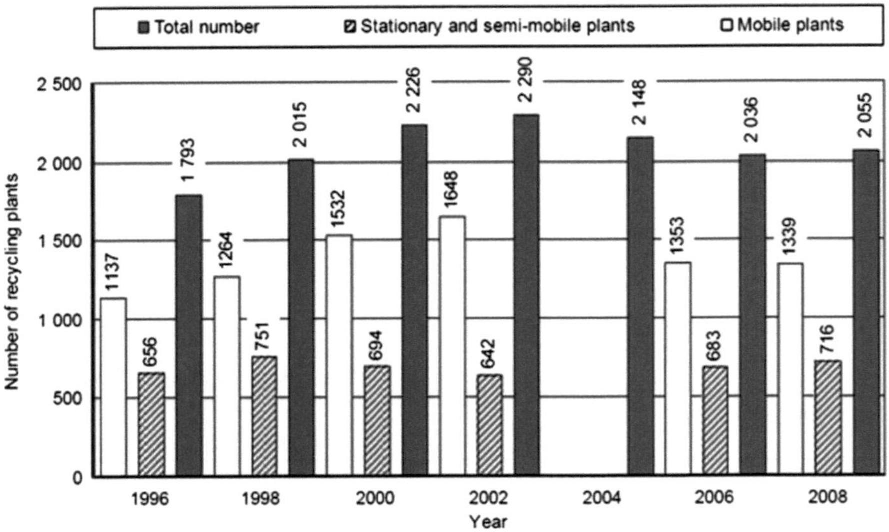

Fig. 3.54 Number of recycling plants for CDW in Germany (Data taken from [126])

In mobile processing plants, the material from the demolition, which must be pre-comminuted, if necessary, is separated first by pre-screening into two factions. The coarse material is fed to the crusher and crushed. An overhead magnet, which is located after the crusher, removed the steel and iron parts. The products created with this treatment technology are the so called pre-screen material and the actual recycling building material. With this variant, essentially only the grain size is influenced. A direct influence on the material composition is not possible.

More sophisticated technologies for the treatment of CDW can be achieved in stationary facilities. Here, these additional processing steps can be carried out:

- Two-stage crusher: Impact crusher, following a jaw crusher
- Sorting of contaminants and impairing substances on a sorting belt and using air-separation or washing in water
- Production of grain fractions by vibration screens.

In connection with the emission control of stationary and mobile recycling facilities schemes of treatment plants and an overview about the state of the art of the used equipment are given [125].

As shown in the diagram in Fig. 3.54, the number of mobile units in Germany is much higher than the number of stationary or semi-mobile plants. The number of stationary plants has changed over the period from 1996 to 2008 only little, while the number of mobile plants has larger fluctuations. The processed amount of Construction and Demolition Waste in mobile units is 30 million tons/year on average (Fig. 3.54). The amount that is processed in stationary and semi-mobile plants is 64 million tons/year. From these values and the number of plants an average capacity of 25.700 tons/year for mobile units and 43.500 tons/year for stationary and semi-mobile plants can be calculated (Fig. 3.55).

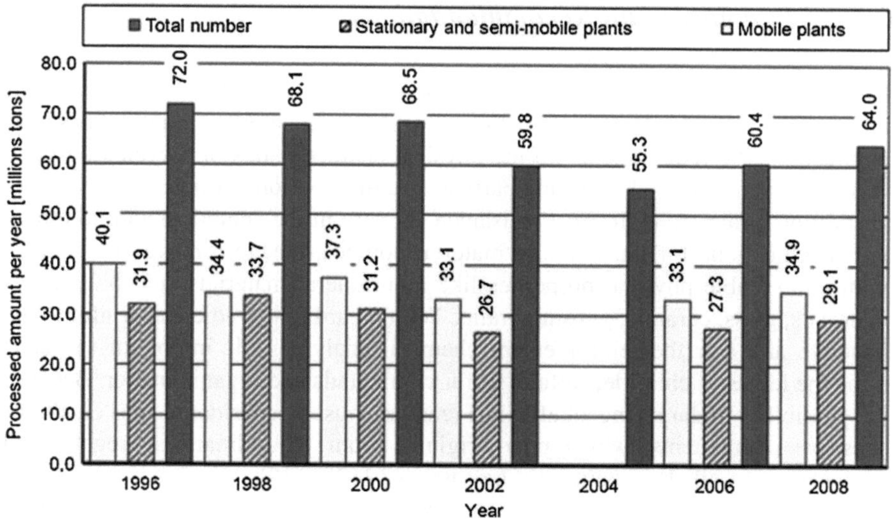

Fig. 3.55 Amount of CDW treated in recycling plants in Germany (Data taken from [126])

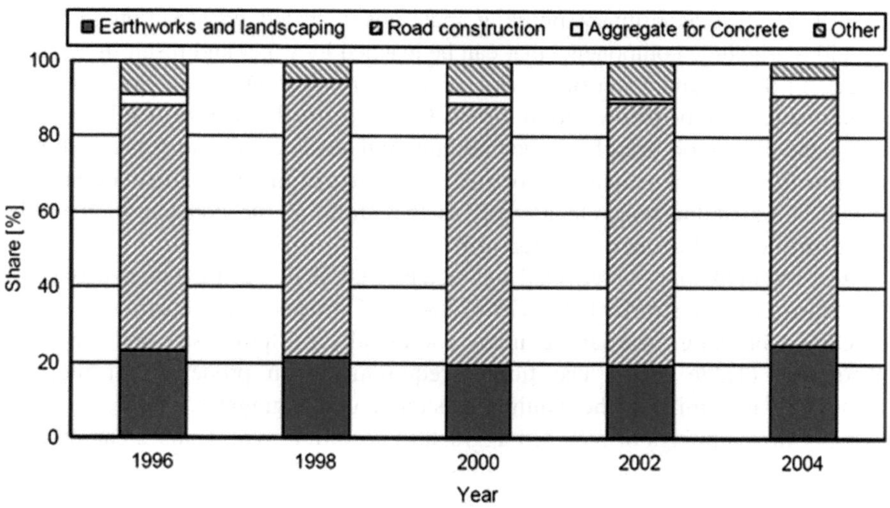

Fig. 3.56 Percentages of the different fields of application of recycling materials from processed CDW [123, 124]

In the areas of application for recycled building materials, made from Construction and Demolition Waste, the road construction dominates, followed by earthworks. Only a small percentage of the material is currently used as aggregate for concrete production (Fig. 3.56). In 2004 2.4 million tons were used for the concrete production. In relation to the production of ready-mixed concrete and concrete products this corresponds to a share of about 4%.

3.5.2 Standards and Recommendations

Rules exist for the required structural properties of recycled building materials, if the material is used in road construction as well as in concrete production. For both applications the requirements on the physical characteristics of recycled building materials such as particle size and particle size distribution, particle density, water absorption, grain strength, frost resistance, etc. are in the center. In addition there are specifications for the building material composition, to remove components with unfavorable physical properties like non-mineral materials, and bituminous materials, glass, ceramic, gypsum grains. With regard to ingredients harmful to the concrete and / or the reinforcement chemical analyses it is important to determine the levels of chloride, sulfate, sulfur compounds and organic matter. Because the possibility of damaging alkali-silica reaction must be considered tests on alkali-sensitive components, which may originate from the primary aggregate, and analyses of the alkali content must be carry out [127–135].

Furthermore, the recycled materials have to meet strict requirements for environmental compatibility. The environmental demands concern both the use of recycled materials in road construction, earthworks and landscaping as well as the use of recycled building materials for the manufacture of construction products [136–138]. Just as primary building materials recycled construction materials should not include components that can be leached by water and result in negative effects of the ground water or the soil. In addition, the contaminant content of the solid matter must be below certain limits. On concrete, the release of substances is evaluated to be mobilized from the building product. The obligatory establishment of procedures for the preparation of aqueous eluates of the recycling materials and the leaching of the concrete as well as the limits of the concentration of the pollutants are still under discussion [139, 140].

The requirements on recycled aggregates that shall be used in concrete are summarized in the standard DIN EN 4226-100. This specification uses the content of certain building materials as index additionally to the density and the water absorption (Table 3.18). The further requirements on properties of recycled aggregates are similar to the requirements on natural aggregates.

Only the chippings of recycled aggregates according type 1 and 2 can be used for the production of concrete according to DIN EN 206-1 and DIN 1045-2. The crusher sand is excluded from the reuse as aggregate. Further rules and limitations are given in the final draft of a guideline of the German Committee for Reinforced Concrete [128]:

- Maximum strength class C30/37
- No lightweight concrete, no prestressed concrete
- Application in dry environments or in environments with low humidity
- Limited portions acc. Table 3.19.

Table 3.18 Quantities of CDW in Germany, fields of application and rates of recycling [134]

Constituents	(% by mass)			
Germany DIN 4226-100: Recycled aggregates	Type 1	Type 2	Type 3	Type 4
	Concrete chippings/ concrete crusher sand	Construction chippings/ construction crusher sand	Masonry chippings/ masonry crusher sand	Mixed chippings/ mixed crusher sand
Concrete and natural aggregates acc. DIN 4226-1	≥90	≥70	≤20	≥80
Clinker, non-pored bricks	≤10	≤30	≥80	
Sand-lime bricks			≤5	
Other mineral materials such as pored brick, lightweight concrete, no-fines concrete, plaster, mortar, porous slag, pumice stone	≤2	≤3	≤5	≤20
Asphalt	≤1	≤1	≤1	
Foreign substances such as glass, non ferrous metal slag, lump gypsum, plastic, metal, wood, plant residue, paper, others	≤0.2	≤0.5	≤0.5	≤1
OD density (oven dry) (kg/m)	≥2,000	≥2,000	≥1,800	≥1,500
Maximum water absorption after 10 min (%)	10	15	20	No requirement

Table 3.19 Permissible portions of recycled aggregates >2 mm related to the total amount of aggregate [134]

Field of application		Recycled aggregates (% by volume)	
ASR-guideline	DIN EN 206–1 and DIN 1045-2	Type 1	Type 2
WO (dry)	Exposure class XC 1	≤45	≤35
WF (humid)	Exposure class X 0		
	Exposure class XC 1 to XC 4		
	Exposure class XF 1 and XF 3	≤35	≤25
	Exposure class XA 1	≤25	≤35

If the portions of recycled aggregates are equal or lower than the required limits the RC-concrete can be designed like concrete made from gravel.

Recycled aggregates from concrete rubble that meet the quality requirements of DIN 4226-100 and that can be used in accordance with the directive of the German Committee for Reinforced Concrete for concrete production can be produced with a minimum technology as well as an advanced technology of processing depending on the quality of the input material:

- Only when an unmixed input material "concrete rubble" is available recycled aggregates according to the requirement can be generated by a treatment that consists only of a comminution in an impact crusher and a subsequent classification.
- If applied an advanced technology, consisting of a two-stage crushing, a sieve classification for setting a defined particle size distribution and a wet scrubber, which can remove light impurities the requirements for the input material can be reduced somewhat.

As part of several projects recycled aggregates from both treatment options were examined and used for the production of concrete [141]. As an example of the achieved quality parameters the properties of the recycled aggregates generated in a stationary system can be named. They showed a high purity (content of concrete and natural aggregates $\geq$ 98.6%) and a low water absorption (fraction 2/8 mm: $\leq$ 3.8%; fraction 8/16 mm: $\leq$ 3.2%). At levels of the recycled aggregates of 25–35 Vol -% of the total aggregates the qualities of the RC concrete were well [142] to very good. The results underscore the importance of a qualified treatment and / or a high purity of the input materials for the manufacture of high quality recycled aggregates.

3.5.3 *New Realizations and Projects*

Periodically increased efforts can be observed to establish closed circuits in the building sector. Here reuse of recycled aggregates for the concrete production are in the focus.

The first demonstration activity was the use of recycled aggregates for the construction of an office building of the Deutsche Bundesstiftung Umwelt DBU in Osnabrueck in 1994. Then a number of buildings follow in connection with the research project "Baustoffkreislauf im Massivbau BiM". Recently, concrete was made of RC aggregate acc. to the current standards. It was used in the context of several flagship projects in residential and administrative buildings. In Table 3.20 some examples of the composition of the concrete with recycled aggregates and the buildings in which it was used are summarized.

All of the flagship projects were carried out in Stuttgart or in the surroundings of Stuttgart. The processing took place in a stationary plant which was equipped with two crushers and a wet scrubber as well as in a mobile plant. In the case of the mobile plant the operator was also responsible for the demolition. Thus it was guaranteed that the input material consists of very clean concrete rubble. An additional advantage were the short transport distances between demolition site, processing plant, ready mix concrete plant and location of the construction project. On base of the positive experience more and more commercial projects using RC-concrete in buildings are in progress.

Table 3.20 Overview on selected projects with RC-concrete [132, 141–149]

Building	Characteristics / mixing design of the concrete	
Administration building Osnabrueck	Natural sand 0/4 mm	669 kg/m^3
Built in 1994/1995 components: Interior walls	RC-aggregates 4/16 mm	669 kg/m^3
Strength class: C 30/37	RC-aggregates 16/32 mm	294 kg/m^3
Amount of RC-concrete: 120 m	Portland cement 45 F	290 kg/m^3
	Fly ash	70 kg/m^3
	Water-cement-ratio	0.69
	Plasticizer	1.8 kg/m^3
	Fresh concrete: Slump after 10 min	410 mm
	Hardened concrete: Compressive strength after 28 days	41 N/mm^2
Residential- and administration building Itzehoe	Natural sand 0/2 mm	36%
Built in 1996/1997	RC-aggregates 2/8 mm	14%
Components: Interior and external walls at moderate moisture	RC-aggregates 8/16 mm	20%
	RC-aggregates 16/32 mm	30%
	Cement CEM II/A-L 32,5 R	350 kg/m^3
	Fly ash	No
	information	
	Water-cement-ratio	
	Plasticizer FM 72	
	Fresh concrete: Slump after 10 min	570 mm
Hardened concrete: Compressive strength after 28 days	45 N/mm^2	
Car park Darmstadt	Natural sand 0/2a mm	585 kg/m^3
Built in 1997/1998	RC-aggregates 2/8 mm	545 kg/m^3
Components: Internal columns, walls, ceilings, exposed concrete	RC-aggregates 8/16 mm	568 kg/m^3
	Cement CEM I 42,5 R	310 kg/m^3
	Fly ash	40 kg/m^3
	Water-cement-ratio	0.55
	Plasticizer FM 26	5–18 ml/kg
	Fresh concrete: Slump after 10 min	550 mm
	Hardened concrete: Compressive strength after 28 days	45 N/mm^2
Residential building "Waldspirale" Darmstadt	Natural sand 0/2a mm	616 kg/m^3
Built in 1999/2000	RC-aggregates 2/8 mm	530 kg/m^3
Components: Interior walls and floor plate	RC-aggregates 8/16 mm	569 kg/m^3
Strength classes: C 20/25, C 30/37	Portland cement CEM I 42,5 R	300 kg/m^3
Amount of RC-concrete: 12,000 m	Fly ash	50 kg/m^3
	Effective water-cement-ratio	0.59
	Plasticizer BV	1.5 kg/m^3
	Consistence of fresh concrete	Regular
	Hardened concrete: Compressive strength after 28 days	42.9 N/mm^2

(continued)

Table 3.20 (continued)

Building	Characteristics / mixing design of the concrete	
Guest house in the residential area on the Rhine Avenue in Ludwigs-hafen	Characteristics of the RC-aggregates 2/8 mm and 8/16 mm	
Built in 2009	Content of concrete and natural rock in the used fractions	99.2% 98.6%
Components: ceilings, walls and elevator shaft		
Quality of RC-concrete: C 30/37, XC 1, F3	Bulk density	2,368 kg/m
Amount of RC-concrete: 600 m		2,394 kg/m
	Water adsorption	3.4%
		2.7%
	Characteristics / mixing design of the concrete	
	Natural sand 0/2 mm	566 kg/m
	RC-aggregates 2/8 mm	149 kg/m^3
	RC-aggregates 8/16 mm	297 kg/m^3
	Natural gravel 2/8 mm	354
	Natural gravel 8/16 mm	318
	Cement CEM II B-V 42,5 R	360 kg/m^3
	Fly ash incorporated in the cement	
	Water-cement-ratio	0.52
	Fresh concrete: Slump after 10 min	480–490 mm
	Hardened concrete: Compressive strength after 28 days	51.8 N/mm
Residential buildings Stuttgart-Ost	Characteristics of the RC-aggregates 2/16 mm	
Built in 2010	Content of concrete and natural rock in the used fraction 2/16 mm	99.4%
Production of concrete of different strength and exposure classes with recycled aggregates		
	Bulk density	2,614 kg/m
	Water adsorption	3.3%
	Fresh concrete: Slump after 10 min	
	C 8/10; X 0	480 mm
	C 12/15, X 0	490 mm
	C 20/25, XC 3	480 mm
	C 25/30; XC 4	490 mm
	C 30/37; XC 1	490 mm
	Hardened concrete: Compressive strength after 28 days	
	C 8/10, X 0	17 N/mm
	C 12/15, X 0	21 N/mm
	C 20/25; XC 3	32 N/mm
	C 25/30, XC 4	38 N/mm
	C 30/37, XC 1	44 N/mm
Residential buildings Stuttgart Rotenbergstraße / Raitelsbergstraße Built in 2010 Components: lean concrete, ceilings, walls Quality of RC-concrete: from C 8/10 to C 30/37 Amount of RC-concrete: 1,500 m, i.e. 68% of the whole concrete quantity		

3.6 Recycling in Italy: Overview Regarding Construction and Demolition Waste

Mario Bassan and Marco Quattrone

3.6.1 Preliminary Remarks About the Method

On the eve of the new European Directive's absorption concerning wastes (2008/98 EC), it is essential to know, in a reliable way, data about construction and demolition waste production, about the number and the location of plants. Without this kind of information, it is very difficult to start new politics about inertwastes' management and recycling.

The European Commission has set the goal, for construction and demolition wastes' recycling, at 70% by 2020. Considering that the actual Italian recycling rate is about 10%, it is clear the necessity of new politics to reach that goal. Regarding C&DW production data, the last "Official Italian Waste Report" indicates a total production of C&D wastes equal to 52,083 million tons (last official available data are dated 2006).

To achieve a correct evaluation of special waste quantities, ISPRA (Environmental Protection and Research Institute) has combined data coming from the waste quantities' declarations through estimate's procedures. In particular, the CDW quantity has been estimated from management operations data, coming from Environmental Declaration Form (MUD) removing the MUD's declarations regarding the midway phases of the management process.

The used methodology sets the total production of CDW equal to the recovered/landfilled quantities. Regarding the recycling material's quantification, the Italian Waste Report presents data concerning not dangerous special wastes; in that Report, there are not specific information about construction and demolition wastes. That depends on the complexity of the waste management and authorization system and, today, a sectorial data collection is impossible.

The data presented in this paper come from the annual report by National Association of Recycled Aggregates Producers (ANPAR). However, this year, cause of the economic situation, the number of plants involved in the study has been lesser than that of the past. It has been chosen to considerer data come from all the plants involved in this analysis during the last 4 years.

M. Bassan (✉) • M. Quattrone
Politecnico di Milano, Milan, Italy

It has been asked, to the plants' manager, to fill a questionnaire with the wastes' quantities relating to the following codes:

EWC 17 01 01	Concrete
EWC 17 01 02	Bricks
EWC 17 01 03	Tiles and ceramics
EWC 17 01 07	Mixture of concrete, bricks, tiles and ceramics other than those mentioned in 17 01 06
EWC 17 03 02	Bituminous mixtures containing other than those mentioned in 17 03 01
EWC 17 04 05	Iron and steel
EWC 17 05 04	Soil and stones other than those mentioned in 17 05 03
EWC 17 05 06	Dredging spoil other than those mentioned 17 05 05
EWC 17 08 02	Gypsum-based construction materials other than those mentioned in 17 08 01
EWC 17 09 04	Mixed construction and demolition wastes other than those mentioned in 17 09 01, 17 09 02 and 17 09 03
Others	

Figure 3.57 shows the geographic location of the plants; they are above all located in northern Italy.

The difficulties to conduct a complete CDW recycling plants census and the consequent lack of data, does not allow to determine clearly how much the analyzed sample can be representative.

A first step in this direction could be the report on the waste treatment plants published by Environmental Services Companies Association (FISE) [150]. The mentioned study has the purpose to arrange a frame concerning the waste treatment plants to understand the actual situation and to arrange corrective measures [151].

3.6.2 The Production of Recycled Aggregates in Italy: Qualitative Aspects

The composition of inert waste is very variable; this is linked to the different wastes' source that depends on local typology and construction technique, climate conditions, economic activities and technologic development of a specific area, furthermore the availability of raw materials.

The analysis of provided data, coming from recycling plants involved in this study, shows that the most of plants receives construction and demolition wastes, including contaminated soil (EWC 17.XX.XX). The ratio between the EWC 17.XX.XX wastes and total waste's quantity is more than 85% except for three cases. The reasons of that difference can be found in the particular conditions of

Fig. 3.57 Location of plants involved in the study (Source: ANPAR)

those places. The first two plants have decided to dedicate their activity on industrial waste while the third one is located near some travertine quarries that send their scraps to the plant. However, the mentioned data are not sufficient to modify the general trend.

Considering the plants in function of their geographic location (North, Centre, South), the percentage of wastes belonged to EWC 17.XX.XX is always higher than 90% of the overall; this confirms that the most CDW comes from construction sector.

Concerning the average composition of treated wastes, Figs. 3.58, 3.59, 3.60, and 3.61 show that, in different proportions, the wastes sent to the plants with important percentages, compared with total, are those belonged to the EWC code shown in Table 3.21.

The "geographic" analysis of these data allows us to notice some significant differences.

Below the subdivision of main waste's typology has reported.

- The code 17 09 04 has a national average equal to 32% that is 17% in the North, 4% in the Centre and 22% in the South.
- The code 17 05 04 has a national average equal to 14% that is 20% in the North, 61% in the Centre and 56% in the South.

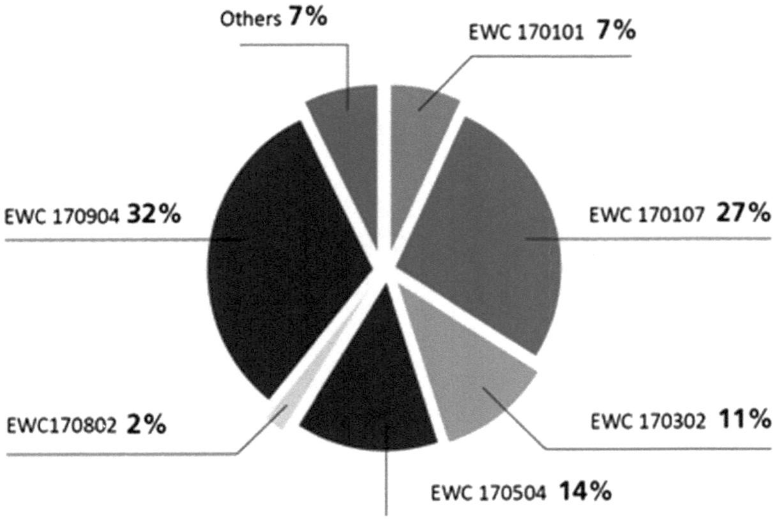

Fig. 3.58 Average composition of wastes treated inside the involved recycling plants

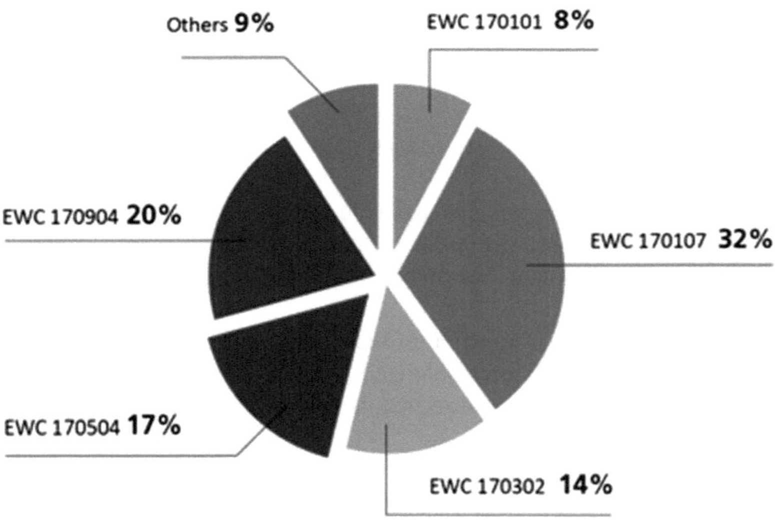

Fig. 3.59 Average composition of wastes treated in the North Italy's recycling plants

- The code 17 01 07 has a national average equal to 27% that is 8% in the North, 15% in the Centre and 1% in the South.

These differences are not very significant and can be justified considering that in the northern plants other kinds of wastes, not only the construction and demolition ones, are sent. Those wastes, after a separation treatment, to eliminate undesirable fractions, and a granulometry sorting became recycled aggregates.

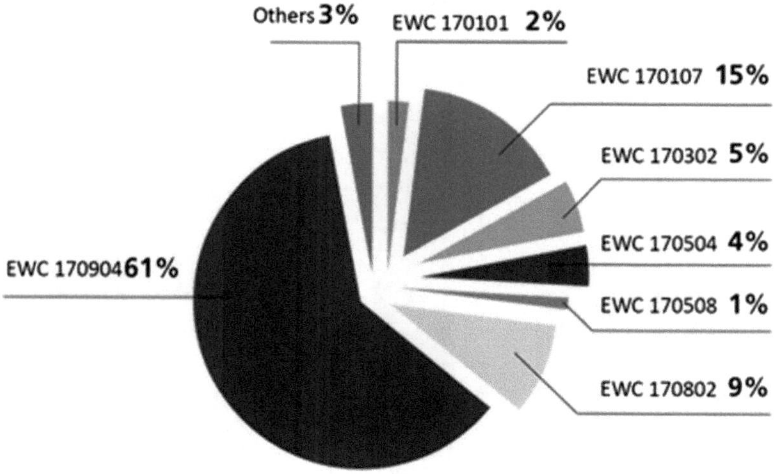

Fig. 3.60 Average composition of wastes treated in centre Italy's recycling plants

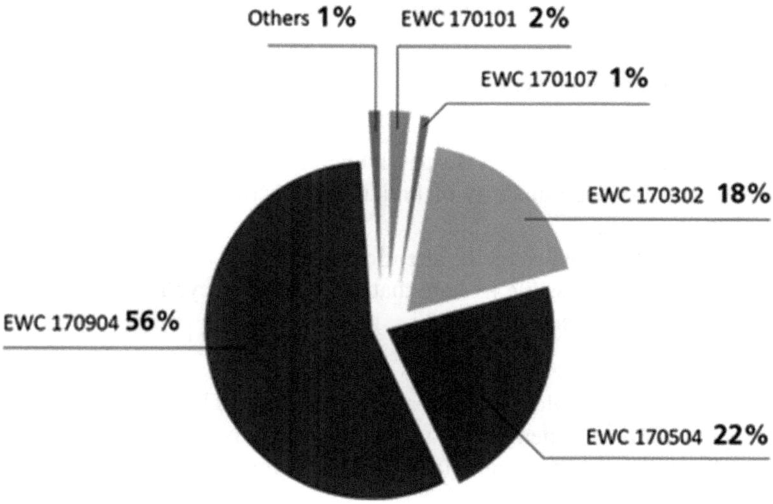

Fig. 3.61 Average composition of wastes treated in South Italy's recycling plants

Table 3.21 Significant percentages of wastes sent to the plants

EW Code	Definition	%
17 09 04	Mixed construction and demolition wastes other than those mentioned in 17 09 01, 17 09 02 and 17 09 03	32
17 05 04	Iron and steel	14
17 01 07	Mixture of concrete, bricks, tiles and ceramics other than those mentioned in 17 01 06	27
17 03 02	Bituminous mixtures containing other than those mentioned in 17 03 01	11

Source: ANPAR

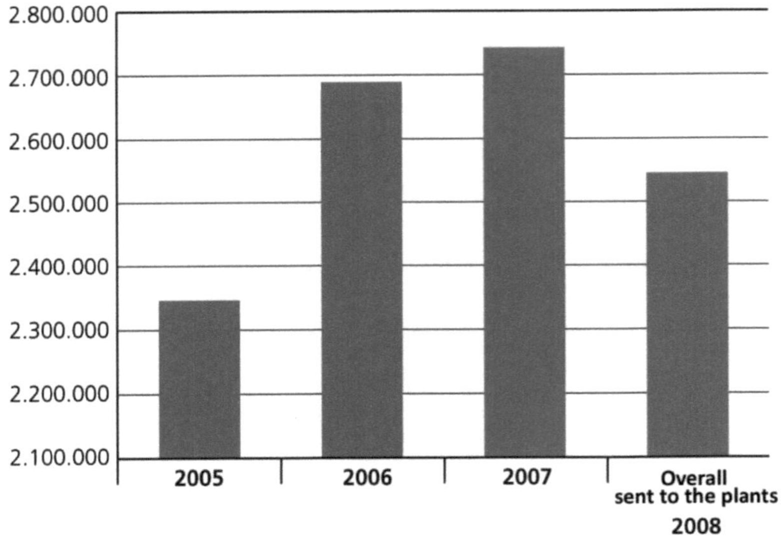

Fig. 3.62 Inert wastes sent to the plants during the period 2005–2008

3.6.3 Quantitative Aspects

Official data, regarding the inert wastes' recycling, even if poorly reliable, indicate that the yearly produced quantity of Italian construction and demolition wastes is about 52 million tons.

Despite during the last years the attention on this kind of waste has considerably risen, nowadays it is possible to state that the C&DW recycling sector is strongly underdeveloped and uncoordinated.

About that:

- An official census of the plants located on the Italian Territory does not exist;
- The construction and demolition waste flow's size is unknown so as its split in the different final destinations (landfill site, recycling plants, volumetric reduction plants, etc.);
- A considerable inert waste's quantity is not recycled/disposed of in a suitable way;
- A homogeneous material's flow, that can be treated and reused inside a construction site, exists but this one has never taken in account from a quantitative point of view.

In 2009, just 49 plants have been involved in this study, 3 plants less of the 2008. However, only data coming from 14 plants have been taken into account. These plants were involved in the study even during the past 4 years as Fig. 3.62 shows.

From the analysis, it is possible to state that after an increase during the 2006, the quantities of C&D wastes sent to the plants have remained constant during the 2007 and have slowly decreased in 2008, according with the trend of the construction activity.

Table 3.22 Percentage of use of recycled aggregates in concrete

Origin of the recycling material	Class of concrete	Rate of use
Demolition of buildings (debris)	= C 8/10	Up to 100%
Demolition of concrete or reinforced concrete	≤C30/37	Up to 30%
	≤C20/25	Up to 60%
Reuse in certified precast concrete industries - any class of concrete	≤C45/55	Up to 15%
Reuse in certified precast concrete industries–class of concrete > C45/55	Same class of original concrete	Up to 5%

If we consider that the official provided data show a positive trend in production of construction and demolition wastes, while, concerning the delivered C&DW quantities, the plants involved in this study recorded an estimated negative trend of about 10%.

3.6.4 Normative Aspects

From 2005 to date, major developments in the use of recycled aggregates have been made. The main reasons can be found in the enacting of two laws: the "Environmental Code" (Decree 152/2006) and the "Construction Regulations".

Regarding the "Environmental Code", which absorbs eight European Directives, the main innovation consists in putting national regulations in line with European standards, especially those concerning the definition of C&D waste, their classification and the test methods adopted to quantify the pollutants.

Previous environmental regulations already promoted the recycling and the use of recycled aggregates for non-structural applications (road sub-layer, embankment etc.). However, these uses are subject to the leaching test performance. In the past, Italian pollutant limits were more restrictive than those of other European Countries were; the outdated Italian test method produced high values of pollutants so the use of recycled aggregates was almost impossible.

Regarding the "Construction Regulations", it is necessary to precise that in 2005 the promulgation of the law's first version allowed the use of recycled aggregates for structural applications. However, this one has been ignored because of a series of mistakes and adverse opinions by some advisory bodies.

Its review has led to the enacting of the aforementioned construction regulations, which are in force since March 2008. Regarding the use of recycled aggregates, the new standard introduces some limit values. These values, expressed in function of concrete's strength class (after 28 days), are listed in Table 3.22.

It is important to highlight that the new regulations do not provide information about the different values of mechanical properties that recycled aggregate concrete could show compared to traditional concrete.

Regarding the technical standards, the new regulations absorb the EN standards, in particular the EN 12620 "Aggregates for concrete". This standard defines the

Table 3.23 Construction and demolition waste production in Europe (tons per capita)

	Year					
Country	2001	2002	2003	2004	2005	2006
Austria	–	–	–	0.81	–	0.81
Belgium	0.81	0.80	1.11	1.06	1.22	1.18
Denmark	0.63	0.75	0.70	0.83	0.97	1.12
Finland	–	–	–	3.99	–	–
France	–	–	–	5.5	–	–
Germany	3.05	2.92	2.71	2.33	2.24	–
Greek	0.41	0.38	0.37	0.37	–	–
Ireland	1.70	–	–	2.74	3.60	3.95
Italy[a]	0.54	0.65	0.74	0.80	0.78	0.88[a]
Luxemburg	–	–	–	5.9	–	–
The Netherlands	1.48	1.47	1.46	1.47	1.58	–
Norway	0.27	0.28	0.27	0.70	0.32	–
Portugal	–	–	–	1.09	–	–
Spain	0.59	0.58	0.66	0.74	0.80	0.88
Sweden	–	–	–	1.14	–	–
United Kingdom	1.74	1.74	1.75	1.66	1.90	1.89

Source: EUROSTAT and ETC/RWM 2008
[a]The datum has been updated and aligned to the last "Italian Waste Report"

properties that aggregates need to have in order to comply with the requirements of EN 206-1 "Concrete – Specification, performance, production and conformity".

The Italian standards UNI 8520-1 and UNI 8520-2 provide the complementary instruction necessary to apply the UNI EN 12620. The former gives prescriptions about designation method and conformity criteria, while the latter specifies requirements of the aggregates for concrete.

Applications of recycled aggregates for civil engineering work and for road construction are provided in the UNI EN 13242 "Aggregates for unbound and hydraulically bound materials for use in civil engineering work and road construction".

3.6.5 A Comparison with European Data

Every year, in Europe, about 850 million tons of construction and demolition wastes are produced. They represent the 31% of the total European wastes' production (source: EUROSTAT and ETC/RWM, 2008). Table 3.23 shows the production's development of C&DW per capita, concerning the old Member Countries and Norway, since 2001. The average for the European Union (27 Countries and Norway) is of 1.74 tons/year per capita.

Analyzing the data it is possible to notice as there are many differences among the old EU configuration's Countries. France and Luxembourg produce 5.5 and 5.9 tons/year per capita respectively. Germany and Ireland produce between 2 and 4 tons/year per capita, while for other Countries the range fluctuates between 0.2 of

Table 3.24 Construction and demolition waste production in Europe (tons per capita)

	Year					
Country	2001	2002	2003	2004	2005	2006
Bulgaria	–	–	–	0.39	–	–
Cyprus	–	–	–	0.58	–	–
Czech Republic	0.85	0.85	1.00	1.44	1.20	1.15
Estonia	0.64	0.94	0.93	1.12	1.61	1.78
Hungary	0.49	0.59	0.51	0.43	0.49	0.54
Latonia	–	0.06	0.03	0.04	0.07	0.05
Lithuania	–		–	0.10	–	0.18
Malta	–	–	–	1.95	–	–
Poland	–	–	–	0.11	0.14	0.44
Romania	0.02	0.03	0.01	0.00	0.02	–
Slovakia	–	–	0.07	0.26	–	–
Slovenia	0.83	0.82	1.00	1.41	1.2	1.18
EU 27+ Norway	–		–	1.74	–	–

Source: EUROSTAT and ETC/RWM, 2008

the Norway and the 2 of the United Kingdom; Italy produce 0.88 tons/year per capita. Even concerning new Member Countries, Table 3.24 shows many differences, however all data are below 1 tons/year per capita except for Malta that produces almost 2 tons/year.

The pro capita production C&D waste index is certainly a data linked to the industrialization level of a Country however, the sensible noticed data difference can be justified because a different data collection's methodology has been used in the different Countries.

To present a clear picture of the problem, it is also important to consider the annual production of aggregates. In 2006, 3.6 billion tons of aggregates have been produced in 21 European Countries. Table 3.25 summarizes these data.

3.6.6 Characteristics of Plants – Technology

The plants' technology, actually popular in Italy, can be grouped in two categories: the stationary and the mobile ones.

Recycling process of C&D waste consists of these stages:

- Coarse separation
- Crushing
- Separation of ferrous elements
- Screening
- Removal of impurity (if necessary)

The coarse separation, made by grab or shears, is the first operation made on the rubble. The purpose of it is to reduce the dimensions of the debris to feed the crusher easily. During the crushing stage, made by the crusher, the dimensions of

Table 3.25 Best estimates of aggregates production data for 2009

	Total number of producers (Companies)	Total number of extraction sites (quarries and pits)	Sand and gravel (Mt)	Crushed rock (Mt)	Marine aggregates (Mt)	Recycled aggregates (Mt)	Manufactured aggregates (Mt)	Total production (Mt)
Austria	1,062	1,362	6	3	0	4	2	97
Belgium	78	104	1	4	3	15	1	78
Bulgaria	190	280	1	1	0	0	0	25
Croatia	260	338	7	2	0	0	0	29
Cyprus	23	23	0	1	0	0	0	12
Czech Rep	198	384	2	4	0	0	0	65
Denmark	350	500	3	0	9	5	0	44
Finland	400	2,091	1	4	0	1	1	65
France	1,428	2,481	14	20	6	15	6	376
Germany	1,280	2,265	23	21	5[a]	61	36	555
Greece	192	213	1	7	0	0	0	71
Hungary	100	100	3	2	0	0	0	57
Ireland	150	600	1	2	0	0	0	39
Italy	1,550	2,460	21	14	0	0	0	350
Netherlands	65	160	4	0	45	22	0	113
Norway	710	927	1	5	0	2	0	66
Poland	2,044	1,786	13	4	0	22	1	203
Portugal	617	1,081	2	2	0	0	0	50
Romania	430	745	2	1	0	1	0	38
Serbia	20	70	1	8	0	0	0	19
Slovakia	180	305	1	1	0	0	0	30
Spain	1,555	1,765	6	17	0	1	0	238
Sweden	985	2,109	2	5	0	1	6	85
Switzerland	540	535	3	5	4	5	0	47
Turkey	770	770	2	29	0	0	0	315
UK	727	1,275	4	8	10	46	10	197
Totals	15,904	24,729	1,259	1,658	82	200	63	3,262

Source: UEPG 2010–2011

[a]Note: As there are less than 3 marine aggregates producers in Germany, the figure given is an estimate, for confidentiality reasons

rubble are reduced to obtain a product of suitable size. This characteristic is obtained by mechanical action as:

- Squeezing
- Impacting
- Grinding

Generally, the crushing cycle is distinguished in primary crushing, secondary crushing and milling to obtain a product with decreasing size.

The classification in different granulometry classes is made by the screening operation. If it is necessary to eliminate impurity like wood, plastic and paper, air separation is the best process to do this: it does not produce washing mud that are hard and expensive to dispose of.

The process ends with the storage of the products.

The main components of a recycling plant are:

- The crusher
- The magnetic separator
- The screen
- The air classifier (if necessary)

Actually the recycled aggregate washing is rare because washing mud disposal is very expensive, and the administrative procedure is complex.

The flowchart in Fig. 3.63 highlights represents the different stages of the material during the recycling process in a typical stationary Italian plant. The productive cycle is organized in order to avoid that unsuitable material is recycled.

Qualitative and quantitative controls are carried out in order to match the information about loaded and truck drivers. The plant is managed by only three operators who use a system for video surveillance. The computerization of the data allows identifying the truck driver if the input material is not conforming to the admittance criterion. The material to process, stored in heaps of same kind, is transferred to the feeding by one mechanical digger thus reducing the probability to introduce unsuitable material. A video camera, videoing the exit mouth of the feeding, holds under control the rubble before it begins the treatment cycle.

In this stage, the operator can stop the feeding and check or put aside the material. To avoid the milling of the fine fraction, a vibrating screen performs the first selection. The coarse material is then sent to the crusher.

After the crusher, an atomized water system allows to eliminate dusts and to pick up them. The output material goes under a magnetic separator and iron elements are stored in a metallic box. Then, a vibrating screen classifies ceramic materials in different granulometry classes (0÷30, 0÷70, >70 mm this one can be crushed once again).

The vibrating screen also separates plastic, wood and paper elements and puts them into a special box. The 30÷70 mm fraction can be stored or sent to another cycle of crushing. At the end of the process the different fractions are stored in heaps by a revolving device which minimizes the dust of fall. Another conveyor belt can feed a further screening station which produces 0÷6, 6÷15 and 15÷30 mm fractions.

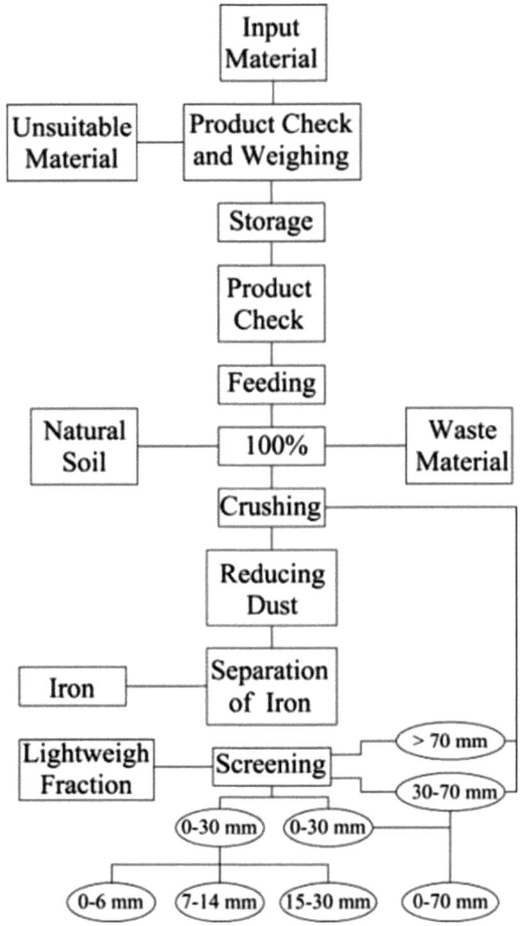

Fig. 3.63 Flowchart of the fixed plant

These granulometry classes are further purified from impurities that are put in special boxes.

Mobile plants, compared to the fixed ones described above, are constitutes by mobile or semi-mobile machines of suitable sizes for road transport. It is less expensive to buy and to manage these machines than the fixed ones, but they are less productive. Their advantage is the possibility to use them inside the demolition site thus avoiding excessive handling of the debris and reducing the high transport costs.

In the field of the recycling of C&DW, mobile plants consist of a crusher, with magnetic separator on the exit mouth, and of a vibrating screen for the granulometry classification. These machines are positioned in series to achieve a recycling station (see Fig. 3.64a). Crusher and screen are usually separated.

Nowadays, it is possible to find a single machine with crusher, magnetic separator and screen (see Fig. 3.64b).

Fig. 3.64 (a) Mobile plant composed by a crusher and a vibrating screen; (b) Mobile plant composed by a single machine with crusher, magnetic separator and vibrating screen

3.6.7 The Administrative Procedure to Install a Recycling Plant and the Complementary Activities

For realizing and managing a new recovered/recycling plants, an authorization is needed; local authorities give that one. Two kinds of authorization are necessary. To achieve the first one it is necessary to present the plant's project and the technical documentation concerning the territorial planning, public welfare/health and the environment one.

The business concerning the recovering of wastes can be done following two procedures: the former follows a simplified administrative procedure while the latter follows the ordinary one. The difference between the two ones concerns the times to achieve the authorizations and the size of the business (for example with the simplified procedure, times are of about 90 days).

The Legislative Decree 4/08, however, has introduced some news; plants' managers that decide to follow the simplified procedure, if the business regards the treatment of not dangerous wastes and the production is higher than 10 tons/day, must necessarily submit the Environmental Impact Evaluation. At this point, there is not a substantial difference between the simplified and the ordinary administrative procedures, concerning the documents to produce for, the only difference (this is the real problem) are the times, always too long.

Often, the construction and demolition waste recycling has carried out parallel to other activities (Fig. 3.65) so that it is possible to optimize the management of wastes and the selling of the products.

In Europe, during the last years, a significant collaboration quarry and recycling activity has developed. On February 2008, the European Aggregates Association and the International Recycling Federation have launched the European Platform for Recycled Aggregates (EPRA). The mission of EPRA is to achieve the best use of recycled aggregates in the most of applications.

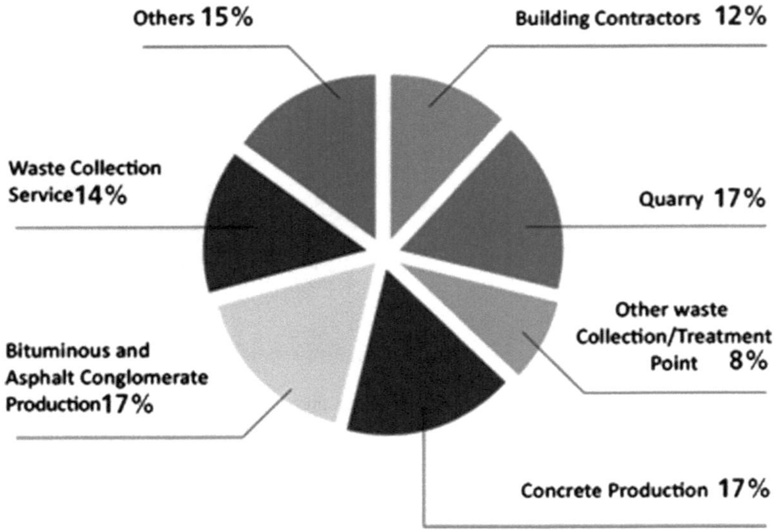

Fig. 3.65 Percentage distribution of integrative activities carried out inside the recycling plants member of National Association of Recycled Aggregates Producers

The match of quarry and recycling activities, although can generate some problems, concerning the administrative procedures, cause many advantages:

- possibility to have in the same production place aggregates from different source; raw material can be used just for high performance applications while the recycled ones for the others;
- reducing the environmental impact linked to the quarry activity on the territory;
- solving the problem of the quarry waste production, they can be recycled inside the plant;
- providing to a third party a recovering service for construction and demolition wastes.

3.6.8 Work Modalities

It is possible to state that every stationary plant, independently if it makes up by stationary or mobile machines, has a quality control production system. This kind of tools allows to the producers to warranty homogeneous materials' stocks. The knowledge of the own materials' characteristics is an unavoidable requirement. The presence of internal laboratories is a proof of that and is the only way to test and know, quickly, the produced recycled material's characteristics.

However, the EC marking on the aggregates, even if mandatory especially for recycled aggregates, is not frequently required from the market when those are used in low performance applications. The enforcement of EU rules, as happen in the structural concrete sector, is producing a significant impulse.

3.6.9 The Recycled Aggregates' Market

Different factors influence the recycled aggregates' market. Among these it is possible to cite the following:

- mining activity taxation;
- landfill cost;
- availability and cost of natural aggregates;
- bias against recycled materials.

Different policies on wastes' management and the availability of natural aggregate are the main cause about this problem. Countries where landfill costs are higher have generally an high recycling rate, on the contrary the availability of raw materials and the low landfill costs, e.g. Italian situation, impede the knowledge's development about recycled aggregates performances and possible applications' fields. For example, concerning landfill cost, prices can be in a range from 5 € per tons in Italy to 50 € per tons in Denmark.

Regarding the availability and the natural aggregate's cost, these factors depend on the Country, by the mines' availability in a specific area and by the transport's cost. Actually in Europe the price per tons fluctuates between a minimum of 2.5 €/tons and a maximum of 12 €/tons. Countries like Italy that have a big availability of natural aggregates tend to sell them at the average price about 6 €/tons. Because of that reason, the price of recycled aggregates should be the 20% lower than the price of the natural ones. Furthermore the concept that recycled aggregates come from wastes meets opposition both from building operators and customers.

3.6.10 Conclusion

According to the ISPRA's estimates, published on Italian Waste Report 2008, about 52 million tons have been produced in Italy [152].

Concerning the qualitative characteristics, in 2008 almost the 100% of the wastes sent to the recycling plants were construction and demolition ones, including contaminated soil.

Among the above-mentioned, the main categories are:

- EWC 17 09 04 mixed construction and demolition wastes other than those mentionedin 17 09 01, 17 09 02 and 17 09 03;
- EWC 17 01 07 mixture of concrete, bricks, tiles and ceramics other than those mentionedin 17 01 06;
- EWC 17 05 04 soil and stones other than those mentioned in 17 05 03.

Regarding the quantitative aspects, despite this year the plants' involvement has been limited because the economic situation, so the study cannot totally be considered representative, the Italian situation is strongly rearward if compared with that of other similar European Countries.

The recycled CDW are about the 10% of the yearly produced inert wastes. The lack of organized measures between public authority and private sectors makes difficult even the achievement of European recycling purpose (the recycling rate equal to 70% in 2020).

Halfway step should be set, for example the achievement of a recycling rate equal to 50% in 2015, to give an impulse to the market.

The failed enforcement of the Ministerial Decree 203/03, concerning the legal requirement for public authorities to use at least the 30% of recycled products, must be report. In this respect, the behaviour of public administrations, especially regarding the use of recycled aggregates in infrastructures and constructions, can be considered disastrous.

3.7 Recycling in the Netherlands

3.7.1 Existing Standards Relevant to CDW

Standards have been developed for application, referring to relevant technical properties, as well as to environmental protection properties.

The specification requirements for recycled aggregates are listed in Tables 3.26 and 3.27.

Table 3.26 PMC recycled concrete aggregate 4/32 as coarse fraction in concrete

Items concerning concrete aggregates 4/32 in concrete	Standards	Requirements	Assessment methods
Composition:			
• Concrete ($\geq$2,100 kg/m^3)	NEN 5905	>90% *(m/m)*	NEN 5942
• Chlorides	NEN-EN 12620	No requirements	NEN-EN 1744-1
• Acid-soluble sulphate content	NEN-EN 12620	More requirements	NEN-EN 1744-1
• Total sulphur content	NEN-EN 12620	$\leq$1% *(m/m)*	NEN-EN 1744-1
• Lightweight organic contaminators	NEN-EN 12620	More requirements	NEN-EN 1744-1
• Constituents causing staining	NEN 5905	No requirements	NEN-EN 1744-1
• Lightweight material	–	No requirements	NEN-EN 1744-1

(continued)

Table 3.26 (continued)

Items concerning concrete aggregates 4/32 in concrete	Standards	Requirements	Assessment methods
Grain distribution	NEN 5905	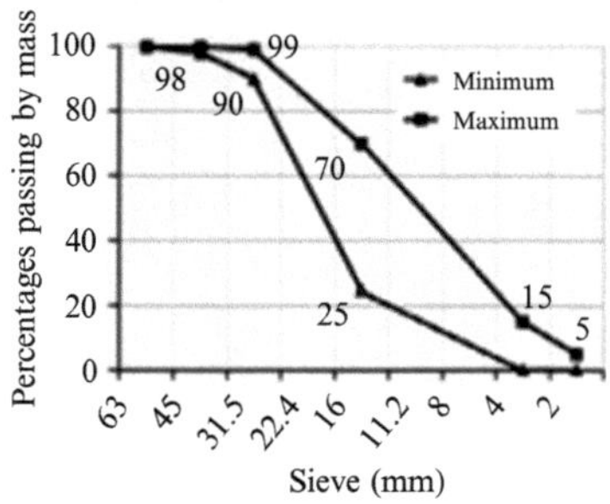	NEN-EN 933-1
Flakiness index	NEN-EN 12620	More requirements	NEN-EN 933-3
Shell content of coarse aggregates	NEN-EN 12620	In practice ≤ 10	NEN-EN 933-7
Fines content	NEN-EN 12620	More requirements	NEN-EN 933-1
Fines quality	NEN-EN 12620	More requirements	NEN-EN 933-8
Los Angeles coefficient	NEN 5905	≤40	NEN-EN 1097-2
Polished stone value	NEN-EN 12620	More requirements	NEN-EN 1097-8
Particle density	NEN-EN 12620	No requirements	NEN-EN 1097-6
Water absorption	NEN-EN 12620	In practice 6% /24 h	NEN-EN 1097-6
Bulk density	NEN-EN 12620	More requirements	NEN-EN 1097-3
Freeze/thaw resistance	NEN-EN 12620	More requirements	NEN-EN 1367-1
Drying shrinkage	NEN-EN 12620	≤0.075%, if required	NEN-EN 1367-4
Alkali-silica reactivity	NEN 5905	No requirements	CUR-recomm. 89
Environment: • Heavy metals (e.g. barium) • Anions (e.g. sulphate) • Organic components (e.g. PAHs)	Building materials decree	No requirements for aggregates, only for the end product, i.e. concrete	Not relevant (no assessment of aggregate)

Table 3.27 PMC recycled mixed aggregate 0/31.5 as sub-base in road construction

Items concerning mixed aggregates 0/31.5 as sub-base in road construction	Standards	Requirements	Assessment methods
Composition:	Standard RAW 2005		Standard RAW 2005
• Concrete and stone (≥2,100 kg/m^3)		≥50% *(m/m)*	
• Concrete (≥2,100 kg/m^3)		≥45% *(m/m)*	
• Masonry (≥1,600 kg/m^3)		≤50% *(m/m)*	
• Other stony materials + asphalt		≤10% *(m/m)*	
• Asphalt		≤5% *(m/m)*	
• Gypsum + non-stony materials		≤1% *(V/V and m/m)*	
• Organic material		≤0.1% *(m/m)*	
• Clumps of clay or soil		No *(visual)*	
Grain distribution	Idem	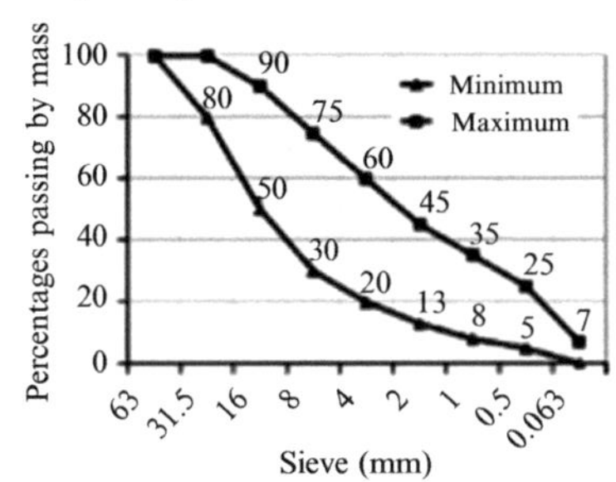	NEN-EN 933-1
Los Angeles coefficient	Idem	≤60	NEN-EN 1097-2
Crushed pieces	Idem	≥90% *(m/m)*	NEN-EN 933-5
Round pieces		≤3% *(m/m)*	NEN-EN 933-5
Flakiness index		≤20	NEN-EN 933-3
$CBR_{0\ days}$	Idem	≥50%	NEN-EN 14227-2,
$CBR_{28\ days}$		≥125% of $CBR_{0\ days}$	Annex D

Environment, heavy metal:	Building materials decree + temporary exemption regulation	Thickness of 70 cm:	NEN 7343 (leaching)
• Barium (Ba) (always critical)		16.45 mg/kg d.m.	NVN 7322
• Copper (Cu) (mostly critical)		0.718 mg/kg d.m.	NVN 7322
Others: Sb, As, Cd, Cr, Co, Hg, Pb, Mo, Ni, Se, Sn, V, Zn		Others: see standard	Others: idem, NVN 7323 + NEN 7324
Environment, anions:	Idem	Thickness of 70 cm:	NEN 7343 (leaching)
• Sulphate (SO_4) (always critical)		3,408 mg/kg d.m.	NNEN-EN-ISO 10304-2
– Outside reserves		1,140 mg/kg d.m.	others: idem + NEN 6589
– In reserves		Others: see standard	
Others: bromide (Br), chloride (Cl), fluoride (F)			
Environment, organic components:	Idem	50.00 mg/kg d.m.	(content)
• PAHs total 10 (always critical)		45.00 mg/kg d.m.	NEN 5731
• PAHs < 8 mm (always critical)		3.000 mg/kg d.m.	NEN 5731
• EOCl (mostly critical)		1.000 mg/kg d.m.	NEN 5735
• Mineral oil (mostly critical)		Others: see standard	NEN 5733
Others: benzene, ethylbenzene, toluene, xylene, phenol, PCBs, (non-)organochlorine pesticides			Others: NVN 5732, NEN 5734, VPR C88-17/-18
Environment, asbestos	Building materials decree (as from 2007)	$\leq$100 mg (serpentine + amphibole $\times$ 10)/kg d.m.	Guideline BRl 2506

3.8 Recycling in Portugal: Overview of CDW

Isabel Martins

3.8.1 *Introduction*

In Portugal, construction and demolition waste (CDW) management was governed by general legislation on waste management but the need of proper managing of this waste stream lead to the publication of specific legislation, Decree Law n. 46/2008 [153], which comes into force in June 2008. This new legal regime for CDW privileges production and hazardousness prevention and appeal to sorting at source, recycling and other forms of recovery instead of landfill disposal.

The new Portuguese framework for CDW particularize the operations concerning the management of waste from construction and demolition and its recent implementation evidenced that unlawful acts like incorrect registration of the CDW at their production place and in the Integrated System of Registration of the Portuguese Environmental Agency (SIRAPA), transport by unlicensed operators and dumping of such wastes still exists as reported by the General Inspectorate of Environment and Territorial Planning (IGAOT).

Several training programs have been performed by public and private institutions within the construction sector in order to contribute to the awareness of various actors regarding the issue of construction and demolition waste management. In spite of the long way that there is still to go, some companies in the construction industry have already put into practice an appropriate management system for their construction and demolition waste.

According to the new legal framework the use of the mineral fraction of CDW in public and private construction works is approved if it comply with the requirements of Portuguese or European standards on this subject and, on the lack of these standards, its use is still allowed if these residues fulfil the technical and environmental requisites established in the specifications for CDW developed by the National Laboratory of Civil Engineering (LNEC).

The LNEC specifications provide guidelines and establish the minimum requirements that recycled material from construction and demolition waste must comply with in order to be used as recycled coarse aggregates for concrete, reclaimed asphalt for hot mix asphalt, aggregates for unbound sub-base and base pavement layers and material for embankment and capping layer of transport infrastructures.

I. Martins (✉)
LNEC, Lisbon, Portugal

Table 3.28 Classification of coarse recycled aggregates in concrete

	Constituents (EN 12620 + A1)					
Classes	RC(%)	RC + RU (%)	RB(%)	RA(%)	X + RG(%)	FL(%)
ARB1	≥ 90		≤10	≤5	≤0.5	≤0,2
ARB2	≥ 70		≤30	≤5	≤1	≤0,5
ARC	≥ 90			≤10	≤2	≤1

RC Concrete and concrete products, mortars, *RU* Unbound aggregates, natural stone, *RA* Bituminous materials, *RB* Masonry, *RG* Glass, *X* Undesirable materials, *FL* Floating material, in volume

3.8.2 Portuguese Specifications for CDW Application

In agreement with Decree-Law 46/2008 the use of construction and demolition waste in Portugal should comply with the requirements of the Portuguese specifications briefly described hereafter:

- LNEC E 471 – Guide for the use of coarse recycled aggregates in concrete [154]
- LNEC E 472 – Guide for the production of recycled hot mix asphalt [155]
- LNEC E 473 – Guide for the use of recycled aggregates in unbound pavement layers [156]
- LNEC E 474 – Guide for the use of recycled materials coming from construction and demolition waste in embankment and capping layer of transport infrastructures [157]

All these specifications present recommendations concerning the practices to be adopted during processing and storage of construction and demolition wastes, requirements for the recycled materials as well as fields of application. A special concern regarding environmental protection is evidenced through the requisite on the release of dangerous substances, which is aligned with the criterion of other European Member States [158].

3.8.2.1 LNEC E 471 – Guide for the Use of Coarse Recycled Aggregates in Concrete

This specification classifies coarse recycled aggregates, within the scope of EN 12620 [159], and defines minimum requirements for their use in concrete. The use of fine recycled aggregates for concrete production is excluded. Coarse recycled aggregates to be used in concrete should conform to the classification of Table 3.28, comply with the properties and minimum requirements of Table 3.29 and for special applications meet the requisites of Table 3.30.

Regarding application of coarse recycled aggregates LNEC E471 states that:

Coarse recycled materials could not be used on concrete that is in contact with water for human consumption.

Table 3.29 Properties and minimum requirements for the use of coarse recycled aggregates in concrete

Properties	Standard	Compliance requirement (according EN 12620 + A1)	Scope
Aggregate sizes	NP EN 933-1	Fulfil § 4.2	All classes
Grading	NP EN 933-1	Fulfil§ 4.3.2	All classes
Constituents	pr EN 933-11	Fulfil one class of Table 1 of this specification	All classes
Shape	NP EN 933-3	FI_{35} (§ 4.4 Table 8)	Aggregates class ARB1
		FI_{50} (§ 4.4 Table 8)	Aggregates class ARB2
		Declared value	Aggregates class ARC
Fines content	NP EN 933-1	f_3 (§ 4.6 Table 11)	Aggregates class ARC
		f_4 (§ 4.6 Table 11)	Aggregates classes ARB1 and ARB2
Resistance to fragmentation	NP EN 1097-2	LA_{50} (§ 5.2 Table 12)	Aggregates class ARB1
		Declared value	Aggregates class ARB2
Particle density	NP EN 1097-6	≥2,000 kg/m^3	Aggregates class ARC
		≥2,200 kg/m^3	Aggregates classes ARB1 and ARB2
Water absorption	NP EN 1097-6	≤7% (§ 5.5)	Aggregates classes ARB1 and ARB2
		Declared value	Aggregates class ARC
Alkali-silica reactivity	see LNEC E 461	Declared value	All classes
Volume stability – Drying shrinkage	NP EN 1367-4 (Annex A)	≤0.075% (§ 5.7.2)	All classes
Acid soluble chloride	EN 1744-5	Declared value	All classes
Water soluble sulphate	NP EN 1744-1, §12	$SS_{0,2}$ (§ 6.3.3)	All classes
Acid soluble sulphate	NP EN 1744-1, §12	$AS_{0,8}$ (§ 6.3.1)	All classes
Total sulphur	NP EN 1744-1, §11	$S \leq 1.0\%$ (§ 6.3.2)	All classes
Organic constituents that affect the rate of setting and hardening of concrete	NP EN 1744-1, §15	Fulfil (a) e (b) §6.4.1	All classes
Other constituents that affect the rate of setting and hardening of concrete	EN 1744-6	Declared value (§6.4.1)	All classes
Release of dangerous substances	EN 12457-4	Classification as waste for disposal in landfill for inert waste	All classes

Table 3.30 Special applications for the use of coarse recycled aggregates in concrete

Properties	Standard	Compliance requirement	Scope	Applications
Alkalis	NP 1382	Value to determine	All classes	Concrete with reactive aggregates
Light organic contaminants	NP EN 1744-1, §14.2	Fulfil § G4 of Annex G of EN 12620	All classes	Architectural concrete

Table 3.31 Fields of application of coarse recycled aggregates in concrete

Class	Strength class	Replacement level (%)	Exposure classes
ARB1	C 40/50	25	X0, XC1 to XC4, XS1 and XA1(2)
ARB2	C 35/45	20	

Application of coarse recycled aggregates, under different conditions, call for specific studies to evaluate their influence on the relevant properties for concrete

ARC coarse recycled materials could only be used for non structural applications (levelling, filling concrete) in non aggressive exposure classes.

ARB1 and ARB2 could be used on concrete elements and on reinforced concrete.

There is no limit for the use of ARB1 and ARB2 coarse recycled aggregates on filling or levelling concrete, in non aggressive exposure classes.

The conditions for application of coarse recycled aggregates in reinforced concrete are expressed below in Table 3.31:

3.8.2.2 LNEC E 472 – Guide for the Production of Recycled Hot Mix Asphalt

Recommendations and requirements for the manufacture and application of reclaimed asphalt, on recycled hot mix asphalt, within the scope of EN 13108-8 [160] are indicated in LNEC E472. The CDW to be recycled in hot mix plants include: asphalt, reclaimed by milling of asphalt road layers; asphalt reclaimed by crushing of slabs ripped up from asphalt pavements; and asphalt from reject and surplus production.

Reclaimed asphalt is identified as MBR1, MBR2 and MBR3 according to the foreign matter and particle size of reclaimed asphalt, maximum size of aggregates and characteristics of the binder of the reclaimed asphalt.

For the production of recycled hot mix asphalt the properties and minimum requirements and the fields of application are indicated in Table 3.32 and in Table 3.33.

3.8.2.3 LNEC E 473 – Guide for the Use of Recycled Aggregates in Unbound Pavement Layers

Recommendations for the use of recycled aggregates, covered by EN 13242 [161] and EN 13285 [162], in unbound layers (sub-base and base) of road pavements are dealt with in LNEC E473.

Table 3.32 Properties and minimum requirements for the reclaimed asphalt to be used in the production of recycled hot mix asphalt

Compliance requirements (EN 13108-8)			Reclaimed asphalt		
Properties		Standard	MBR1	MBR2	MBR3
Foreign matter on the reclaimed asphalt		EN 12697-42	F1	F5	F5
Binder of the reclaimed asphalt	Type of binder	–	Bitumen (traditional) or Modified Bitumen	Bitumen (traditional) or Modified Bitumen	Bitumen (traditional), Modified Bitumen or Hard Bitumen
	Binder properties	EN 12697-3 or EN 12697-4 EN 1426 or EN 1427	P_{15} or S_{70} or $P \geq 15 \times 10^{-1}$ mm or $S \leq 70°C$	P_{15} or S_{70} or $P \geq 15 \times 10^{-1}$ mm or $S \leq 70°C$	P_{dec} or S_{dec} or Declare source and properties of the binder
Particle size reclaimed asphalt		NP EN 933-1	32 mm	32 mm	32 mm
Aggregate properties	Mean grading	NP EN 933-1	Declared value	Declared value	Declared value
	Maximum size, D	NP EN 933-1	Declared value	Declared value	Declared value
Binder content of reclaimed asphalt		EN 12697-1	Declared value	Declared value	Declared value
Maximum water content of reclaimed asphalt		EN 12697-14	5%	5%	5%

Table 3.33 Applications reclaimed asphalt to be used in the production of recycled hot mix asphalt

Classes of reclaimed asphalt (EN 13108-8)	Application	Maximum incorporation (%)
MBR1	Wearing course	10
	Base and levelling course	50
MBR2	Base and levelling course	25
MBR3	Base and levelling course	10

Table 3.34 Classification of recycled aggregates for use in unbound granular pavement layers

	Constituents					
	EN 13242 + A1					
Classes	RC + RU + RG (%)	RG (%)	RA (%)	RA (%)	FL (%)	X (%)
B	≥90	≤5	≤10	≤5	≤5	≤1
C	≥50	≤5	≤10	≤30	≤5	≤1

Recycled aggregates in unbound granular pavement layers obeys to the classification of Table 3.34, and the properties and minimum requirements for the classes B and C, are indicated on Table 3.35. Regarding their application, data is presented in Table 3.36.

3.8.2.4 LNEC E 474 – Guide for the Use of Recycled Materials Coming from Construction and Demolition Waste in Embankment and Capping Layer of Transport Infrastructures

For use in embankment and capping layer of transport infrastructure, construction and demolition waste covered by this specification are grouped into three classes (B, C or MB), defined based on the relative proportions of each constituents as shown in Table 3.37. According to geotechnical characteristics these materials are categorized as MAT1 and MAT 2 obeying the properties and minimum requirements of Table 3.38. Data in Table 3.39 indicate the corresponding fields of application

It must be emphasized that materials type MAT1 are not allowed in the lower part of embankments built in floodable areas and specific studies are foreseen for materials treated with hydraulic binders.

3.8.3 Recycling Plants

In Portugal there are between 15 and 20 stationary recycling plants, licensed by the Portuguese Environmental Agency (APA), for recovery of the mineral fraction of CDW and production of materials to be used in the construction sector. Most of these recycling plants are located in the metropolitan areas surrounding the main

Table 3.35 Properties and minimum requirements for recycled aggregates to be used in unbound granular pavement layers

Compliance requirements			Recycled aggregates		
			AGER1	AGER2	AGER3
Parameters	Properties	Standard	B or C	B or C	B
Geometrical and nature	Mixture designation	EN 13285	0/31.5	0/31.5	0/31.5
	Oversize (EN 933-1 + A1)	EN 13285	OC_{75}	OC_{80}	OC_{85}
	Grading class (EN 933-1 + A1)	EN 13285	GB	GB	GA
	Fines content (EN 933-1 + A1)	EN 13285	UF_9 LF_2	UF_9 LF_2	UF_9 LF_2
	Fines quality (EN 933-9)	EN 13242 + A1	$MB_{0/D} \leq 1.0$	$MB_{0/D} \leq 0.8$	$MB_{0/D} \leq 0.8$
	Percentage of crushed or broken particles and totally round particles in coarse aggregates (EN 933-5)	EN 13242 + A1	C50/30	C50/10	C90/3
Mechanical	Resistance to fragmentation and to wear (EN 1097-2 and EN 1097-1)	EN 13242 + A1	LA_{45} MDE_{45} $LA + MDE \leq 85$	LA_{40} MDE_{40} $LA + MDE \leq 75$	LA_{40} MDE_{35} $LA + MDE \leq 70$
Chemical	Water soluble sulphate (EN 1744-1)	EN 13242 + A1	$SS_{0,7}$	$SS_{0,7}$	$SS_{0,7}$
	Release of dangerous substances (EN 12457-4)	–	Classification as wastes for deposition in landfill for inert waste		

Table 3.36 Application of recycled aggregates in unbound pavement layers

Category	AGER1		AGER2		AGER3
Classes	C	B	C	B	B
Application in sub-base course – ADTt	≤50	≤150	≤150	≤300	≤300
Application in base course – ADTt	NR	≤150	≤150	≤150	≤300

ADTt Average daily truck traffic, *NR* Not recommendable

Table 3.37 Classification of recycled materials for embankment and capping layer of transport infrastructure

	Constituents					
Class	RC + RU + RG	RG (%)	RA (%)	RB + RS	FL (%)	X (%)
B	≥90%	≤10	≤5	≤10%	≤5	≤1
MB	≤70%	≤25	≥30	≤70%	≤5	≤1
C	No limit	≤25	≤30	No limit	≤5	≤1

RS Soils

Table 3.38 Properties and minimum requirements for recycled materials for embankment and capping layer of transport infrastructure

Compliance requirements			Recycled materials		
			MAT 1	MAT 2	
Parameters	Properties	Standard	B, MB and C	B and C	MB
Geometrical and nature	Maximum particle size	–	D_{max} ≤150 mm	D_{max} ≤80 mm	
	Maximum fines content (below 80 μm)	LNEC E 196	10%	10%	
	Fines quality	EN 933-9	$MB_{0/D}$ ≤2	$MB_{0/D}$ ≤1	
Mechanical	Resistance to fragmentation and	EN 1097-2 + A1	–	LA_{45}	–
	to wear	EN 1097-1 + A1		MDE_{45}	
Chemical	Water soluble sulphate	EN 1744-1	0.7%	0.7%	
	Release of dangerous substances	EN 12457-4-	Classification as waste for disposal in landfill for inert waste		

Table 3.39 Application of recycled materials for embankment and capping layer of transport infrastructure

Category	MAT1			MAT2		
Class	B	MB	C	B	MB	C
Capping layer	X	X	X	✔	X	✔
Embankment	✔	✔	✔	✔	✔	✔

cities, like Lisbon and Oporto, as denoted by the black dots in the Portugal map, Fig. 3.66, which match the locals of higher construction activity.

A study carried out in 2008 evidenced that in the city of Lisbon the production of CDW wastes is mainly due to remodelling works and not to new constructions in opposition to what is observed in the neighbouring cities [163].

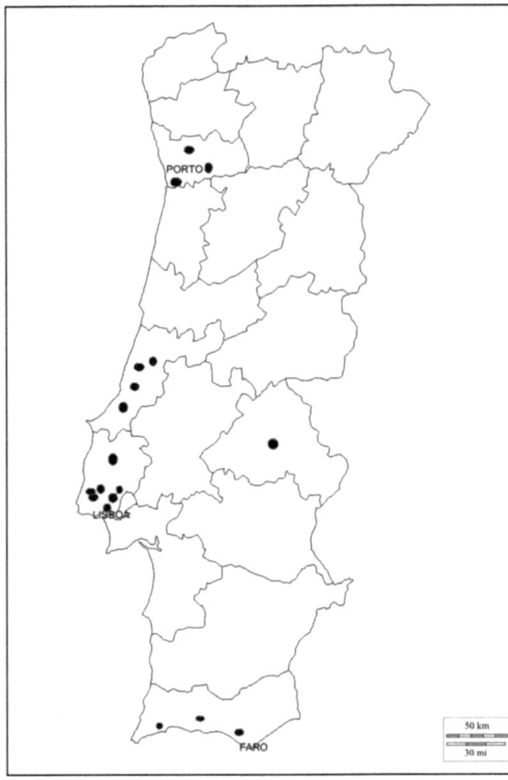

Fig. 3.66 Portuguese stationary recycling plants for the mineral fraction of C&DW

Mobile recycling plants are also used to recovery of the mineral fraction of CDW but there is no available information about them.

Actually the recycling plants have had much difficulty in putting on the market their recycled materials. Barriers to their use are mainly dictated by economic aspects, lack of experience, assurance of quality of recycled materials, existence of regular supply/demand systems and environmental concerns.

3.8.4 Recycling Rates

The existing data regarding construction and demolition waste production in Portugal is provisional. According to the Portuguese Environmental Agency the total waste, listed in chapter 17 of the List of Wastes, received at the recycling plans in 2009 was approximately two million tons. However CDW generation must be higher since the CDW reused or recycled on site in civil works is not registered on the Portuguese electronic database.

Figure 3.67 presents the total production of waste within the chapter 17 of the List of Waste, expressed as percentage, including soils and excluding dangerous

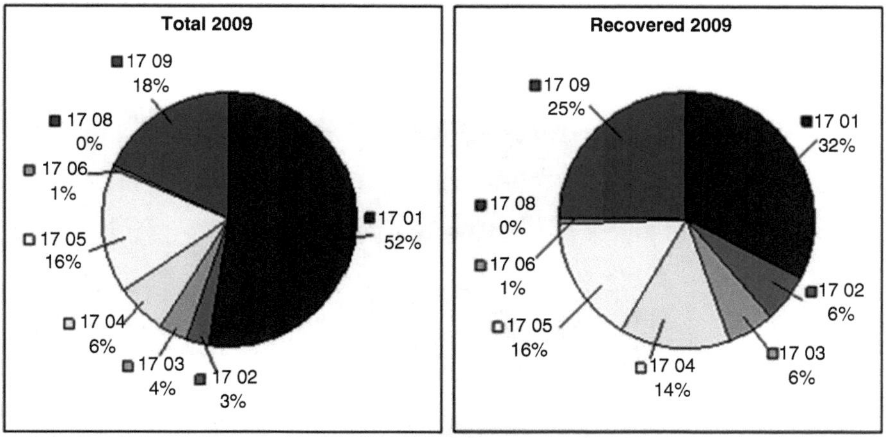

Fig. 3.67 Production and recovery of waste within the chapter 17 of the list of wastes in 2009

wastes, received at recycling plants. CDW classified as 17 01 of List of Wastes – concrete, bricks, tiles, ceramics, and gypsum-based materials – constitute about half of the total waste from chapter 17 and 32% of it was subject to recovery operations. Nevertheless the available data show that only 4% has been effectively processed and is available to be used as inorganic material in the construction sector; most of the remainder was stored for further processing.

In spite of the impossibility of knowing the total production of CDW in Portugal and the difficulty to point a value or a range of values for the recycling rate it is necessary a better marketing of recycled materials from CDW and the adoption of additional measures in order to increase the recycling and meet the European goals for 2020 for this waste flow. In 17th June 2011 a new Portuguese legislation, Decree-Law 73/2011 [164] transposing the Waste Framework Directive 2008/98/EC [165], foresees that, whenever technically feasible, it is mandatory the use of at least 5% of recycled materials in the hiring of construction works and maintenance of infrastructure under the Code of Public Contracts, with a view to preserve natural resources and promote waste recovery.

3.8.5 Case Studies

Regardless of incomplete implementation of the new law for CDW management successful recycling cases in Portugal could be mentioned. All these cases involve previous development of appropriate environmental management plans, investment on employee training and extensive sorting of waste in situ in order to minimize the costs of waste management (Fig. 3.68).

Fig. 3.68 Construction of pier north of Lisbon airport

Between 2008 and 2010 the construction of the new pier north of Lisbon airport, with a gross area of 13,000 m^2, produced a total of 15,650 tons of CDW and the concrete slabs of the existing pavement after grinding and characterisation were used in situ as recycled aggregates for the layer beneath the wearing course of the new pier. A recycling rate of 96% was achieved in this project.

In 2009/2010 during construction of Estoril-Sol Residence, with a gross floor area of buildings around 30,000 m^2, about 4,250 tons of CDW was generated, being 2,400 tons of concrete, classified within the code 17 01 01 of the List of Waste, that were recovered by licensed operators. The total recycling rate of 99% evidenced the adequate CDW management.

Also in 2009 the demolition of a set of industrial and office buildings in Lisbon, occupying an area of 7,000 m^2, were entirely demolished resulting in the production of 32,700 tons of inert waste, partially recycled in situ. About 25,000 tons of recycled material were transported to other works to be used in infrastructures. In this process only 0.3% of CDW were eliminated, corresponding to a recycling rate of 99.7%.

Acknowledgements The author wishes to thank the National Laboratory for Civil Engineering (LNEC), the Portuguese Environmental Agency (APA), the Construction Group EDIFER and Demolition Enterprise DEMOTRI, for their collaboration on the data presented.

3.9 Recycling in Spain: Overview of CDW

Enric Vázquez

3.9.1 Introduction

The information presented in this overview is mostly obtained with the GEAR Project and complemented with the projects RECNHOR and CLEAM. The Project "Spanish Guide on Recycled Aggregates" (GEAR Project) is a technical and scientific project subsidized by the Ministerio de Medio Ambiente, y Medio Rural y Marino, of the Spanish Government, with 2.5 M€ within the framework of the National Plan of Scientific Research, Development and Technological Innovation.

Coordinated by the GERD (Spanish Association of the Recycling Industry), the GEAR Project, possesses the direct participation of 73 CDW recycling plants of Spain, 4 Universities:

- Universitat Politècnica de Catalunya (UPC)
- Universidad de Oviedo (UNIOVI)
- Universidad d'A Coruña (UDC)
- Universidad Politécnica de Valencia (UPV)

and 3 Technological Centers:

- Instituto Tecnológico de la Construcción (AIDICO), Valencia
- Instituto Tecnológico de Rocas Ornamentales y Materiales de Construcción (INTROMAC), Caceres
- Asociación para la Investigación y Desarrollo Industrial de los Recursos Naturales (AITEMIN), Madrid

Under the scientific direction of Prof. E. Vázquez (UPC, Barcelona), the technical direction of E. Varela (UPC) and the Management of A. Güell (GERD), the GEAR Project aims the normalization of CDW recycled products and its uses in civil construction across the elaboration of the "Spanish Guide on Recycled Aggregates". The first stage of the project consists of the accomplishment of a technical and statistical field study to characterize the situation of the Spanish Recycling Plants. This diagnosis study consists on the analysis of data from 73 Spanish recycling plants, chosen by their importance and location, production systems and uses. The aim is that the final document reflects the reality of the Spanish recycling market, considering all its technical, scientific and environmental

E. Vázquez (✉)
Materials Section of the Department of Construction Engineering, Universitat Politecnica de Catalunya, Barcelona, Spain

Table 3.40 Characterization tests adopted for the diagnosis stage

Property	Control test	Standard
Geometrical	Granulometry	UNE EN 933-1:1998
	Particle shape/flakiness index	UNE EN 933-3
Physical mechanical and physical	Absorption/density	UNE EN 1097-6
	Composition	EN 933-11:2009
	Sand equivalent test	UNE EN 933-8
	Methylene blue test (when applicable)	UNE-EN 933-9:1999
	Cleaning (organic impurity content)	UNE 146130
	Resistance to fragmentation (Los ángeles test)	UNE EN 1097-2
	Plasticity (Atterberg limits)	UNE 103104
Chemical	Total sulfur content	UNE EN 1744-1:1999
	Organic contaminants	UNE EN 1744-1:1999 (APDO 15.1)
	Salts	NLT 114
	Water Soluble Sulfate content	UNE EN 1744-1:1999
	Acid Soluble Sulfate content	UNE EN 1744-1:1999
	Gypsum	NLT 115

aspects, presents several real controlled applications, analyses future possibilities and offer recommendations and limits based on European standards and the Spanish research projects.

3.9.2 Diagnosis Method

The diagnosis consisted in collecting technical information from the Spanish CDW recycling plants, mainly considering the methods of production of recycled aggregates, the results of control tests conducted and the uses proposed by the companies. The collection of this information has been done during technical visits, applying specification sheets.

In order to complement the information, the recycling companies have executed a specific plan of tests for the Project defined by the Coordination. The planning of the tests and their frequency of accomplishments has been defined by the needs of the project and the conditions of production of each plant. In order to fulfil this plan, all participants have received indications and specifications on how to correctly carry out tests plan so to guarantee its representativeness and strictness. The Table 3.40 presents the laboratory tests carried out in this stage of the project.

Every University and Technological Centre has verified the results obtained in the assigned correspondent territory by the accredited laboratories used by the companies, and executed a random part of the tests presented to control the good execution.

In the second stage of the Project, the tests plan will be applied with more specific tests for the analysis of the recycled aggregate application. Based on the

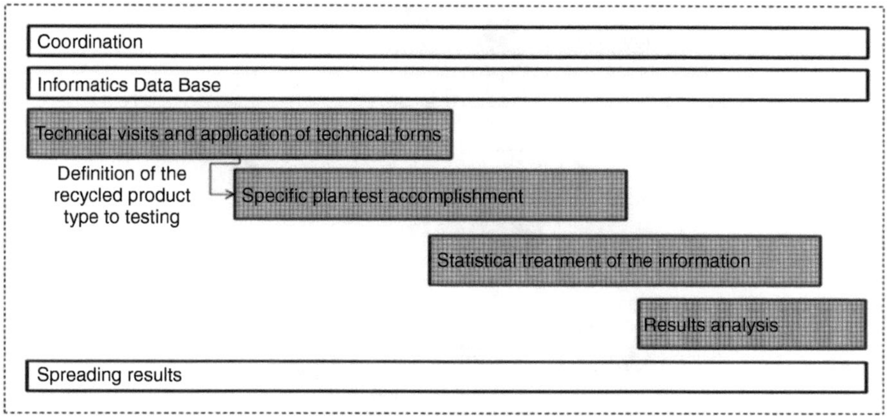

Fig. 3.69 Task flow of GEAR project's diagnosis stage

present diagnose results, the project adopted the following research lines for the next stage: roller compacted concretes, non-structural concretes, pre fabricated elements, fillings / drainages, aggregates for roads bases and gravel and soil cement.

All the obtained information has been registered in the project database. The Fig. 3.69 presents the task flow of the activities that correspond to the stage of diagnosis of the GEAR Project.

3.9.3 Production of Recycled Aggregates in Spain

In Fig. 3.70 we present the distribution of the analyzed plants in the Spanish Autonomies.

From 73 plants investigated from 2008 to 2011, 35 were direct participants of the Project and 38 were incorporated later as collaborators. The total number of plants in Spain in 2011 was 156.

The total amount of CDW estimated in 2008 by the "*Ministerio de Medio Ambiente*" was about 38 million tons. This amount has diminished dramatically in the last 2 years and we don't have precise data. In 2008, 15% of waste was processed. In 2008, 1.5 million tons were produced and commercialized. The most used recycled product has been the mixed recycled aggregate, in the all-in-one fraction (47.1% of total commercialized tons in 2008, object of this analysis), followed by concrete recycled aggregate, also in the all-in-one fraction (17.95% of total). The distribution in different types of aggregates produced can be seen in the Fig. 3.71.

Only a small proportion of concrete coarse aggregates >4 mm, were used in structural concrete in emblematic constructions as the bridges in Barcelona and Valencia and in housing near Madrid.

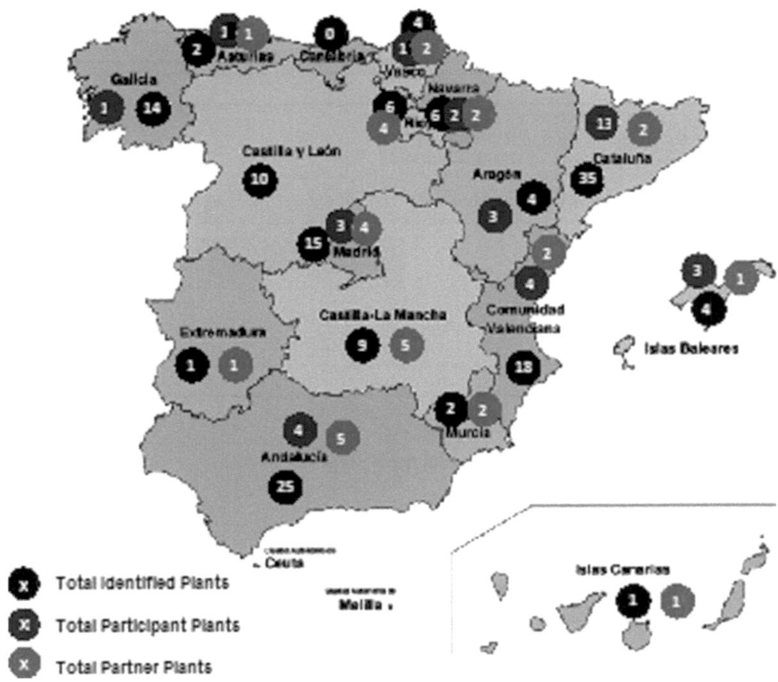

Fig. 3.70 Map of participation plants

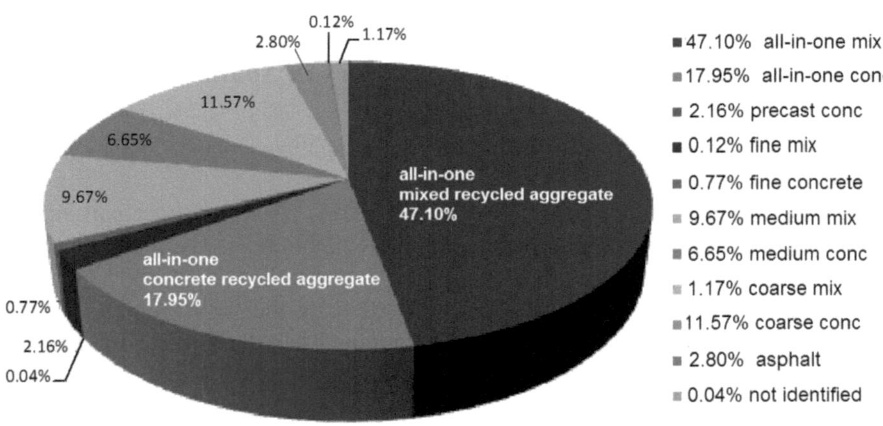

Fig. 3.71 Types of aggregates produced

The 59% of the 0/40 mm size of **Concrete Recycled Aggregates** was used in road sub base unbounded or as soil cement mixtures. Most of the 0/40 mm size of **Mixed Aggregate** is used in environmentally controlled fills and small country roads but there are some cases of use in soil cement. The fine fraction of those aggregates was mostly used as fill or non-identified applications. The 42% of the aggregates >40 mm, concrete and mixed were used in fills and 35% in country roads.

Fig. 3.72 Types of plants

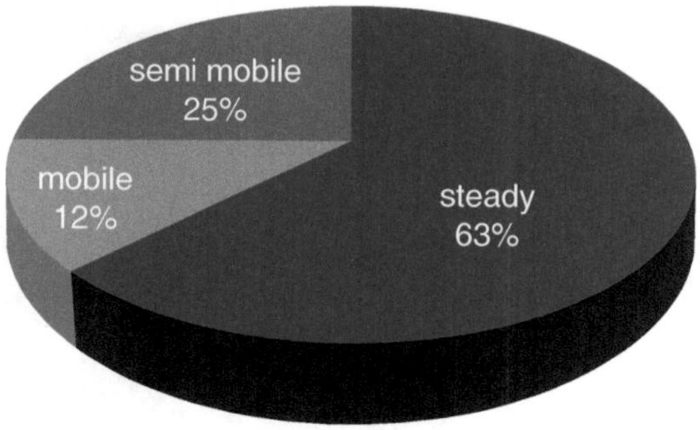

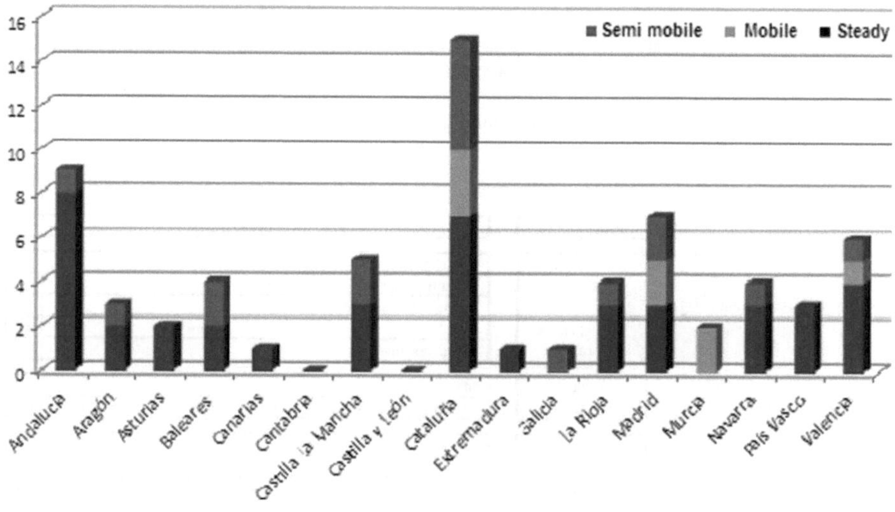

Fig. 3.73 Distribution of plants

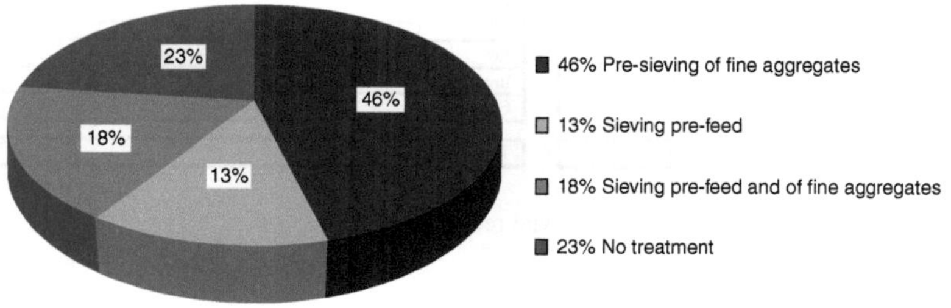

Fig. 3.74 Plants with pre treatment and pre elimination of fines

The different types of plants are presented in Fig. 3.72.

Mobile plants are only present in Catalonia, Madrid, Murcia and Valencia. General distribution of mobile, semi mobile and steady plants can be seen in Fig. 3.73.

The proportion of plants with pre treatment and pre elimination of fines can be seen in Fig. 3.74.

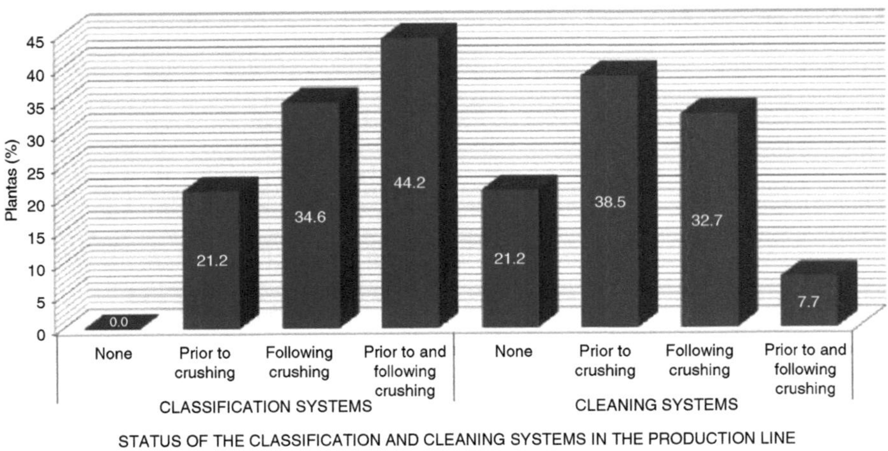

Fig. 3.75 Plants with classification and cleaning

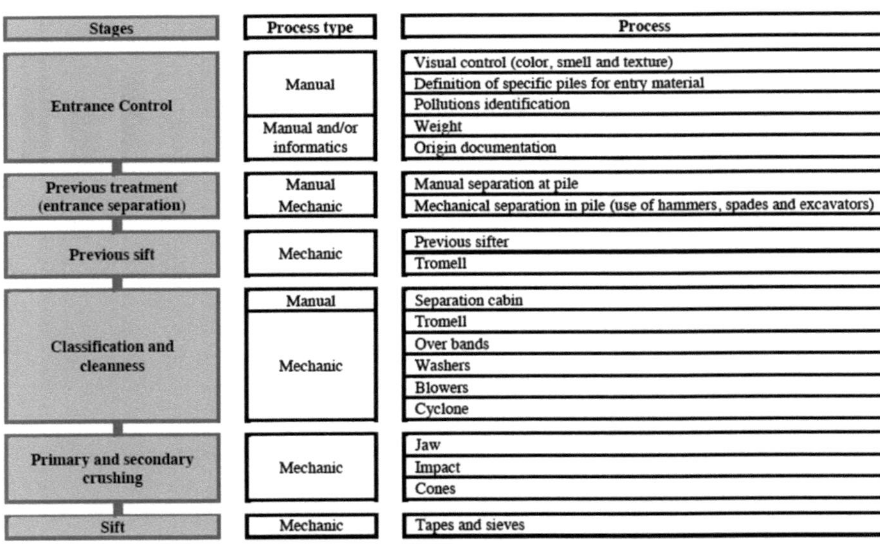

Fig. 3.76 Production flow of a Spanish type recycling plant

In Fig. 3.75 we can see the situation of plants with classification and cleaning.

In general as we can deduce from the obtained data the production scheme of a recycling plant in Spain generally consists of all or some of the stages presented. The design criteria of a recycling plant can change a lot for each plant depending of the place and demand of products. In the most complete plants the production flow is as can be seen in Fig. 3.76.

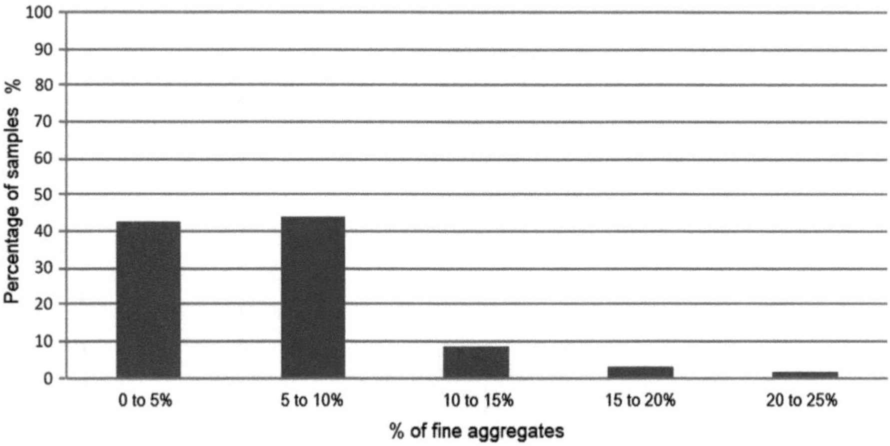

Fig. 3.77 Fines in recycled aggregates

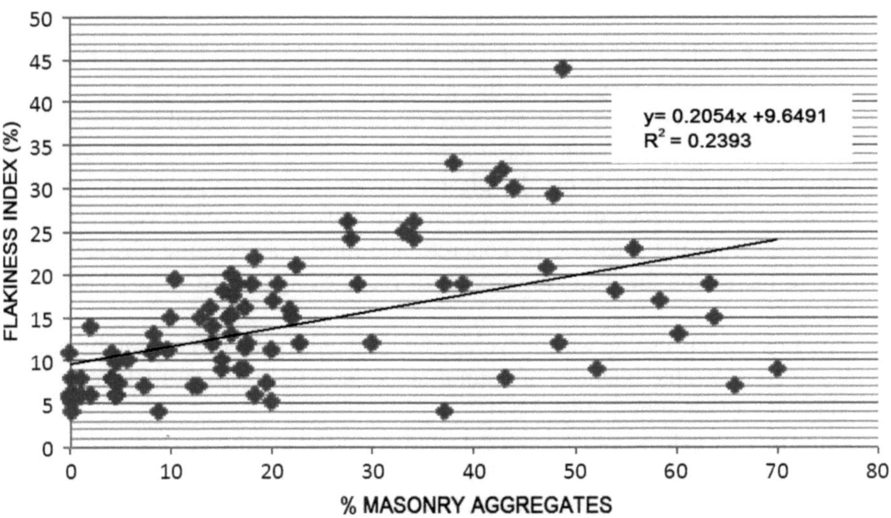

Fig. 3.78 Flakiness index related with % of masonry aggregates

3.9.4 Properties of Spanish Recycled Aggregates

We include in this overview only the most relevant properties. The analysis of the **size distribution** detected excess of fines in an important proportion of the recycled aggregates commercialized (Fig. 3.77).

The **flakiness index** was clearly related with the proportion of masonry elements present (Fig. 3.78).

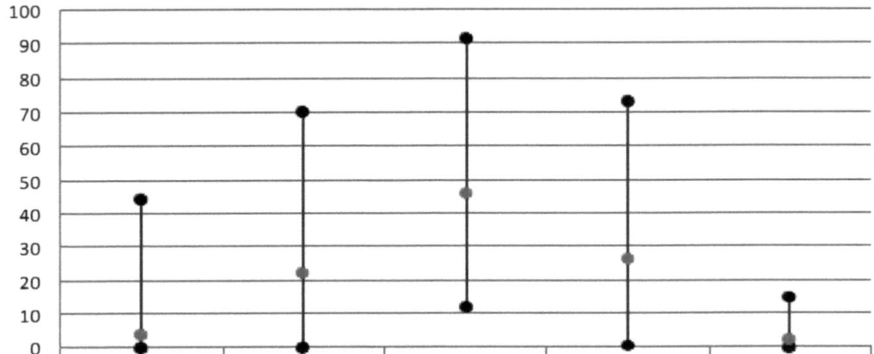

	Asphalt	Ceramics and masonry	Concrete aggregates	Natural stones	Others
● min	0.00	0.00	12.00	0.19	0.00
● med	3.51	22.00	45.86	25.98	1.95
● max	44.21	70.00	91.39	73.00	14.50

Fig. 3.79 Composition of mixed aggregates

The **composition** was evaluated with EN 933-11. Figure 3.79 shows the minimum and maximum ranges, and the average found for each type of material identified in recycled aggregates analyzed during the experimental campaign.

Absorption is an important parameter of quality of material in aggregates. Although ceramic materials usually present higher water absorption than concrete materials, is the presence of stone materials that most influence the values of these parameters. Both the presence of concrete and mortar as ceramic elements contribute to increasing the water absorption of recycled aggregate. (Figs. 3.80, 3.81, and 3.82).

The **Los Angeles** value was in average better than expected. This can be related with the considerable quantity of natural stone present in Spanish CDW and the good quality of ceramics. (Fig. 3.83)

Only 16% of values >40% were detected considering all types and samples.

Sufficient good values were obtained in **sand equivalent** for mixed aggregates 0/40 mm, (Fig. 3.84) for his use in sub bases and soil cement.

The cleanness coefficient of Spanish recycled aggregates is in average a weak point. Only 52% of the products and samples are under 2, the limit established in the Spanish Road Code PG3 for all kind of aggregates.

The good values are related with plants that select the material before processing or plants provided with air and water cleaning systems.

The proportion of **soluble salts** in the recycled aggregates is another weak point considering all the compositions. The limit established for the use in contact with soil is <1% of soluble salts different from gypsum. Only 43% are under this limit (Fig. 3.85).

As in the case of cleanness coefficient, the good values are related with plants that select carefully the material before processing or plants provided with air and water cleaning systems.

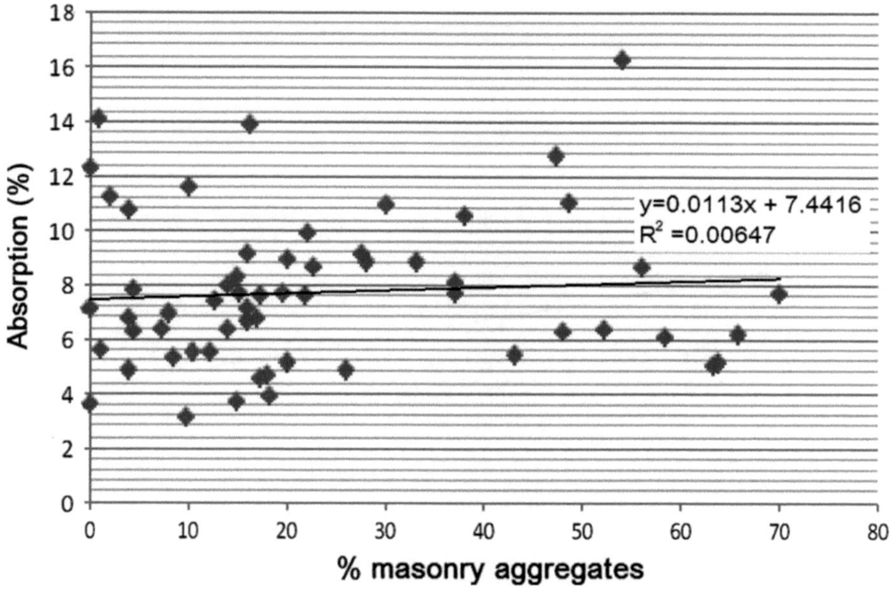

Fig. 3.80 Absorption related with % masonry aggregates

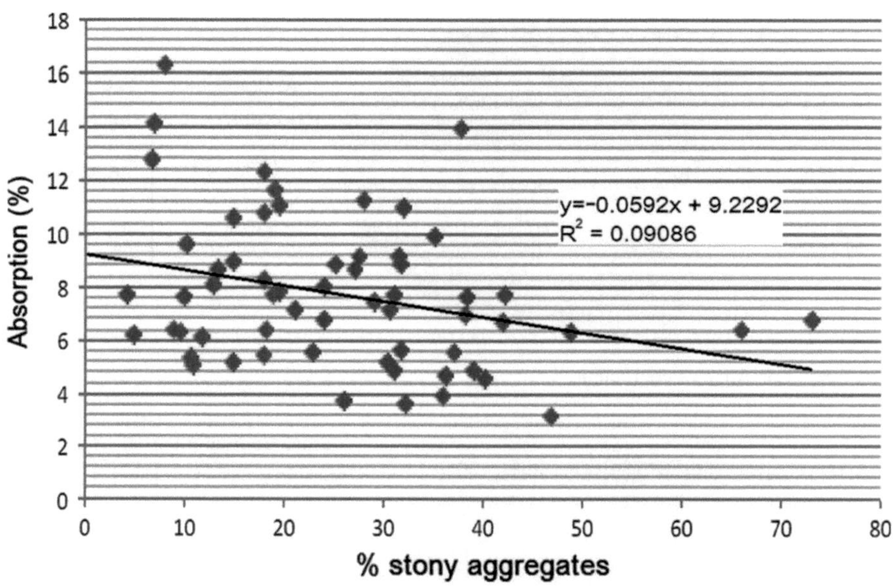

Fig. 3.81 Absorption related with % concrete aggregates

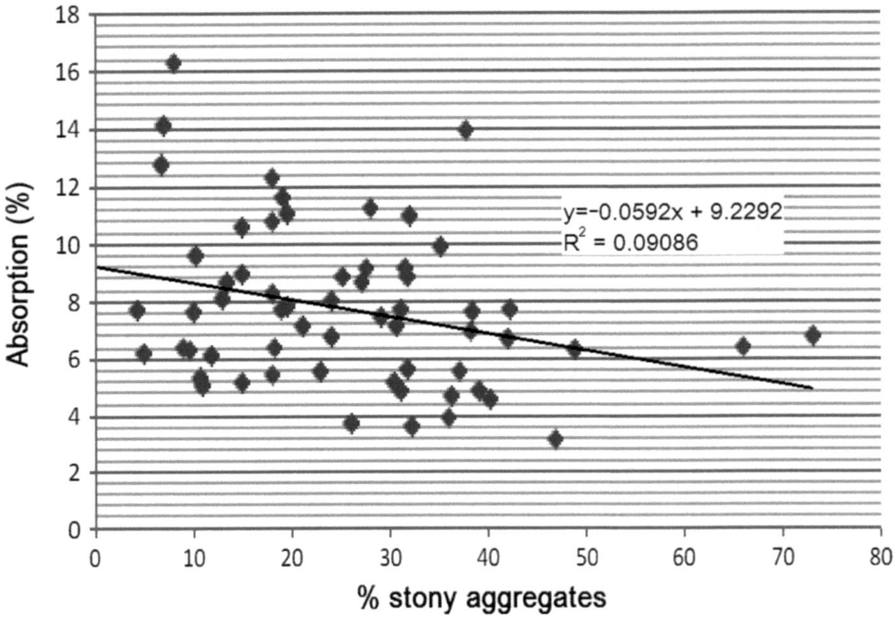

Fig. 3.82 Absorption related with % of stony aggregates

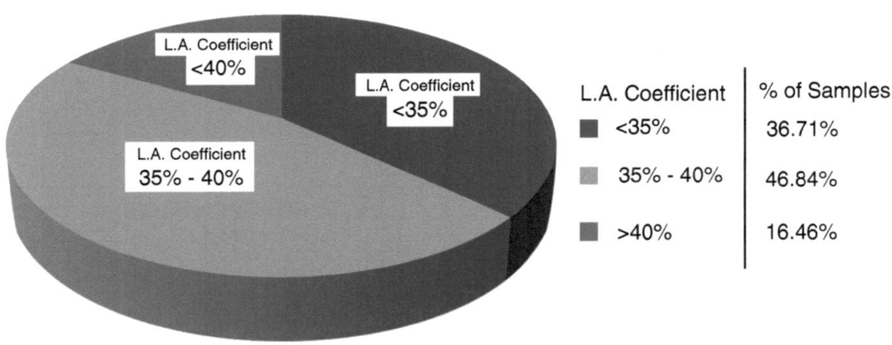

Fig. 3.83 Los Angeles coefficient

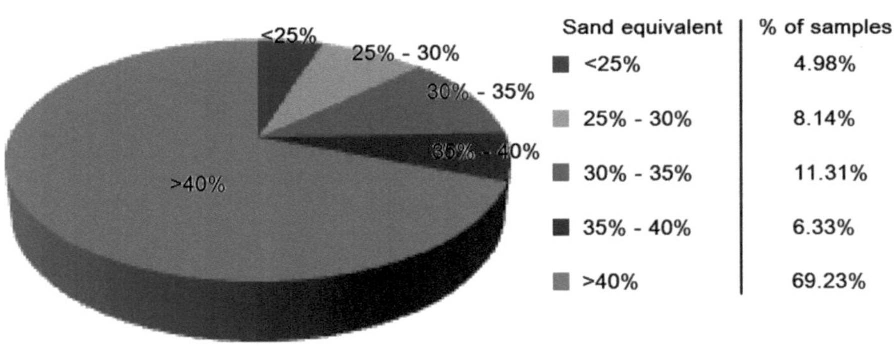

Fig. 3.84 Sand equivalent

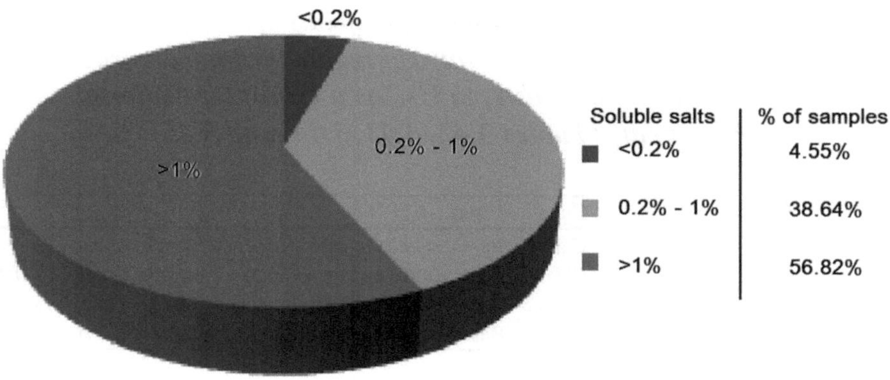

Fig. 3.85 Soluble salts

Table 3.41 Classification of recycled aggregates

Category	Description	Proportion of constituent (% total weight) Ru (unbound aggregate, natural stone) (%)	Rc (concrete and mortar)	Rb (Ceramic) (%)	Ra (Bituminous materials) (%)	X (Other) (%)
Ar H	Recycled concrete aggregate	≥90		≤10	≤5	≤1
ArM H	Mixed recycled concrete aggregate	≥70		≤30	≤5	≤1
ArM C	Mixed recycled ceramic aggregate	<70		>30	≤5	≤1
ArM A	Mixed recycled asphalt aggregate	–		–	5–30	≤1

The presence of **gypsum** is a weak point in relation with the use of the recycled aggregate in concrete or soil cement, because In the average gypsum content in recycled concrete aggregate (with minimum 90% concrete) is of 0.6% only, with maximum values of 2%, which is produced in fewer plants. In contrast, the mixed type aggregate, with a minimum concrete content of 70%, which is the most widespread, has an average of 2.5% and varies between 0.5 and 6.7%, always in weight. In general the plants cleaning with water have the lowest gypsum content.

3.9.5 The Proposed Classification of the Recycled Aggregates

The Gear Project proposed a classification of the recycled aggregates based in the predominant components (Table 3.41).

3.9.6 *GEAR-PRT-02: Technical Specifications and Recommendations for Aggregates from Construction and Demolition Waste, to Use as a Granular Material in Pavements (Tables 3.42, 3.43, 3.44, and 3.45)*

Table 3.42 Usage classes for recycled graded-aggregate, according to technical specifications degree

Class	Usage
Class1	Bases, subbases and shoulders for traffic routes T0
Class 2	Bases, subbases and shoulders for traffic routes T1 y T2
Class 3	Bases, subbases and shoulders for traffic routes T3 y T4
Class 4	Bases and subbases for roads with lower traffic category T4

Table 3.43 Viability of application by category of aggregate and usage class

		Recycled aggregate category				
Viability of application		Ar H	ArM H	ArM C	ArM A	MI
Usage class	Class 1	X				
	Class 2	X	X			
	Class 3	X	X	X	X	
	Class 4	X	X	X	X	

Table 3.44 Limits of the size distribution for recycled aggregates

	Cumulative percent of aggregate sieved (% mass min-max)[b]								
Category[a]	40	25	20	8	4	2	0.500	0.250	0.063
ZA25	100	75–100	65–90	40–63	26–45	15–32	7–21	4–16	0–9
ZA20	–	100	75–100	45–73	31–54	20–40	9–24	5–18	0–9
ZAD20	–	100	65–100	30–58	14–37	0–15	0–6	0–4	0–2

[a]The graded-aggregate type designation is based on the nominal maximum size, according to the range of % passing, set in the corresponding Limits of the size distribution
[b]The sieve sizing is defined by UNE-EN 933-2

Table 3.45 Summary of recommended technical requirements

	Class/usage			
Technical requirements	Class 1	Class 2	Class 3	Class 4
Granulometry	Table 3.44	Table 3.44	Table 3.44	Table 3.44
Flakiness index (%)	<35	<35	<35	<35
Grinded particles (%)	100	≥75	≥50	≥75
Composition[a] categories	ArH ó ArM A	ArH, ArM A ó ArM C	All categories	All
Los Ángeles coefficient (%)	≤35	≤35	≤40	≤45
Liquid limit	Non-plastic	Non-plastic	Non-plastic	<25
Plasticity	Non-plastic	Non-plastic	Non-plastic	<6
Sand equivalent (%)	>40	>40	>35	>30
Cleanness coefficient[a] (%)	<2	<2	<2	N.A.
Clay lumps (%)	<1	<1	<1	N.A.
Organic matter (%)	≤0.2	≤0.2	≤0.2	N.A.
Sulphur compounds (%)	≤1	≤1	≤1	≤1
Water soluble sulphate (%)	≤0.5	≤0.5	≤0.5	≤0.5

[a]Orientative and not restrictive requirement

3.9.7 GEAR-PRT-04: Technical Specifications and Recommendations for Aggregates from Construction and Demolition Waste, to Use as a Granular Material in Pavement Surfaces, Treated with Hydraulic Binders (Tables 3.46, 3.47, 3.48, 3.49, and 3.50)

Table 3.46 Usage classes in the current technical recommendations, according to technical specifications degree

Classes		Application
Soil cement	SC20	Soil-cement applicable to pavements and shoulders T3 y T4
	SC40	Soil-cement applicable to pavements and shoulders T00 a T2
Gravel cement	GC20	Gravel-cement applicable to pavements T00 a T2
	GC25	Gravel-cement applicable to pavements T3 y T4 and shoulders, replacing SC40

Table 3.47 Viability of application by category of aggregate and usage class

			Recycled aggregate category				
Viability of application			Ar H	ArM H	ArM C	ArM A	MI
Usage class	Soil cement	SC20					
		SC40	X	X	X		
	Gravel cement	GC20	X	X			
		GC25	X	X	X	X	

Table 3.48 Limits of the size distribution for soil cement application

	Cumulative percent of aggregate sieved (% mass min-max)[a]								
Class	50	40	25	20	12.5	8	4	2	0.500	0.063
SC40	100	80–100	67–100	62–100	53–100	45–89	30–65	17–52	5–37	2–20
SC20	–	–	100	92–100	75–100	63–100	48–100	36–94	18–65	2–35

[a]The sieve sizing is defined by UNE-EN 933–2 (mm)

Table 3.49 Limits of the size distribution for gravel cement application

	Cumulative percent of aggregate sieved (% mass min-max)[a]							
Class	40	25	20	8	4	2	0.500	0.063
GC25	100	76–100	67–91	38–63	25–48	16–37	6–21	1–7
GC20	–	100	80–100	44–68	28–51	19–39	7–22	1–7

[a]The sieve sizing is defined by UNE-EN 933–2 (mm)

Table 3.50 Summary of recommended technical requirements

		Class/usage			
Technical requirements		SC20	SC40	GC20	GC25
Granulometry		Table 3.48	Table 3.48	Table 3.49	Table 3.49
Flakiness index		–	–	≤30%	≤35 o ≤40%
Grinded particles		–	–	≥50 o ≥75%	≥30 o ≥50%
Composition[a]		–	ArH, ArM H ó ArM C	Ar H ó ArM H	All categories
Los Ángeles coefficient		–	–	≤35%	≥45 o ≥ 40%
Liquid limit		<30	<30	No plasticity	<25
Plasticity		<15	<15	No plasticity	<6
Sand Equivalent (S.E.)		–	–	≥ 40%	≥35%
Methylene blue index		–	–	<10 if does not comply S.E. >30%	
Clay lumps	Coarse aggregate	–	–	≤1%	≤1%
	Fine aggregate			≤0.25%	≤0.25%
Compressive strength 7d		2.5–5 Mpa	2.5–5 Mpa	4.5–7 Mpa	4.5–6 Mpa
Workability term t_{pm}	Compacted the whole width of roadway	180 min	180 min	180 min	180 min
	Compacted by strips of roadway	240 min	240 min	240 min	240 min
Organic matter (%)		≤1	≤1	≤1	≤1
Sulphur compounds (%)		≤1	≤1	≤1	≤1
Acid soluble sulphate (%)		≤0.8	≤0.8	≤0.8	≤0.8
Potential alkali reactivity		Non-reactive	Non-reactive	Non-reactive	Non-reactive

[a]Orientative and not restrictive requirement

3.9.8 GEAR-PRT-05: Technical Specifications and Recommendations for Aggregates from Construction and Demolition Waste, to Use as a Granular Material in Precast Concrete Units, Treated with Hydraulic Binders

The applications considered are the following:

- Kerbs (BD)
- Concrete blocks/concrete hollow bricks (BV)
- Benches (BNC)
- Blocks (BLQ)

This proposal is in accordance with the following technical standards, and adapts those requirements to the recycled aggregates specifications.

- EN 1340:2004 Concrete kerb units. Requirements and testing methods.
- EN 127340 Concrete kerb units. Requirements and testing methods. National complement to the standard UNE-EN 1340
- prEN 15037-2 Precast concrete products Beams and floor systems Part 2: Concrete blocks.

Table 3.51 Usage classes in the current technical recommendations, according to technical specifications degree

Class			Application	Characteristics
Class 1	Kerbs	BD.-1	Kerbs type R6, R5 y R3,5	Made from Ar H or ArM H
	Concrete blocks/ concrete hollow bricks	BV.-1	Concrete blocks/concrete hollow bricks for floor slabs	Percentage of recycled aggregate used ≤20%
			Flexural strength ≥ 4KN	Good surface finish
	Benches	BNC.-1	Benches for seating on public roads and parks	Adverse weather conditions
	Blocks	BLQ.-1	Concrete blocks/concrete hollow bricks for walling	
Class 2	Kerbs	BD.-2	Kerbs type R5 y R3,5	Made from Ar H or ArM H
	Concrete blocks/ concrete hollow bricks	BV.-2	Concrete blocks/concrete hollow bricks for floor slabs	Percentage of recycled aggregate used: 20–100%
			Flexural strength <4KN	It does not matter the surface finish
	Benches	BNC.-2	Benches for seating on public roads and parks	It is not subject to adverse weather or freeze-thaw cycles
	Blocks	BLQ.-2	Concrete blocks/concrete hollow bricks for walling	
Class 3	Kerbs	BD.-3	Kerbs type R3,5	Made from ArM C
	Concrete blocks/ concrete hollow bricks	BV.-3	Concrete blocks/concrete hollow bricks for floor slabs	Percentage of recycled aggregate used ≤20%
			Flexural strength ≥ 3KN	It does not matter the surface finish
	Benches	BNC.-3	Benches for seating on public roads and parks	It is not subject to adverse weather or freeze-thaw cycles
	Blocks	BLQ.-3	Concrete blocks/concrete hollow bricks for walling	
Class 4	Kerbs	BD.-4	Kerbs type R3,5	Made from ArM C
	Concrete blocks/ concrete hollow bricks	BV.-4	Concrete blocks/concrete hollow bricks for floor slabs	Percentage of recycled aggregate used: 20–100%
			Flexural strength <3KN	It does not matter the surface finish
	Benches	BNC.-4	Benches for seating on public roads and parks	It is not subject to adverse weather or freeze-thaw cycles
	Blocks	BLQ.-4	Concrete blocks/concrete hollow bricks for walling	

- DIN 483 Terms of delivery and testing standards of concrete kerb units
- NFP-98302 precast concrete kerb units and channels
- BS 340 British Specification for "Precast concrete: Kerbs, channels, edgings and quadrants" (Tables 3.51, 3.52, 3.53, 3.54, 3.55, and 3.56)

Table 3.52 Limits of the size distribution for recycled aggregates

Application		Granulometry – EN 933–1													
		Cumulative percent of aggregate sieved (% mass min-max)													
		40	25	20	12.5	10	8	6.3	4	2	1	0.500	0.250	0.125	0.063
Class 1	BD.-1 BV.-1	–	–	–	95–100	80–95	60–90	60–85	50–75	35–60	25–40	14–25	9–20	4–15	0–15
	BNC.-1 BLQ.-1	95–100	88–97	80–90	60–75	50–65	40–60	37–55	25–48	17–38	13–27	10–21	6–15	3–12	0–8
Class 2	BD.-2 BV.-2	–	–	–	95–100	80–95	60–90	60–85	50–75	35–60	25–40	14–25	9–20	4–15	0–15
	BNC.-2 BLQ.-2	95–100	88–97	80–90	60–75	50–65	40–60	37–55	25–48	17–38	13–27	10–21	6–15	3–12	0–8
Class 3	BD.-3 BV.-3	–	–	–	95–100	80–95	62–90	58–85	50–75	35–62	25–45	14–35	9–27	4–22	0–18
	BNC.-3 BLQ.-3	95–100	88–97	80–90	62–75	52–65	42–60	35–55	25–48	17–40	13–30	10–25	6–18	3–14	0–10
Class 4	BD.-4 BV.-4	–	–	–	95–100	80–95	62–90	58–85	50–75	35–62	25–45	14–35	9–27	4–22	0–18
	BNC.-4 BLQ.-4	95–100	88–97	80–90	62–75	52–65	42–60	35–55	25–48	17–40	13–30	10–25	6–18	3–14	0–10

Table 3.53 Composition of recycled aggregates

	% Substitution percentage of recycled aggregates			
Material	≤50%		51–100%	
Asphalt	≤1%	Or the sum of them three does not exceed 5%	≤1%	Or the sum of them three does not exceed 3%
Lightweight particles (L)	≤3%		≤2%	
Other materials (X)	≤3%		≤2%	

Table 3.54 Summary of recommended technical requirements

	Class/applicación			
Technical requirements	Class 1	Class 2	Class 3	Class 4
Granulometry	Table 3.52	Table 3.52	Table 3.52	Table 3.52
Flakiness index (%)	<35	<35	<35	<35
Percentage of 0/4 mm particle size distribution	≤75% (for BD and BV) ≤48% (for BNC and BLQ)	≤75% (for BD and BV) ≤48% (for BNC and BLQ)	≤75% (for BD and BV) ≤48% (for BNC and BLQ)	≤75% (for BD and BV) ≤48% (for BNC and BLQ)
Percentage of fine aggregates (<0.063 mm)	≤15% (for BD and BV) ≤8% (for BNC and BLQ)	≤15% (for BD and BV) ≤8% (for BNC and BLQ)	≤18% (for BD and BV) ≤10% (for BNC and BLQ)	≤18% (for BD and BV) ≤10% (for BNC and BLQ)
Absorption (%)	≤11	≤11	≤8	≤8
Los Ángeles coefficient (%)	<50	<50	<45	<40
Sand equivalent (S.E.)	≥50% ó ≥20% y M.B. <6.5	≥50% ó ≥20% y M.B. <6.5	≥50% ó ≥20% y M.B. <4.5	≥50% ó ≥20% y M.B. <4.5
Composition[a]	Table 3.53	Table 3.53	Table 3.53	Table 3.53
Organic matter	No organic matter	No organic matter	No organic matter	No organic matter
Sulphur compounds (%)	≤1	≤1	≤1	≤1
Acid soluble sulphate (%)	≤0.8	≤0.8	≤0.8	≤0.8
Total chlorides (%)	≤0.05	≤0.05	≤0.05	≤0.05
Gypsum (%)	≤1	≤1	≤1	≤1

[a]Orientative and not restrictive requirement

Table 3.55 Viability of application by category of aggregate and usage class

		Recycled aggregate category			
Viability of application		ArH	ArM H	ArM C	ArM A
Usage class	Class 1	X	X		
	Class 2	X	X		
	Class 3			X	
	Class 4			X	

Table 3.56 Viability of application by category of aggregate and usage class

			Recycled aggregate category				
Viability of application			Ar H	ArM H	ArM C	ArM A	MI
Usage class	Soil cement	SC20					
		SC40	X	X	X		
	Gravel cement	GC20	X	X			
		GC25	X	X	X	X	

Table 3.57 Viability of application by category of aggregate and usage class

		Recycled aggregate category				
Viability of application		Ar H	ArM H	ArM C	ArM A	MI
Usage class	Class 1	X	X	X		
	Class 2	X	X	X		

Table 3.58 Limits of the size distribution for recycled aggregates

	Granulometry – EN 933-1									
	Cumulative percent of aggregate sieved (% mass min-max)[a]									
Class	32	25	20	16	12.5	8	4	2	0.5	0.063
Class 1	100	90–100	78–100	67–100	55–85	44–68	28–51	19–39	7–22	1–7
Class 2	100	90–100	78–100	67–100	55–85	44–68	28–51	19–39	7–22	1–7

[a]The sieve sizing is defined by UNE-EN 933-2 (mm)

Table 3.59 Composition of coarse recycled aggregate

Componente	Porcentaje en masa (%)	Categoría según UNE 12620:09
Rc (hormigón, productos de hormigón, mortero, elementos de albañilería de hormigón)	Rc + Ru >50	R_{cu50}
Ru (áridos no ligados, piedra natural, árido ligado hidráulicamente)		
Rb (elementos de albañilería de arcilla, de silicato cálcico y hormigón aireado no flotante)	Rb <50	R_{b50-}
Ra (materiales bituminosos)	Ra <5	R_{a5-}
X (arcilla, tierra, yeso, plástico, madera y otros)	X + Rg <1	XR_{g1-}
Rg (vidrio)		
FL (materiales flotantes)	<2 cm^3/kg	FL_{2-}

3.9.9 *GEAR-PRT-07: Technical Specifications and Recommendations for Aggregates from Construction and Demolition Waste, to Use as a Granular Material Treated with Hydraulic Binders in Roller Compacted Concrete*

The applications considered in these technical specifications and recommendations, refer to binded structural layers of pavements with roller compacted concretes, using recycled aggregates. The applications are the following:

- Pavement layers
- Pavement bases

This proposal is in accordance with the following technical standards, and adapts those requirements to the recycled aggregates specifications.

- UNE-EN 12620:2009: Aggregates for concrete
- EHE-08, anejo 15: Hormigones con árido reciclado (recycled aggregate concretes), and anejo 18: Hormigones de uso no estructural (non-structural concretes) (Tables 3.57, 3.58, 3.59, 3.60).

Table 3.60 Summary of recommended technical requirements

Technical requirements		Class/application Class 1	Class 2
Granulometry	Minimum particle size	4 mm	4 mm
	Spindle	Table 3.58	Table 3.58
Flakiness index (%)		≤30	≤35
Composition[a]		Table 3.59	Table 3.59
Absorption (%)		≤10	≤10
Density (kg/m^3)		≥2,100	≥2,100
Los Ángeles coefficient (%)		≤35	≤40
Cleanness coefficient (%)		≤2	≤2
Organic matter (%)		≤1	≤1
Sulphur compounds (%)		≤1	≤1
Acid soluble sulphate (%)		≤0.8	≤0.8
Oxidizable sulphurs		Must not contain	Must not contain
Water soluble compounds (%)		≤1	≤1

[a]Orientative and not restrictive requirement

3.9.10 GEAR-PRT-06: Technical Specifications and Recommendations for Aggregates from Construction and Demolition Waste, to Use as a Granular Material Treated with Hydraulic Binders in Mass Concrete

The applications considered in these technical specifications and recommendations are the following:

- Structural concretes (HE) up characteristic compressive strength (fck) = 30 MPa.
- Non-Structural concretes (HNE) up characteristic compressive strength (fck) = 20 MPa
- Lean Concrete (HL) with a minimum cement content = 150 kg/m^3.

This proposal is in accordance with the following technical standards, and adapts those requirements to the recycled aggregates specifications.

- EHE-08, anejo 15: Hormigones con árido reciclado (recycled aggregate concretes), y anejo 18: Hormigones de uso no estructural (non-structural concretes).
- UNE-EN 12620:2009: Aggregates for concrete

3.9.10.1 Structural Concretes (HE)

Are subjected to Structural Concrete Instruction EHE-08 and especially its Annex 15. This Annex is the result of the Group's investigations GT2 / 5, for RECNHOR project coordinated by Dr. Pilar Alaejos (from CEDEX), and also involving the CEDEX, polytechnic universities of Catalonia (Universitat Politècnica de Catalunya), and Madrid (Universidad Politécnica de Madrid), the universities of A Coruña

Table 3.61 Maximum content of impurities

Element	Max. cont of impurities % total weight of the sample
Ceramic material	5
Light particles	1
Asphalt	1
Others materials (glass, plastic, metal, etc)	1

Table 3.62 Frequency of testing production control

Property	Standard	Frequency	
Granulometry	UNE-EN 933-1:98	1/week	Each 2.000 tons
Flakiness index	UNE-EN 933-4:00	1/month	Each 10.000 tons
Fines content	UNE-EN 933-2:96	1/week	Each 2.000 tons
Los Angeles coefficient	UNE-EN 1097-2:99	1/month	Each 2.000 tons
Absorption	UNE-EN 1097-6:01	1/week	Each 2.000 tons
Estability in front of $MgSO_4$[a]	UNE-EN 1367-2:99	1/6 months	Each 10.000 tons
Soft particles	UNE 4134:58	1/month	Each 10.000 tons
Clay lumps	UNE 7133:58	1/week	Each 2.000 tons
Light particles	UNE 7244	1/month	Each 10.000 tons
Sulphur compounds	UNE-EN 1744-1	1/3 months	Each 10.000 tons
Acid soluble sulphate	UNE-EN 1744-1	1/3 months	Each 10.000 tons
Total chlorides	UNE-EN 1744-1	1/3 months	Each 10.000 tons
Impurities	PrEN 933-11	1/week	Each 2.000 tons

[a]Only applicable in frost environment and deicing salts

(Universidade da Coruña), the Spanish Institute of cement and its applications (Instituto Español del Cemento y sus Aplicaciones-IECA) and the University of Cantabria (Universidad de Cantabria).

The most important points regarding materials are reflected in the Tables 3.61 and 3.62

3.9.10.2 Non-structural Concretes (HNE)

According to Structural Concrete Instruction EHE-08, Annex 18, for the manufacture of the non-structural concrete may be used up 100% of coarse recycled aggregate, provided it meets the specifications defined for it in Annex No. 15 of this Instruction.

The minimum characteristic compressive strength (f_{ck}) of non-structural concretes will be 15 N/mm^2. Due to the low resistance which requires these concretes and consequently low content of cement, inside its requirements do not appear any reference to exposure conditions.

3.9.11 Associated Works to GEAR Project ("Pre-normative Research on the Use of Recycled Aggregate")

Unbound recycled aggregate
Rural roads, forestry, bases for sports fields: 17 works

Bases and sub-bases in urban paving: 7 works
Bases and sub-bases in different traffic roads: 19 works

Recycled aggregates for soil-cement: 5 works
Recycled aggregates for precast concretes: 5 companies

Kerbs
Concrete blocks/ concrete hollow bricks
Blocks for walls
Benches (urban furniture)

Mass concrete:4 works

Sidewalks
Foundations
Non-structural concrete in buildings

Recycled aggregate for dry compacted concrete: 1 paved road.
Demonstration projects associated with other projects

1 bridge in Valencia (CEDEX)
1 bridge in Barcelona (UPC) – (board)

References

1. Mueller, A.: Closed loop of concrete rubble. In: Bauhaus Universität Weimar. Available at: http://www.uni-weimar.de/Bauing/aufber/Lehre/Gastvorlesung/Barcelona/Lecture07.pdf. Accessed 2007
2. Conselho Nacional Do Meio Ambiente (Conama): Resolução nº 307. Brasília (2002)
3. NBR 15.116 (ABNT): NBR 15.116: agregados de resíduos sólidos da construção civil: utilização em pavimentação e preparo de concreto sem função estrutural – requisitos. Standard, ASSOCIAÇÃO BRASILEIRA DE NORMAS TÉCNICAS (ABNT), Rio de Janeiro (2004a)
4. NBR 15.113 (ABNT): NBR 15.113: resíduos sólidos da construção civil e resíduos inertes: aterros: diretrizes para projeto, implantação e operação. Standard, ASSOCIAÇÃO BRASILEIRA DE NORMAS TÉCNICAS (ABNT), Rio de Janeiro (2004b)
5. Pinto, T.P., Rodrigo, J.L.: Manejo e gestão de resíduos da construção civil: como implantar um sistema de manejo e gestão nos municípios 1. Ministério Das Cidades - Ministério Do Meio Ambiente – Caixa Econômica Federal, Brasília (2005)
6. Pinto, T.: Gestão ambiental de resíduos da construção civil: a experiência do SINDUSCON-SP. SINDICATO DA INDÚSTRIA DA CONSTRUÇÃO DO ESTADO DE SÃO PAULO (SINDUSCON-SP), São Paulo (2005)

7. Marques, J.: Estudo da gestão municipal de resíduos de construção e demolição na bacia hidrográfica do Turvo Grande (UGRHI-15). Dissertation, Escola de Engenharia de São Carlos, São Carlos (2009)
8. Miranda, L., Angulo, S.C., Careli, E.D.: A reciclagem de resíduos de construção e demolição no Brasil: 1986–2008. Ambiente Construído **9**(1), 57–71 (2009)
9. Angulo, S.C., et al.: On the classification of mixed construction and demolition waste aggregate by porosity and its impact on the mechanical performance of concrete. Mater. Struct. **43**, 519–528 (2010)
10. ASTM C 33–82: ASTM C 33-82-Standard Specification for Concrete Aggregates. Standard, ASTM-American Standards for Testing and Materials, Philadelphia (1982)
11. ASTM 125-79a: ASTM 125-79a-Standard Definitions of Terms Relating to Concrete and Concrete Aggregates. Standard, ASTM-American Standards for Testing and Materials, Philadelphia (1979)
12. FHWA-Federal Highway Administration: Transportation Applications of Recycled Concrete Aggregate: FHWA State of the Practice National Review September 2004. National Review. US Department of Transportation and Federal Highway Administration, Washington, DC (2004)
13. Li, S., Nantung, T., Jiang, Y.: Assessing issues, technologies and data needs to meet traffic input requirements by mechanistic-empirical pavement design guides: implementation initiatives. Transportation research record. J. Transp Res. Board **1917**, 141–148 (2005)
14. Mishulovich, A.: Used concrete/construction debris in Portland cement manufacturing. Portland Cement Association Research and Development Report Serial No. 2635 (2003)
15. Taschereau, D.: Moving mountains. Solid Waste and Recycling. 3–6 December/January (2001)
16. Wilson, A.: Concrete: even smarter than you think. Available at www.cement.ca/cement.nsf/e/5707B2CB7E10E24585256E1C007D7C9D? (2003). Accessed 2003
17. Ontario Ministry of Natural Resources: The state of the aggregate resource in Ontario study. Consolidated report, Government of Ontario, Ministry of Natural Resources, Ontario (2010)
18. Lamond, J., Campbell Sr., R., Campbell, T., Cazares, J., Giraldi, A., Halczak, W., Hale Jr., H.C., Jenkins, N.J.T., Miller, R., Seabrook, P.: Removal and reuse of hardened concrete. ACI Mater. J. **99**(3), 300–325 (2002)
19. Fathifazl, G., Abbas, A., Razaqpur, A.G., Isgor, O.B., Fournier, B., Foo, S.: New mixture proportioning method for concrete made with coarse recycled concrete aggregate. ASCE-J. Mater. Civ. Eng. **21**(10), 601–611 (2009)
20. Abbas, A., Fathifazl, G., Isgor, O.B., Razaqpur, A.G., Fournier, B., Foo, S.: Proposed method for determining the residual mortar content of recycled concrete aggregates. J. ASTM Int. **5**(1), 12 (2008)
21. Abbas, A., Fathifazl, G., Fournier, B., Isgor, O.B., Zavadil, R., Razaqpur, A.G., Foo, S.: Quantification of the residual mortar content in recycled concrete aggregates by image analysis. J. Mater. Charact. **60**(7), 716–772 (2009)
22. Fathifazl, G., Razaqpur, A.G., Isgor, O.B., Abbas, A., Fournier, B., Foo, S.: Creep and drying shrinkage characteristics of concrete produced with coarse recycled concrete aggregate (RCA). Cem. Concr. Compos. **33**(10), 1026–1037 (2011)
23. Banthia, N., Chan, C.: Use of recycled aggregate in plain and fiber-reinforced shotcrete. ACI Concr. Int. **22**(6), 41–45 (2000)
24. Fathifazl, G., Razaqpur, A.G., Isgor, O.B., Abbas, A., Fournier, B., Foo, S.: Flexural performance of steel reinforced recycled concrete (RRC) beams. ACI Struct. J. **106**(6), 858–867 (2009)
25. CSA A23.3-04: A23.3-04-Design of Concrete Structures. Standard, Canadian Standard Association (CSA), Rexdale, Ontario (2004)
26. ACI 318–05: ACI 318–05 – Building Code Requirements for Structural Concrete. Standard, American Concrete Institute, Farmington Hills (2005)

27. Fathifazl, G., Razaqpur, A.G., Isgor, O.B., Abbas, A., Fournier, B., Foo, S.: Shear strength of steel reinforced recycled concrete beams with stirrups. Mag. Concr. Res. **62**(10), 685–699 (2010)
28. Fathifazl, G., Razaqpur, A.G., Isgor, O.B., Abbas, A., Fournier, B., Foo, S.: Shear strength of steel reinforced recycled concrete beams without stirrups. Mag. Concr. Res. **61**(7), 387–400 (2009)
29. Fathifazl, G., Razaqpur, A.G., Isgor, O.B., Abbas, A., Fournier, B., Foo, S.: Shear capacity evaluation of steel reinforced recycled concrete (RRC) beams. Eng. Struct. J. **33**(3), 1025–1033 (2011)
30. ASTM A 944–99: ASTM A 944-99-Standard Test Method for Comparing Bond Strength of Steel Reinforcing Bars to Concrete Using Beam-End Specimens. Standard, ASTM-American Standards for Testing and Materials (1999 (Reapproved 2004))
31. Fathifazl, G.: Structural performance of steel reinforced recycled concrete members. Dissertation, Department of Civil and Environmental Engineering, Carleton University, Carleton, Canada (2008)
32. Butler, L., West, J.S., Tighe, S.L.: The effect of recycled concrete aggregate properties on the bond strength between RCA concrete and steel reinforcement. J. Cem. Concr. Res. **41**(10), 1037–1049 (2011). doi:10.1016/j.cemconres.2011.06.004
33. Abbas, A., Fathifazl, G., Isgor, O.B., Razaqpur, A.G., Fournier, B., Foo, S.: Durability of recycled aggregate concrete designed with equivalent mortar volume method. Special issue of the journal of cement and concrete composites on sustainability of civil engineering structures – durability of concrete. J. Cem. Concr. Compos. **31**(8), 555–563 (2009)
34. ASTM: Concrete and Concrete Aggregates. Annual Book of American Society for Testing and Materials, vol. 4.02. Annual Book, American Society for Testing and Materials, Philadelphia (2003)
35. ACI 222R- 96: ACI 222R- 96 – Corrosion of Metal in Concrete. Standard, ACI-American Concrete Institute, Farmington Hills (1996)
36. RILEM Recommendations CPC-18: Measurement of hardened concrete carbonation depth. Mater. Struct. **21**(6), 453–455 (1988)
37. Salem, R.M., Burdette, E.G., Jackson, N.M.: Resistance to freeze and tawing of recycled aggregate concrete. ACI Mater. J. **100**(3), 216–221 (2003)
38. Salem, R.M., Burdette, E.G.: Role of chemical and mineral admixtures on physical properties and frost resistance of recycled aggregate concrete. ACI Mater. J. **95**(5), 558–563 (1998)
39. Shehata, M.H., Christidis, C., Mikhaiel, W., Rogers, C., Lachemi, M.: Reactivity of reclaimed concrete aggregate produced from concrete affected by alkali–silica reaction. Cem. Concr. Res. **40**(4), 575–582 (2010)
40. Shehata, M., Mikhaiel, W., Lachemi, M., Rogers, C.: Mitigating the expansion in concrete containing reclaimed aggregate produced from demolished concrete affected by alkali-silica reaction EM-039. In: Engineering, C. (ed.) 2nd International Engineering Mechanics and Materials Specialty Conference, Ottawa, ON, pp. 01–08, 14–18 June 2011
41. Adams, M.P., Gray, B., Ideker, J.H., Tanner, J.E., Jones, A., Fournier, B., Beauchemin, S., Shehata, M., Johnson, R.: Applicability of standard alkali-silica reactivity testing methods for recycled concrete aggregate. Paper submitted for the 14th International Conference on AAR in Concrete, Austin, TX, 2012
42. Beauchemin, S., Fournier, B.: Influence of the nature and composition of recycled aggregate material on their performance in the accelerated mortar bar testing. In: Concrete, I. (ed.) 14th International Conference on AAR in Concrete, Austin, TX, 2011
43. Smith, J., Tighe, S.: Moving towards an environmental sustainable concrete pavement using recycled concrete aggregate. In: Transportation Association of Canada – Conference Proceedings, Toronto, ON, 2008
44. Smith, J., Tighe, S.: Recycled concrete aggregate coefficient of thermal expansion: characterization, variability, and impacts on pavement performance. Natl. Acad. Sci. Transp. Res. Rec. **2113**, 53–61 (2009)

45. Rizvi, R., Henderson, V., Tighe, S., Norris, J.: Evaluating the use of recycled concrete aggregate in pervious concrete pavement. Transp. Res. Board **2164**, 132–140 (Accepted for Publication in the Transportation Research Record) (2010)
46. Chan, P., Tighe, S.: Quantifying Pavement Sustainability. Final Report. Ministry of Transportation, Toronto (2010)
47. Xiao, J.: Recycled Concrete. Chinese Building Construction Publishing Press, Beijing (2008)
48. Deng, X.: Study on effect of compressive strength of recycled aggregate concrete with water cement ratio (only available in Chinese). Chin. Concr. J. **2**, 46–48 (2005)
49. Li, J.: Study on mechanical behavior of recycled aggregate concrete. Dissertation, Tongji University (2004)
50. Liu, Y.P., Chen, Z.F.: Research of application and basic mechanical properties on recycled concrete (only available in Chinese). Chin. Concr. J. **12**, 43–45 (2009)
51. Xiao, J.Z., Li, J.B., Huang, J.: Influence of recycled coarse aggregate replacement percentage on compressive strength of concrete (only available in Chinese. J. Build. Mater. **9**(3), 297–301 (2006)
52. Tang, J.: Preliminary study on compressive strength of recycled aggregate concrete (only available in Chinese). Sichuan Build. Sci. **33**(4), 183–186 (2007)
53. Jin, C., Wang, X.P., Akinkurolere, O.O., Jiang, C.R.: Experimental research on the conversion relationships between the mechanical performance indexes of recycled concrete (only available in Chinese. Chin. Concr. J. **49**(11), 37 (2008)
54. Xiao, J.Z., Liu, Q., Li, W.G., Tam, V.: On the micro- and meso-structure and failure mechanism of recycled concrete (only available in Chinese). J. Qingdao Technol. Univ. **30**(2), 24–30 (2009)
55. Xiao, J.Z., Li, J.B., Sun, Z.P., Hao, X.M.: Study on compressive strength of recycled aggregate concrete (only available in Chinese). J. Tongji Univ. (Nat. Sci.) **32**(12), 1558–1561 (2004)
56. Buck, A.D.: Recycled concrete as a source of aggregate. ACI J. **74**, 212–219 (1977)
57. Xiao, J.Z., Li, J.B., Zhang, C.Z.: On relationships between the mechanical properties of recycled aggregate concrete: an overview. Mater. Struct. **39**, 655–664 (2006)
58. Li, W.X., Zhang, X., Liu, X.: Mechanical properties of recycled aggregate concrete study of the impact of factors (only available in Chinese). Chin. Concr. J. **10**, 60–63 (2009)
59. Li, J.B., Xiao, J.Z., Sun, Z.P.: Properties of recycled coarse aggregate and its influence on recycled concrete (only available in Chinese). Chin. J. Build. Mater. **7**(4), 390–395 (2004)
60. Zhang, Y.M., Qin, H.G., Sun, W., Hao, D.M., Ning, Z.: Preliminary study on the proportion design of recycled aggregate concrete (only available in Chinese). China Concr. Cem. Prod. **1**, 7–9 (2002)
61. Poon, C.S., Shui, Z., Lam, L.: Effect of microstructure of ITZ on compressive strength of concrete prepared with recycled aggregates. Constr. Build Mater. **18**(6), 46–468 (2004)
62. Du, T.: Experimental study on the microstructure and basic behaviors of recycled high performance concrete. Dissertation, Huazhong University of Science and Technology. (only available in Chinese) Huazhong China (2006)
63. Shui, Z.H., Pan, Z.S., Zhu, W.Q., Zhan, B.H.: Microscopic structural features of the recycled aggregate concrete (only available in Chinese). J. Wuhan Univ. Technol. **25**(12), 99–102 (2003)
64. Xiao, J.Z., Liu, Q., Du, J.T., Zhang, C.Z.: Micro-damage mechanisms and property fluctuation of recycled aggregate concrete. Key Eng. Mater. **348–349**, 61–64 (2007)
65. Xiao, J.Z., Lan, Y.: Investigation on the tensile behavior of recycled aggregate concrete (only available in Chinese). Chin. J. Build. Mater. **8**(2), 154–158 (2006)
66. Yang, X., Wu, J., Liang, J.G.: Experimental study on relationship between tensile strength and compressive strength of recycled aggregate concrete (only available in Chinese). Sichuan Build. Sci. **35**(5), 190–192 (2009)
67. Cheng, G.Y.: Experimental study on the basic performance of recycled aggregate concrete with different displacement ratio (only available in Chinese). Chin. Concr. J. **11**, 67–70 (2005)

68. Ge, X.L., Zeng, L.: Study on influence of recycled coarse aggregate on the strength of recycled concrete (only available in Chinese). J. Shaanxi Architect. Build. Mater. **114**, 36–38 (2004)
69. Xiao, J.Z., Lei, B., Yuan, B.: Splitting tensile strength distribution of concrete with different recycled coarse aggregates (only available in Chinese). Chin. J. Build. Mater. **11**(2), 223–229 (2008)
70. Xiao, J.Z., Li, J.B.: Study on relationships between strength indexes of recycled concrete (only available in Chinese). Chin. J. Build. Mater. **9**, 197–201 (2005)
71. Zhou, H., Liu, B.K., Lu, G.: Experimental research on the basic mechanical properties of recycled aggregate concrete (only available in Chinese). J. Anhui Inst. Architect. Ind. **16**(6), 4–8 (2008)
72. Hu, M.P.: Mechanical properties of concrete prepared with different recycled coarse aggregates replacement rate (only available in Chinese). Concr. J. **2**, 52–54 (2007)
73. Huang, Y., Deng, Z.H., Luo, Y.M., Yang, H.F.: Experimental study on shear strength of recycled aggregate concrete (only available in Chinese). Chin. Concr. J. **2**(31), 14–17 (2010)
74. Guo, Z.H., Shi, X.D.: Reinforced Concrete Theory and Analysis. Tsinghua University Publishing Press, Beijing (only available in Chinese) (2003)
75. Wu, F.Y., Yang, Y.F.: Preliminary research on behaviors of recycled concrete-filled stub columns (only available in Chinese). J. Fuzhou Univ. **33**(315), 305–308 (2005)
76. Yang, Y.F.: Theoretical research on load-deformation of recycled aggregate-filled steel tubular members (only available in Chinese). Chin. J. Ind. Constr. **37**(12), 1–6 (2007)
77. Xiao, J.Z., Yang, J.: On recycled concrete confined by GFRP tube under axial compression (only available in Chinese). J. Tongji Univ. (Nat. Sci.) **37**(12), 1586 (2009)
78. Xiao, J.Z., Li, P.S., Qin, W.: Study on bond-slip between recycled concrete and rebars (only available in Chinese). J. Tongji Univ. (Nat. Sci.) **34**(1), 13–16 (2006)
79. Xiao, J.Z., Huang, Y.B.: Residual compressive strength of recycled concrete after high temperature (only available in Chinese). Chin. J. Build. Mater. **9**(3), 255–259 (2006)
80. GB50010: Code for Design of Concrete Structures. Beijing (2002) only available in Chinese
81. Xiao, J.Z., Huang, Y.B., Zheng, Y.C.: Residual flexural strength of recycled concrete after elevated-temperatures (only available in Chinese). Chin. J. Build. Mater. **26**(3), 32–36 (2009)
82. Xiao, J.Z.: Experimental investigation on complete stress-strain curve of recycled concrete under uniaxial loading (only available in Chinese). J. Tongji Univ. **35**(11), 1445–1449 (2007)
83. Hu, Q., Song, C., Zou, C.Y.: Experimental research on the mechanical properties of recycled concrete (only available in Chinese). J. Harbin Inst. Technol. **41**(4), 33–36 (2009)
84. Xiao, J.Z., Du, J.T.: Complete stress-strain curve of concrete with different recycled coarse aggregates under uniaxial compression (only available in Chinese). Chin. J. Build. Mater. **11**(1), 111–115 (2008)
85. Ravindrarajah, R.S., Tam, C.T.: Properties of concrete made with crushed concrete as coarse aggregate. Mag. Concr. Res. **37**(130), 29–38 (1985)
86. Song, C.: Analysis on mechanical properties and microstructure of recycled concrete. Dissertation, Harbin Institute of Technology. (only available in Chinese), Harbin-China (2003)
87. Xiao, J.Z., Li, J.B., Zhang, C.Z.: Mechanical properties of recycled aggregate concrete under uniaxial loading. Cem. Concr. Res. **35**(6), 1187–1194 (2005)
88. Xu, W.: Properties of recycled aggregate concrete under uniaxial loading (only available in Chinese). Chin. Concr. J. **10**, 21–23 (2006)
89. Deng, Z.H., Yang, H.F., Lin, J., Wen, S.H.: Experimental study on the stress-strain curve of recycled concrete (only available in Chinese). Chin. Concr. J. **11**, 22–24 (2008)
90. Xing, Z.X., Zhou, Y.N.: Study on the main performance of regenerated concrete. J. North China Inst. Water Conserv. Hydroelectr Power **19**(2), 30–32 (1998)
91. Xiao, J.-Z., Lan, Y.: Experimental study on flexural performance of recycled concrete beams (only available in Chinese). Spec. Struct. **9**(1), 9–12 (2006)
92. Xiao, J.Z., Li, J.B., Lan, Y.: Research on recycled aggregate concrete-a review (only available in Chinese. Chin. Concr. J. **10**(57), 17–20 (2003)

93. Xu, Y.Z., Shi, J.G.: Analyses and evaluation of the behaviour of recycled aggregate and recycled concrete (only available in Chinese). Chin. Concr. J. **7**, 41–46 (2006)
94. RILEM: Specifications for concrete with recycled aggregate. Mater. Struct. **27**(9), 557–559 (1994)
95. Ding, S., Sun, W.M., Guo, Z.-G., Dai, W.Y., Ni, T.Y., Shen, D.: Experiment on deformation behaviour of recycled concrete beams (only available in Chinese). Ind. Constr. **39**(Supplement), 864–867 (2009)
96. Li PX, S.: Research on serviceable properties of reinforced recycled concrete beams. (only available in Chinese). J. Build. Struct. S1(Supplementary Issue), 27–31 (2008)
97. Hu, Q., Huang, Q., Zou, C.-Y.: Experimental study on partial recycled concrete beams (only available in Chinese). J. Herbin Inst. Technol. **41**(6), 38–42 (2009)
98. EN 1992-1-1: Eurocode 2-Design of Concrete Structures. (2004)
99. GB50068: Unified Standard for Reliability Design of Building Structures Beijing (only available in Chinese) (2001)
100. Xiao, J.-Z., Lan, Y.: Experimental study on shear behaviour of recycled concrete beams (only available in Chinese). Struct. Eng. **20**(6), 54–58 (2004)
101. Zhou, J.H., Jiang, H.: Shear behaviour of recycled coarse aggregate concrete beam (only available in Chinese). J. Shenyang Jianzhu Univ. (Nat. Sci. Ed.) **24**(5), 683–688 (2009)
102. Zhang, L., Zhang, X., Yan, G.: Experimental research on the shearing capacity of recycled aggregate concrete beams without stirrups (only available in Chinese). J. Zhengzhou Univ. (Eng. Sci.) **27**, 18–23 (2006)
103. DG/TJ08-2018-2007: Technical Code for Application of Recycled Aggregate Concrete (only available in Chinese) (2007)
104. Zhou, J.H., Yang, Y.-S., Jiao, X.: Experimental study on axial pressure bearing capacity of recycled concrete column (only available in Chinese). J. Shenyang Jianzhu Univ. (Nat. Sci. Ed.) **24**(4), 572–576 (2008)
105. Xiao, J.-Z., Shen, H.B., Huang, Y.B.: Test on compression performance of recycled concrete columns (only available in Chinese). Struct. Eng. **22**(6), 73–77 (2006)
106. Yang, Y.F., Han, L.H.: Experimental behaviour of recycled aggregate concrete filled steel tubular columns. J. Constr. Steel Res. **62**(12), 1310–1324 (2006)
107. Yang, Y. F.: Performance of recycled aggregate concrete-filled steel tubular members under various loadings. In: Proceedings of the 2nd International Conference on Waste Engineering and Management, Shanghai, China, pp. 475–484 (2010)
108. ACI Committee 318: Building Code Requirements for Structural Concrete and Commentary. Detroit (1999)
109. AIJ: Recommendations for Design and Construction of Concrete Filled Steel Tubular Structures. Architectural Institute of Japan, Tokyo (1997)
110. Zhou, J.H., Wang, X.B., Yu, T.H.: Mechanic behaviour test on recycled concrete simply-supported rectangular slabs (only available in Chinese). J. Shenyang Jianzhu Univ. (Nat. Sci. Ed.) **24**(3), 411–415 (2008)
111. Xiao, J.-Z., Li, H., Jin, S.C., Li, B.: Longitudinal shear test on steel deck recycled aggregate concrete composite slabs (only available in Chinese). Struct. Eng. **26**(4), 91–95 (2010)
112. Xiao, J.-Z., Zhu, X.H.: Study on seismic behaviour of recycled concrete frame joints (only available in Chinese). J. Tongji Univ. (Nat. Sci.) **33**(4), 436–440 (2005)
113. Xiao, J.-Z., Tawana, M.M., Wang, P.J.: Test on the seismic performance of frame joints with pre-cast recycled concrete beams and columns. In: Proceedings of the 2nd International Conference on Waste Engineering and Management, Shanghai, China, pp. 773–786 (2010)
114. Li, J.B., Xiao, J.Z., Huang, J.: Influence of recycled coarse aggregate replacement percentage on compressive strength of concrete (only available in Chinese). Chin. J. Build. Mater. **9**(3), 297–301 (2006)
115. Cao, W.L., Xu, T.G., Liu, Q., Zhang, J.W., Zhang, Y.Q.: Experimental study on seismic performance of high-rise recycled aggregate concrete shear wall (only available in Chinese). World Earthq. Eng. **25**(2), 18–23 (2009)

116. Cao, W.L., Dong, H.Y., Zhang, J.W.: Study on seismic performance of RAC shear wall with different shear-span ratio. In: Proceedings of the 2nd International Conference on Waste Engineering and Management, Shanghai, China, pp. 652–660 (2010)
117. Xiao, J.-Z., Huang, J.D., Yao, Y.: Experimental study on seismic behaviour of recycled concrete small-sized block walls (only available in Chinese). China Civ. Eng. J. (2010)
118. Xiao, J.-Z., Sun, Y.D., Falkner, H.: Seismic performance of frame structures with recycled aggregate concrete. Eng. Struct. **28**(1), 1–8 (2006)
119. GB50011: Code for Seismic Design of Buildings, Beijing (only available in Chinese) (2001)
120. Sun, Y.D., Xiao, J.-Z., Zhou, D.Y.: Experimental research on seismic behaviour of recycled concrete frame filled with recycled lightweight masonry blocks (only available in Chinese). Earthq. Eng. Eng. Vib. **25**, 124–131 (2005)
121. Xiao, J.-Z., Wang, C.-Q., Zhu, B.R., Huang, J.D.: Shake table test on recycled concrete block masonry buildings (only available in Chinese). J. Sichuan Univ. (Nat. Sci.) **42**(5), 120–126 (2010)
122. Xiao, J.-Z., Wang, C.-Q., Li, J., Tawana, M.M.: Shaking table model tests on recycled aggregate concrete frame structure. ACI Struct. J. (2010)
123. Arbeitsgemeinschaft Kreislaufwirtschaftsträger Bau (KWTB): Monitoring-Bericht Bauabfälle., Berlin/ Düsseldorf/ Duisburg (2000, 2001, 2003, 2005, 2007)
124. Arbeitsgemeinschaft Kreislaufwirtschaftsträger Bau (KWTB): Monitoring-Bericht Bauabfälle. Erhebung 2006, Berlin (2008) unveröffentlicht
125. VDI-Richtlinie 2095: VDI-Richtlinie 2095: Emission control. Treatment of mineral construction-site and demolition waste. Stationary and mobile demolition waste recycling facilities. Part 1. Richtlinie, VEREIN DEUTSCHER INGEINEURE (2011)
126. Umweltstatistische Erhebnungen, Abfallwirtschaft. In: Statistisches Bundesamt Deutschland. Available at www.destatis.de (2011). Accessed 2011
127. TL Gestein-StB 04: Technische Lieferbedingungen für Gesteinskörnungen im Straßenbau. FGSVNr. 613, FGSV Verlag GmbH, Köln (2007) Ausgabe 2004, Fassung 2007
128. DIN EN 13043:2002–12: Gesteinskörnungen für Asphalte und Oberflächenbehandlungen für Straßen, Flugplätze und andere Verkehrsflächen. Standard, Berlin (2002)
129. DIN EN 206–1:2001:07/A1:2004-10/A2:2005–09: Beton – Teil 1: Festlegung, Eigenschaften, Herstellung und Konformität. Standard, Berlin (2001)
130. DIN 1045–2:2008–08: Tragwerke aus Beton, Stahlbeton und Spannbeton – Teil 2: Beton – Festlegung, Eigenschaften, Herstellung und Konformität – Anwendungsregeln zu DIN EN 2206–1. Standard, Berlin (2008)
131. DIN EN 12620:2008–07: Gesteinskörnungen für Beton. Standard, Berlin (2008)
132. DIN EN 13055–1:2002-08/Berichtigung1:2004–12: Leichte Gesteinskörnungen für Beton, Mörtel und Einpressmörtel. Standard, Berlin (2002)
133. DIN 4226–100:2002–02: Gesteinskörnungen für Beton und Mörtel. Teil 100: Rezyklierte Gesteinskörnungen. Aggregates for Mortar and Concrete, Part 100: Recycled Aggregates. Germany. Standard, Berlin (2002)
134. Deutscher Ausschuss für Stahlbeton: DAfStb-Richtlinie: Beton nach DIN EN 206–1 und DIN 1045–2 mit rezyklierten Gesteinskörnungen nach DIN 4226–100 - Teil 1: Anforderungen an den Beton für die Bemessung nach DIN 1045–1. Standard, Berlin (2004)
135. Deutscher Ausschuss für Stahlbeton: DAfStb-Richtlinie: Vorbeugende Maßnahmen gegen schädigende Alkalireaktionen im Beton (Alkali-Richtlinie), Berlin (2007)
136. LAGA-Mitteilung 20: Anforderungen an die stoffliche Verwertung von mineralischen Abfällen – Technische Regeln, in der jeweils aktuellen Fassung. Available at: http://www.bafg.de/nn_161560/Baggergut/DE/04__Richtlinien/LAGA__TR__Boden,templateId=raw, property=publicationFile.pdf. Accessed 2012
137. Deutsches Institut für Bautechnik (DIBt): Zulassungsgrundsätze für die Bewertung der Auswirkungen von Bauprodukten auf Boden und Grundwasser. Teile 1 bis 3, Berlin (2009)

138. Pawel, A.: Nachweis der Umweltverträglichkeit von Bauprodukten im Rahmen von allgemeinen bauaufsichtlichen Zulassungen. In: Neue Entwicklungen im Betonbau, Fachtagung des Deutschen Ausschuss für Stahlbeton, Berlin (2007)
139. Susset, B., Leuchs, W.: Ableitung von Materialwerten im Eluat und Einbaumöglichkeiten mineralischer Ersatzbaustoffe. Umsetzung der Ergebnisse des BMBF-Verbundes "Sickerwasserprognose" in konkrete Vorschläge zur Harmonisierung von Methoden. Abschlussbericht, Auftrag des Umweltbundesamtes. Forschungsprojekt im Auftrag des Umweltbundesamtes. FuE-Vorhaben Förderkennzeichen 205 74 251 (2008)
140. Gäth, S., Luckner, L.: Wissenschaftliche Bewertung des UBA-Abschlussberichtes "Ableitung von Materialwerten im Eluat und Einbaumöglichkeiten mineralischer Ersatzbaustoffe" von Susset&Leuchs. Justus-Liebig-Universität Gießen, Berlin (2008)
141. Ifeu-Institut Heidelberg: Hochwertige Verwertung von Bauschutt als Zuschlag für die Betonherstellung, im Auftrag des Ministeriums für Umwelt, Naturschutz und Verkehr Baden-Württemberg, Zwischenbericht. Zwischenbericht, Ministeriums für Umwelt, Naturschutz und Verkehr. http://www.rc-beton.de/rc-beton-projekte/stuttgart-ost/stuttgart-ost-download.html (2010)
142. Mettke, A.: RC-Beton – Qualität und Qualitätssicherung. Aufbereitung von Baustoffen und Wiederverwertung. In: Fachtagung Recycling R'10, Weimar (2010)
143. Vyncke, J.: Use of recycled materials as aggregates in the construction industrie. Half-yearly publication Combined Volume **2** Issue 3 and 4. European Thematic Network on Recycling in Construction – DG XII (March/September 2000)
144. Wöhnl, U.: Recyclingbeton für Bauteile im Hochbau. Sonderdruck aus Beton **44**(9), 499–503 (1994)
145. In: BiM-Online Homepage. Available at http://www.b-i-m.de/projekte/projframe.htm (2011). Accessed 2011
146. Haase, R., Dahms, J.: Baustoffkreislauf am besonderem Beispiel von Beton im norddeutschen Raum. Beton **48**(6), 350–355 (1998)
147. Grübl, P., Rühl, M.: Beton unter Verwendung von rezyklierter Gesteinskörnung. Beton **52**(5), 250–254 (2002)
148. Philipp Holzmann A.G.: Haus komplett aus Recyclingbeton – Recyclingbeton für eine Autobahnbrücke. Tiefbau **112**(4), 234–235 (Haus komplett aus Recyclingbeton – 1999)
149. Mettke, A.: Der Einsatz von Recycling-Beton in der Praxis, Fallbeispiel Ludwigshafen. Berlin (November 2010)
150. AA. VV.: Gli impianti per il trattamento dei rifiuti in Italia. Rapporto, FISE Assoambiente (2009)
151. AA. VV.: L'Italia del recupero. rapporto FISE UNIRE sul riciclo dei rifiuti. rapporto, FISE UNIRE (2010)
152. AA. VV.: Rapporto rifiuti. Istituto Superiore per la Protezione e la Ricerca Ambientale (2008)
153. PORTUGAL-Ministério da Ambiente, do Ordenamento do Território e do Desenvolvimento Regional.: Decreto-Lei 46/2008, de 12 de Março (2008)
154. LNEC-E471: Guia para a utilização de agregados reciclados grossos em betões de ligantes hidráulicos. Laboratório Nacional de Engenharia Civil (2009)
155. LNEC-E472: Guia para a reciclagem de misturas betuminosas a quente em central. Laboratório Nacional de Engenharia Civil (2009)
156. LNEC-E473: Guia para a utilização de agregados reciclados em camadas não ligadas de pavimentos. Laboratório Nacional de Engenharia Civil (2009)
157. LNEC-E474: Guia para a utilização de materiais reciclados provenientes de resíduos de construção e demolição em aterro e camada de leito de infra-estruturas de transporte. Laboratório Nacional de Engenharia Civil (2009)
158. Böhmer, S., Moser, M., Neubauer, C., Peltoniemi, M., Schachermayer, E., Tesar, M., Walter, B., Winter, B.: Aggregates case study. Final Report referring to contract n° 150787–2007 f1sc-at, aggregates case study – data gathering' (2008)

159. CEN. EN 12620: 2002 + A1: Aggregates for Concrete. Comité Européen de Normalisation (2008)
160. CEN. EN 13108–8: 2005.: Bituminous Mixtures – Material Specifications – Part 8: Reclaimed asphalt. Comité Européen de Normalisation (2005)
161. CEN. EN 13242:2002 + A1:2007.: Aggregates for unbound and hydraulically bound materials for use in civil engineering work and road construction. Comité Européen de Normalisation, (2007) Incorporating corrigendum May 2004
162. CEN. EN 13285:2003.: Unbound mixtures – Specification., Comité Européen de Normalisation (2003)
163. Braz de Melo, A.B.., Gonçalves, A.F., Martins, I.M.: Construction and demolition waste generation and management in Lisbon (Portugal). Res. Conserv. Recycl. **55**(12), 1252–1264 (2011)
164. PORTUGAL-Ministério da Ambiente e do Ordenamento do Território.: Decreto-Lei 73/ 2011, de 17 de Junho. Ministério da Ambiente e do Ordenamento do Território (2011)
165. OJEU. 2008/98/CE.: Directive 2008/98/EC of the European Parliament and the Council of 19 November 2008 on Waste and Repealing certain Directives (Text with EEA relevance) (2008)

Chapter 4
Recycled Concrete

Abstract A new approach to the proportioning of concrete mixtures with recycled aggregate is presented. This procedure requires less cement, which makes it more economical and environmentally attractive. Recent publications, opinions and discrepancies in relation to the durability of recycled concrete, such as freeze-thaw standards, studies of chloride penetration and diffusion, and sulfate attacks, are examined. The removal of gypsum constituents from recycled aggregates during their processing seems indispensable. This can be achieved with jigging machines, in which asymmetrical jig stroke diagrams can be implemented. Another effective technique for reducing gypsum content is selection by means of near infrared spectroscopy. It is discussed if appropriate test methods exist to evaluate the specific characteristics of recycled concrete aggregates with respect to frost durability. The miscellany of recently published papers, with results based in laboratory studies with accelerated tests led to conclusions relatively optimistic about the use of concrete with recycled concrete aggregates in environments with chlorides.

Keywords Recycled concrete • Proportioning • Freeze and thaw • Gypsum • Chlorides

4.1 Methods in Mixture Proportioning and His Influence in the Properties

Enric Vázquez

In the 2nd International Conference on Waste Engineering and Management held in Shanghai I 2010, A.G. Razaqpur et al. [1] presented a new method of concrete mix design for concrete made with recycled aggregate. The method is based on

E. Vázquez (✉)
Materials Section of the Department of Construction Engineering, Universitat Politecnica de Catalunya, Barcelona, Spain
e-mail: enric.vazquez@upc.edu

E. Vázquez (ed.), *Progress of Recycling in the Built Environment: Final Report of the RILEM Technical Committee 217-PRE*, RILEM State-of-the-Art Reports 8, DOI 10.1007/978-94-007-4908-5_4, © RILEM 2013

considering the concrete recycled aggregate as a two phase material, the old natural aggregate and the old mortar, and taken the old mortar as a part of the total mortar content of the new concrete. The key to the system is to design a reference concrete with natural aggregate, having the desired compressive strength (ACI method), and design the new recycled concrete with the same total mortar volume as the reference concrete. After an extensive experimental verification, the authors conclude that the designed new recycled concrete can satisfy the strength, the durability and all the requirements of a structural concrete with less cement.

The first step is the determination of the volume fractions of the original natural aggregate, and the residual mortar in the RCA (recycled coarse aggregate). Knowing the RMC (residual mortar content), and the bulk specific gravity of the RCA (recycled coarse aggregate) and the original natural aggregate, can be determined the bulk specific gravity of the RMC, and consequently, the volume fractions of residual mortars and original natural aggregate [1].

$$V_{RCA}^{RCA-concrete} = \frac{V_{NA}^{NA-concrete} \times (1 - R)}{1 - RMC \times \dfrac{SG_b^{RCA}}{SG_b^{OVA}}} \tag{4.1}$$

Where, $V_{RCA}^{RCA-concrete}$ is the volume fraction of coarse RCA in RCA-concrete; $V_{NA}^{NA-concrete}$ is the volume fraction of fresh natural aggregate in the companion normal concrete; RMC is the residual mortar content of RCA; R is the replacement ratio which represents the volumetric ratio of the fresh natural aggregate in RCA-concrete to fresh natural aggregate in the companion NA-concrete; SG_b^{RCA} and SG_b^{OVA} are the bulk specific gravities of RCA and original virgin aggregate, respectively.

As second step a reference concrete mix proportions, with natural aggregates only, are determined using the ACI method to satisfy a specified strength and other properties. Based on the composition of the RCA, the minimum amount of coarse natural aggregate required in the RCA concrete is determined.

Third step is the determination of the required volume of RCA and NA in RCA concrete.

Fourth step is to calculate the required oven dry weight of RCA and NA in RCA concrete.

Fifth step is to calculate the required new mortar volume in RCA concrete.

Sixth step is to calculate the required water, cement and fine aggregate proportions in RCA concrete.

The validity is verified with two RCA sources and using the same OPC and sand in all mixes. The w/c ratio and air content of all mixes were held at 0.45 and 6%.

Using the new proportioning method, resulted 6 and 13% **higher compressive strength** for all the mixes with recycled aggregate, compared with the reference concretes. Using the conventional method of mix proportioning resulted lower elastic modulus for the concretes with RCA.

Using the new method resulted 11 and 14% **higher elastic moduli** than the reference concretes. The observed higher moduli can be attributed to the equality of the total natural aggregate volumes of the recycled and reference concretes.

Creep at 305 days was 3 and 11% lower in the recycled concretes proportioned with the new method than the reference concretes. The RCA mixes proportioned with the conventional method had higher creep than the reference concretes. Since all the factors affecting creep where kept constant, the higher total mortar content, compared with those proportioned by the new method are identified as the main reason for the higher creep.

Shrinkage at the age of 224 days with RCA, concretes proportioned with the new method were lower as the shrinkage of the reference concretes. As in creep, the total mortar content can be considered the main reason.

The Freeze and Thaw resistance is related to the difference in the proportioning of RCA in the two types of mixes, and the RCA concrete mixes proportioned with the new method had higher resistances to freeze and thaw than the mixes proportioned with the conventional method.

The Chloride Penetration was tested with the ASTM C 1556–06. The Chloride concentration on the exposed surface were below the limit of the ACI in all the concretes, but differences between the TWO RCA concretes can be related with different salt contaminations of the old concrete.

The RCA concrete specimen proportioned with the new method had level **carbonation** limit then the concrete proportioned with the conventional method, and similar level then the reference concrete.

Other tests showed that there was no major difference between the observed **flexural b**ehaviour of reinforced RCA concrete beams and reference concrete beams. The same was observed for **shear** performance. In the **Bond** performance the RAC concrete designed by the new method presented the best results.

Jimenez et al. [2] have worked with the EMV method and Spanish recycled aggregates in the materials laboratory of UPC (Universitat Politècnica de Catalunya). An experimental campaign has been performed in order to examine different concrete mix proportioning methods, taking into account the use of recycled concrete aggregates. With this purpose, two different mixes, each one of them using 0.45 and 0.6 water/cement ratios (w/c), were prepared. The mixes were prepared using ACI proportioning methodology [3] and the design method proposed by Razaqpur and Fathifazl, based on the fact that the recycled concrete aggregate (RA) is a two phase material comprising old mortar on the one hand, and natural aggregate on the other. The first of these two mixes was prepared using clean recycled concrete aggregates (A) and the second one was prepared using mixed recycled aggregate with predominance of concrete (B). The chosen cement was CEM I 42.5 R.

Other mixes are being studied, after the verification of the novel method on the initial mixes, using Bolomey mix proportioning method [4] and an adaptation of the novel method to Bolomey. The results of this part will be published later.

In order to prepare the different mixes, a characterization campaign over the aggregates was performed. The recycled aggregates bulk densities were 2.31 and

Table 4.1 Mix proportions of concrete per m^3

Identification	w/c	RA [%]	Water [kg]	Cement [kg]	Sand [kg]	Limestone [kg]	RA (A) [kg]	RA (B) [kg]	HRWRA [g]
EMV-A	0.45	65	140	311	577	421	782	–	2,176
ACI-A	0.45	100	184	409	752	–	786	–	1,227
EMV-B	0.45	68	145	323	600	369	–	788	1,939
ACI-B	0.45	100	184	409	746	–	–	798	1,227
EMV-A	0.60	65	140	233	644	421	782	–	1,632
ACI-A	0.60	100	184	307	839	–	786	–	613
EMV-B	0.60	68	145	242	669	369	–	788	1,454
ACI-B	0.60	100	184	307	833	–	–	798	613

Table 4.2 Average values of compressive strengths and modulus of elasticity of ACI and EMV mixes

	Compressive strength [N/mm^2]		Modulus of elasticity [N/mm^2]	
	w/c ratio		w/c ratio	
Concrete mix	0.45	0.6	0.45	0.6
ACI-A	31.6	20.7	21,286	18,649
EMV-A	32.9	21.1	26,641	22,946
ACI-B	33.1	22.1	21,854	19,382
EMV-B	39.2	28.5	25,924	25,336

2.33 [g/cm^3], and the attached mortar amounts of them were 39 and 32% for aggregates A and B respectively. The natural aggregates bulk densities were 2.67 and 2.65 [g/cm^3] for limestone and sand respectively.

To obtain the mortar content of the RA, a thermal shock methodology was used [5]. This consists on subjecting the recycling aggregate to cycles of water submersion and high temperature shock, in order to create stresses by the rapid evaporation of the absorbed water of the aggregate and also by throwing the heated aggregate back to the water. Its purpose is to crumble down the attached mortar by the appearance of cracks, which will debilitate the bond between the natural aggregate and the mortar.

After the aggregates characterization, the mix proportions were calculated. The results are shown in Table 4.1.

The designation of the mixes corresponding to EMV, are the ones using the Razaqpur method, which was named after Equivalent Mortar Volume. The letter after, corresponds to the type of aggregate used.

A number of different tests were performed in order to verify the quality and characteristics of the concrete. In Table 4.2 the results of the compressive strength and elasticity modulus are shown. The compressive strength tests were performed over cubic specimens (10 × 10 cm) and the elasticity modulus over cylindrical specimens (10 × 20 cm). For the compressive strengths, a conversion factor has been used in order to present the results based on a cylindrical specimen of 15 × 30 cm [4]

It can be noticed that, in general terms, the results of the compressive strengths of the novel method are similar or better than the conventional methods, in both

0.45 and 0.6 w/c ratios. This was achieved using less cement in all of the cases, going up to around 24% reduction in weight when comparing ACI-A with EMV-A mixes, and around 21% reduction when comparing ACI-B with EMV-B mixes.

The elasticity modulus showed the same tendency of the compressive strength results, but with higher differences in both 0.45 and 0.6 w/c ratios. This higher difference is a result of the total natural aggregate in the mixes, because this property is based on this issue, and EMV method achieves higher amounts by replacing the old mortar by natural aggregates.

4.2 Durability

Enric Vázquez and Marilda Barra

4.2.1 Sulfates

The four most significant applications of recycled aggregate are in unbound road base layers, bounded overlays with a low lime or cement content, in mass concrete and in various structural concrete applications. Their necessities, from a durability standpoint, are different in each application. Although they share some characteristics, they do it with a different level of demand.

A **sulfate attack** can originate from the same components of the recycled aggregate. The sulfates that will act are of the sum of all sulfates present in the system. In all cases, there is a common necessity to prevent the formation of ettringite and thaumausite. From an environmental point of view, it is also important to prevent the excessive leaching of the sulfates. The need to limit the presence of sulfates in recycled aggregate is a common necessity; nonetheless, the environment surrounding the material is of particular importance as it varies from place to place affecting the overall product and leaching release.

In recycled concrete aggregates, sulfates can be present as gypsum and as sulfates from admixtures or binders. In mixed aggregates and CDW in general, their presence is attributed to residues of plasterwork, insulation prefabricated pieces, drywall, etc.

This contamination of final products is of particular concern in countries with wide use of gypsum, such as Spain [6] and Germany [7]. As a construction material, gypsum is good, efficient and economic, which means that the percentage of

E. Vázquez (✉) • M. Barra
Materials Section of the Department of Construction Engineering, Universitat Politecnica de Catalunya, Barcelona, Spain
e-mail: enric.vazquez@upc.edu

gypsum in building rubble can be expected to increase in the future. But this can make the re-use of building rubble difficult, and even call it into question altogether. Data from Germany and Spain confirm this trend.

In eastern Germany [7] the average gypsum content in recycled aggregates is 2.73 mass % and varies between 1.8 and 5.35 mass %. In Spain [6], the average gypsum content in recycled concrete aggregate (with minimum 90% concrete) is of 0.6% only, with maximum values of 2%, which is produced in fewer plants. In contrast, the mixed type aggregate, with a minimum concrete content of 70%, which is the most widespread, has an average of 2.5% and varies between 0.5 and 6.7%, always in weight.

Regulation limits for sulfates vary according to application and are virtually the same within the EU. Yet, there has been pressure in changing them coming from two sides: some researchers are of the opinion that tolerance of sulfate in recycled aggregates should be lower, while complaints to the contrary were heard from producers in the EQAR 2011 congress [8]. In any case, the removal of gypsum constituents from recycled aggregates during their processing seems indispensable. This can be achieved with jigging machines, in which asymmetrical jig stroke diagrams can be implemented [7]. A significant reduction in the gypsum content in processed crushed concrete can be achieved by jig sorting, fulfilling requirements even with input material that contains gypsum. As A. Müller et al. have proved, with medium and high input levels of gypsum, two passes through the jigging machine are needed to comply with the requirements specified by the DIN 4226–100. This can be achieved either by intermediate stockpiling of the material and feeding it through the machine again or by two jigging machines connected in series.

Another effective technique for reducing gypsum content is selection by means of near infrared spectroscopy. This newly developed technique has been successful in Spanish Basque processing plants, among others.

Whatever the technique chosen, the improvement of gypsum removal in processing facilities is for many a must if they want to survive in today's market.

4.2.2 Frost-Susceptible Recycled Concrete Aggregates

Frost resistance appears as another durability issue discussed in both concrete applications and road base. Demolished concretes to be recycled show great diversity in properties with regard to the potential to resist environmental effects, such as frost action, when they are used in a new concrete as aggregate.

Ch. Hendriks [9] was of the opinion that recycled aggregate concrete is frost resistant unless moisture content is extremely high. The combination of frost and the use of de-icing salts (chlorides) also result in damage when the porosity of recycled aggregate is very high.

It should be discussed if appropriate test methods exist to evaluate the specific characteristics of recycled concrete aggregates with respect to frost durability.

Table 4.3 Properties of source concretes

Type	Code	W/C	Air content (%)	Compressive strength at 28 days (MPa)
Air-entrained	HQC-AE	0.35	4.4	60.7
	MQC-AE	0.45	4.0	49.0
	LQC-AE	0.63	4.5	28.3
Non-air-entrained	MQC-NON-AE	0.45	1.2	55.8

Many authors think that this fact raises another important prejudice against the performance of recycled concrete aggregates compared to natural aggregates.

Gokce et al. [10] concluded after a long series of tests, that the direct unbound frost soundness test offers more realistic testing conditions to judge the soundness of recycled concrete aggregates. In this test, there is no chemical process decomposing the adhered mortar. In comparison with existing standard procedures, which condition the aggregate at 110°C, a moderate pre-drying process at 50°C is proposed. This seems to be more appropriate to the cement-bonded characteristic of the recycled concrete aggregates, as it eliminates the uncontrolled damage of test material that the usual conditioning at 110°C causes.

The sulfate soundness test fails to predict the frost susceptibility of the recycled concrete aggregate. This is because the procedure of the test severely decomposes the mortar adhering to the original coarse aggregate regardless of its quality or air void system. Thus, the sulfate soundness test would be applicable for only natural (inert) aggregates as a rapid indication of durability.

Due to the presence of adhered mortar, recycled concrete aggregate has a significantly higher absorption compared to natural coarse aggregate. In general, the frost susceptibility of an ordinary concrete aggregate is judged by its physical properties. Although high moisture absorption is often used as an index for unsoundness, many aggregates can absorb large amounts of water but remain sound. Unsoundness is therefore related to pore size distribution, rather than to the total porosity of the aggregate [11].

In Tables 4.3 and 4.4 we reproduce a part of the important study of Gokce. In his research, Gokce used three air-entrained (AE) type concretes with low, medium and high water–cement ratios; and a non-air-entrained (NON-AE) type concrete with a medium water–cement ratio; as the starting material to produce recycled coarse aggregate.

The concrete mixtures were proportioned to give a slump of 80 ± 20 mm and air content of 4.5 ± 1.5% with water–cement ratio of 0.55. The mixture proportions and measured fresh concrete properties are shown in Table 4.5. The freeze-thaw test procedure was performed in accordance with JSCE-G 501, corresponding to ASTM C 666 Procedure A. The durability factor (DF) was calculated at the end of 300 cycles. The weight, dynamic modulus of elasticity and length change measurements continued up to 500 cycles.

The following experimental procedure was implemented in the direct testing of the freeze-thaw resistance of recycled coarse aggregates and natural crushed stone:

1. Drying of the sample at 50°C and 60% R.H. until reaching constant weight.
2. Saturation of the sample with 3% NaCl solution for 24 h.

Table 4.4 Properties of aggregates

Source	Process level	Code	Adhered mortar content (%)	Density S.S.D. (g/cm^3)	Absorption (%)	JIS crushing value (%)
Sandstone	–	VC	0	2.65	0.68	3.0
HQC-AE	1	A1	52.3	2.42	4.88	3.8
	3	A3	30.2	2.51	3.14	1.5
MQC-AE	1	B1	55.0	2.41	5.58	5.2
	3	B3	32.4	2.50	3.19	1.7
LQC-AE	1	C1	52.3	2.37	6.27	6.3
	3	C3	32.3	2.48	3.76	2.3
MQC-NON-AE	1	B'1	55.7	2.44	5.57	3.7
	3	B'3	45.5	2.47	5.07	2.1

Table 4.5 Mixture proportions and fresh properties of concretes

		Materials (kg/m^3)								
						Admixture				
Mixture code	W/C	Water	Cement	Fine agg	Coarse agg	AE-WR	SP	Air cont (%)	Slump (mm)	T (°C)
VC-VS	0.55	165	300	813	1,039	0.72	0.94	4.3	90	17.0
A1-VS	0.55	160	291	819	959	0.47	0.91	4.5	90	16.8
A3-VS	0.55	157	285	825	1,002	0.46	0.89	4.0	65	16.0
B1-VS	0.55	160	291	819	955	0.47	0.91	4.1	80	16.1
B3-VS	0.55	157	285	825	998	0.46	0.89	3.8	65	16.0
C1-VS	0.55	160	291	822	940	0.47	0.91	4.5	65	19.4
C3-VS	0.55	158	287	826	988	0.46	0.90	4.7	75	20.3
B'1-VS	0.55	160	291	851	939	0.47	0.91	4.7	85	25.5
B'3-VS	0.55	156	284	817	996	0.57	0.89	4.1	80	21.5

3. Removing the saturated sample from the solution and wiping the particles with a towel to bring them into SSD condition.
4. Weighing the sample prior to frost test (W_0).
5. Re-immersion of the sample into the NaCl solution for 30 min due to the immediate loss of solution in the aggregate surface after weighing.
6. Removing the saturated sample from the solution without allowing the surface to lose wetness.

Next, the testing sample, in saturated and slightly surface wet condition, was placed into a double wall bag made of two plies of poly. Then, all the bags containing the aggregate samples were sunk into the freezing pool.

One freezing and thawing cycle took about 5 h. The freeze-thaw equipment was set to ensure a temperature not lower than −20°C at freezing, and not higher than 15°C at thawing, at any point of the chamber. The saturated samples were subjected to up to 90 freeze-thaw cycles. Every 30 cycles, the weight loss of the test samples was checked according to the following procedure:

1. Roughly sieving the wet sample to remove the particles under 5 mm as much as possible.

Table 4.6 Freezing and thawing test results of the concretes

Concrete mixture	No. of cycles at which the exposure ended	300 cycles DF	300 cycles Weight loss (%)	300 cycles Length change ($\times 10^{-6}$)	End of cycles Relative d. modulus of elasticity (%)	End of cycles Weight loss (%)	End of cycles Length change ($\times 10^{-6}$)
VC-VS	500	87	2.10	199	76	4.39	371
A1-VS	500	91	1.39	150	82	3.25	235
A3-VS	500	92	1.41	156	89	2.48	240
B1-VS	500	96	2.68	130	94	4.70	237
B3-VS	500	97	2.07	106	92	4.00	157
C1-VS	500	92	2.58	205	85	5.12	349
C3-VS	500	93	3.10	196	80	5.37	397
B'1-VS	90	17	–	–	56	0.81	1074
B'3-VS	180	35	–	–	58	1.47	862

2. Wiping the sample with a towel without causing a further weight loss.
3. Sieving the surface dry sample again to remove the remaining particles finer than 5 mm.
4. Determining the new weight of the sample over 5 mm at the end of *N* cycles.
5. Re-immersion of the sample into the NaCl solution for 30 min, due to the loss of the solution on the aggregate surface, before starting the test for the next 30 cycles.

The calculation of the frost soundness (FS) after *N* cycle was done as follows:

$$FS_N = \frac{W_o - W_N}{W_o} \times 100(\%)$$

where FS = freeze-thaw soundness loss of the aggregate tested after N cycles (%), W_0 = weight of the saturated-surface dry sample prior to the frost test (g) and W_N = weight of the saturated-surface dry sample after *N* cycles (g).

The results summarized in Table 4.6 reveal that the concretes produced with the recycled coarse aggregates belonging to air-entrained type of source concretes were highly durable against freezing and thawing (DF > 91), and were even slightly superior than the reference concrete produced with virgin aggregates.

While the frost soundness loss of the non-air-entrained type recycled aggregate ranged between 34.2 and 37.1%. The results for the virgin aggregate and air-entrained type recycled aggregates were below 10%, except for one with 16.4%, originated from low quality source concrete with a high adhered mortar content (C1). In the opinion of the author it is not possible to propose a certain limit value for the frost soundness loss with the limited data of this study. A further quantitative assessment based on the statistical interpretation of a large number of test results, to be obtained from a broad range of demolished concrete sources in terms of quality, air-entrainment characteristics and damage accumulation during past service life, would be favorable to establish an exact durability criteria.

J. Mehus et al. [12], from the Norwegian Building Research Institute, present an overview of research conducted as a part of the Norwegian Roads Recycled Materials R&D Program. A proposed adjustment of the test method specified in EN 1367–1 avoiding pre-drying of the recycled aggregate, has been evaluated. Furthermore, various exposure conditions with recycled aggregate, including submersion and saturation in deicing salts, have also been investigated. This has resulted in the proposal of a revised test method based on EN 1367–1, for documentation of freeze-thaw properties of recycled aggregate. The specimen preparation procedure in EN 1367–1 specifying pre-drying affects the recycled aggregate freeze-thaw durability.

A. Richardson et al. [13] have shown that the use of air entrainment and polypropylene fibers in concrete made with recycled aggregate are equally effective at providing freeze-thaw durability when compared with concrete made with natural aggregates, air entrainment and polypropylene fibers.

The results show that the concrete samples made with recycled aggregate were slightly more durable than those made with virgin aggregate. The samples made with recycled aggregate and an addition of polypropylene fibers were found to have an average compressive strength of 13.8 N/mm^2, while the samples made with natural aggregate with polypropylene fibers had an average compressive strength of 12.9 N/mm^2 -which is a difference of 7% and within normal batching tolerances. The concrete samples made with recycled aggregates were more durable than plain samples made with natural aggregate by 68%. This data may be explained by the variability of the good quality recycled aggregate and curing procedure using soaked aggregate. Overall, the results show that recycled aggregates, with the aid of additives, can be used for applications where concrete is subject to freezing and thawing while still providing the durability that natural aggregates offer.

4.2.3 Chloride Penetration

As is well known, corrosion of reinforcing steel as consequence of penetration of chlorides through concrete, from sea water, marine environment, de-icing salts, have serious implications for structures and represents important economic damages for many countries. Should not be ruled out the presence of chlorides in some components of concrete, but this is easily avoidable with a correct control of materials.

Different physical mechanisms may be distinguished with respect to the driving force. These mechanisms may be referred to the permeation of a salt solution, the capillary absorption or chloride-containing liquids, and the diffusion of free chloride ions. J. Kropp [14] describes the transport of chloride ion as the result of the action of three mechanisms: permeation, capillary absorption and diffusion. The diffusion will exist only if there is not flow of water in the concrete. In reality it is very difficult, if not impossible, quantify the partial contributions of each mechanism and the combined effects of them. In the determination of diffusion coefficients by some rapid methods, overlaps an external electric potential that causes a migration transport.

All these mechanisms act only on the free chlorides. The chlorides can be chemically binded by the C_3A present in the hydrated cement, forming the chloro-aluminate or Friedel's salt ($3CaO.Al_2O_3.CaCl_2.10H_2O$) and can be binded too by adsorption by the CSH. Blast furnace slag and fly ash in blended cements, reduce the concentration of C_3A and generally chlorides binding capacity [14].

Other publications suggest that the cements with high slag content, increases the ability of chlorides to bind [14]. There are factors that can affect retention, such as type of chlorides and moisture. Tuutti [15] indicates that chlorides from the addition of $CaCl_2$ are combined in larger quantities than those from KCl. The pH of the liquid in the pores, the water/cement ratio and especially the type of test used, should also be taken into account.

It is considered that only the free chlorides are able to reach the reinforcing steel, and therefore, bound chlorides are innocuous with respect to corrosion. It should be noted that physically bound chlorides, can be released by carbonation or entry of sulphates [16]. Likewise, the chloroaluminate can be decomposed by the same agents, producing calcium carbonate, aluminum oxide and releasing chloride. This may mean that in a carbonate matrix, could be free the most of chlorides. Despite this argument, be bound, strongly retards the transport and the ability of chloride to reach the reinforcing steel, and therefore, the effect should always be considered when determining the transport of chlorides in the concrete. The complete mechanism requires even more research and experimental determinations.

These processes also apply to recycled aggregate concretes. They must be take into account the presence of adhered mortar from the original concrete, and even aggregates that are pure mortar. Thus, it is necessary consider not only the porosity of the new mortar, but also the possible contribution from the mortar from the recycled aggregate. A common simplification of the problem is to consider only the possible contribution of recycled aggregate porosity, ignoring the contribution of other factors that may influence the final result of changing behavior in chlorides.

These factors include prior contamination of recycled aggregates by chlorides and/or sulfates, the water/cement ratio in mortar of the new concrete, pre-wetting, the ability of binding chlorides, and the test used to determine the diffusion coefficient.

A review of recent publications about the transport of chloride in recycled concrete, allows an assessment of the importance of each one of these factors. If the recycled concrete aggregate is contaminated by chlorides, the new concrete will have an additional quantity of chlorides.

L. Friedl et al. [17] established that the pretreatment of recycled aggregate (e.g. moistening), has a large effect on the quantity and the speed of redistribution of chlorides between recycled aggregate and the new matrix. Additional tests with conventional concrete corrosion cells, where individual chloride contaminated recycled aggregates were placed near the embedded steel, showed intense pitting corrosion after 90 days. For these investigations the composition of the recycled aggregate concrete, especially the type and the proportion of recycled aggregate, were varied. The results indicate that the quantity of recycled aggregate has a minor influence on chloride ingress. However, the concrete composition and the type of recycled aggregate can enhance the chloride ingress into the recycled aggregate concrete.

Y.A. Villagran et al. [18] analyzed in his work, the influence of coarse recycled concrete aggregate on the chloride penetration rate and binding capacity, of concrete under marine exposure. Concrete specimens were molded and exposed to a natural marine atmosphere. After exposure periods of 6, 12, and 18 months, ingress profiles of total and water-soluble chlorides were measured. Results indicate that recycled aggregate incorporation had two opposed effects in concrete:

- It increases its chloride penetration rate
- The chloride binding capacity. Therefore, water-soluble free chloride contents in conventional and recycled concrete series had similar values.

The diffusion coefficient D is a chloride parameter considered useful to evaluate the transport of chloride in a concrete. Tests are available to accelerate the transport of chloride, either by means of an external electric field, a concentrated solution of chloride or with rising temperature

The immersion test NT BUILD 443, as for non-steady-state diffusion, the RCM test, as for non-steady-state migration, the UNE 83987, as for steady state and non-steady-state migration and the resistivity test were selected for a final evaluation in the European project CHLORTEST. In parallel an international inter-laboratory comparison was organized by RILEM TC 178-TMC [19], and other comparisons were made by Tang Luping et al. [20]. The values obtained of the non-steady-state migration from the UNE 83987 are comparable with those obtained with NT BUILD 443 for relative porous concretes, but deviate for dense concrete. For the same 2 tests the steady state diffusion coefficient obtained from UNE 83987 is smaller then the non-steady-state coefficient from the NT BUILD 443. These and other relations must be taken into account if we want to draw valid conclusions, comparing studies on chloride penetration into concrete with recycled aggregates made with immersion tests, dissemination and migration respectively.

V.A. Ulloa et al. [21] studied the influence of the use of recycled aggregates in the chloride penetration coefficient, determined according to UNE 83987. The type of aggregates, the water/cement ratio in new concrete and the age of this at the beginning of the attack, are the three variables. For the part of the study focused on the use of recycled concrete aggregate (ArH), were used recycled aggregates from a plant in Valencia, with absorptions of 5.26 and 4.51% respectively. The effective w/c ratio of the concretes were 0.40 and 0.50, and dosages for cement CEM I 42,5/SR were 380 y 330 kg/m^3

Non-standard concretes with mixed recycled aggregates were studied, with 21% ceramic aggregate and absorption of 8.6%. The dosages shown in Table 4.7, were carried out by the method of Bolomey. For the study of the penetration of chlorides, diffusion coefficients for the hardened concrete for stationary and nonstationary states are determined. Cores of 7.5 $\times$ 3 cm obtained from cylindrical specimens of 15 $\times$ 30 cm were tested. The results obtained for the compressive strength and the diffusion coefficients are shown in the graphs (Figs. 4.1, 4.2, and 4.3).

Table 4.7 Quantities of material per m^3, for different dosages

Dosage/m^3						
w/c ratio	0.4			0.5		
Type of recycled aggregate	AN	ArH	ArMh	AN	ArH	ArMh
Water	152	152	152	165	165	165
Cement	380	380	380	330	330	330
Sand 0/4 mm	920	751	852	933	788	885
Medium aggregate 7/12 mm	418	0	0	399	0	0
Coarse aggregate 12/20 mm	591	0	0	585	0	0
Recycled aggregate GRB		1,076	875	0.00	1,050	855
Recycled aggregate TECREC						
Theoretical Density (kg/m^3)	2,461	2,359	2,259	2,415	2,333	2,235

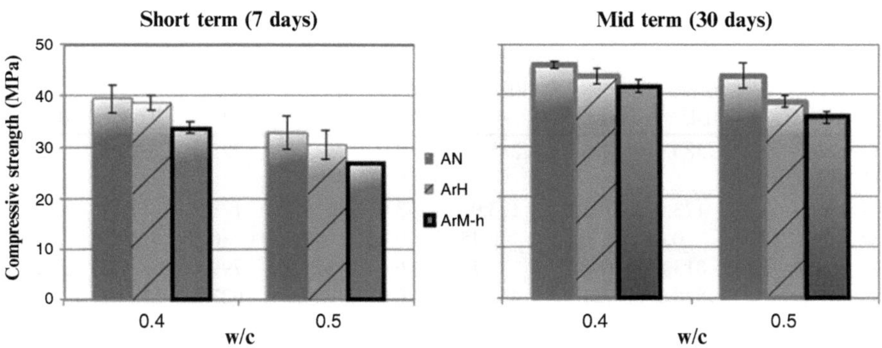

Fig. 4.1 Compressive strength in the short and mid-term, for concretes performed with different types of coarse aggregates and two water–cement ratios

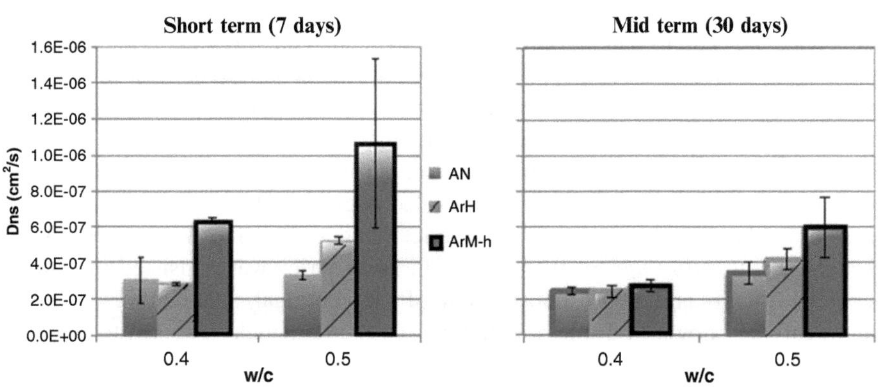

Fig. 4.2 Non-Stationary Chloride Diffusion Coefficient Dns, analyzed for concretes in the short and mid-term

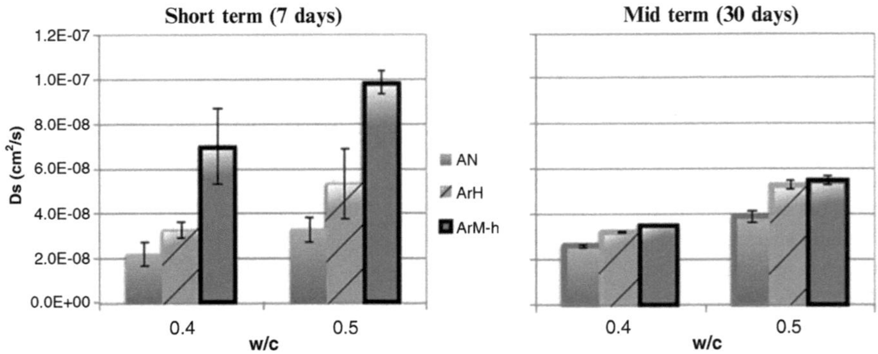

Fig. 4.3 Stationary Chloride Diffusion Coefficient Ds, analyzed for concretes in the short and mid-term

Table 4.8 Concrete mixes

	Concrete mixes (kg/m^3)							
Materials	0%-0.45	20%-0.45	50%-0.45	100%-0.45	0%-0.60	20%-0.60	50%-0.60	100%-0.60
Cement CEM I 52.5 SR	385.0	385.0	385.0	385.0	325.0	325.0	325.0	325.0
Water	173.0	173.0	173.0	173.0	178.7	178.7	178.7	178.7
w/c	0.45	0.45	0.45	0.45	0.55	0.55	0.55	0.55
Fine agg.	813.8	698.2	741.7	746.1	874.9	799.4	831.3	824.9
Coarse agg.	870.9	747.9	431.7	0.0	754.9	627.5	367.1	0.0
Medium agg.	167.6	143.9	83.1	0.0	253.9	211.0	123.4	0.0
Recycled agg.	0.0	222.9	514.8	963.2	0.0	209.6	490.5	923.4
Superplas.	3.85	3.85	3.85	3.85	3.25	3.25	3.25	3.25
Total	2414.2	2374.8	2333.3	2271.2	2390.8	2354.7	2319.2	2255.3

For recycled aggregates ArH, with the above-mentioned porosities, the chloride penetration is affected by the water-cement ratio into the cement matrix of new concrete. To water-cement ratio of 0.4, resultant concretes do not experiment significant losses of durability, in terms of chloride penetration, even in this case, which ArH aggregate is used in the entire fraction >4 mm. For concretes with ARMH, chloride penetration values are significantly higher.

M. Barra et al. [22], have performed an extensive experimental program in order to evaluate the repercussions of the recycled aggregates content in the properties and in the durability of concrete. Two series of concrete mixes were tested, with a water-cement ratio (w/c) of 0.45 and 0.50. Each series was consisted of 4 mixes with 0, 20, 50 and 100% substitution of conventional aggregate by recycled aggregate, as shown in Table 4.8. Notice that the nomenclature used to represent each mix indicates the percentage of substitutions of conventional aggregate by recycled aggregate in volume and the water-cement ratio (w/c).

Calcareous aggregate (coarse, medium and fine) and one type of coarse recycled aggregate were used. The bulk density and water absorption of recycled aggregate

Table 4.9 Compressive strength at 28 days

Concrete	Compressive strength (N/mm^2)
0%-0.45	71
20%-0.45	61
50%-0.45	56
100%-0.45	50
0%-0.55	62
20%-0.55	55
50%-0.55	48
100%-0.55	41

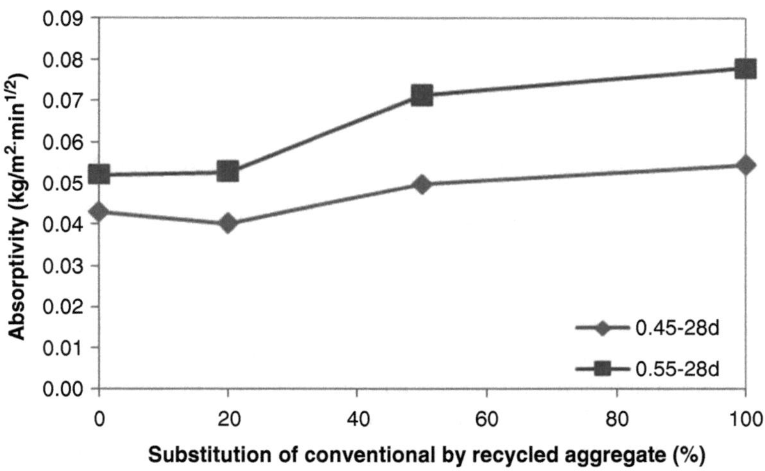

Fig. 4.4 Absorptivity assessed for different mixes

was 2.33 g/cm^3 and 4.6%, respectively. The visual characterization (UNE EN 933–11) of the former showed the following results: concrete 55.9%, aggregates 34.5%, bituminous material 8.5%, masonry 0.7%, gypsum 0.4% and others 0.005%.

The average compressive strength at 28 days measured in cylindrical specimens with 150 mm of diameter and 300 mm of height are summarized in Table 4.9 for all mixes.

Additional tests were conducted to assess the capillary suction, the penetration depth and profile (NTBuild 443:1995).The capillary suction test was performed at 28 days from the casting of the specimens. The results obtained were then used to assess the absorptivity of the different concrete mixes, as illustrated in Fig. 4.4.

For the same w/c, the absorptivity estimated for the mixes with low content of recycled aggregate (0 and 20%) are similar. Higher percentages of substitutions of conventional aggregate (50 and 100%) lead to an increase in the absorptivity estimated. For instance, the difference between the absorptivity of the mix without recycled aggregate and with 100% of substitution are 27.5% for a w/c of 0.45 and 49.1% for a w/c of 0.55. Such result may be attributed to the higher porosity of the recycled aggregate used that facilitates the entrance of liquids in the microstructure.

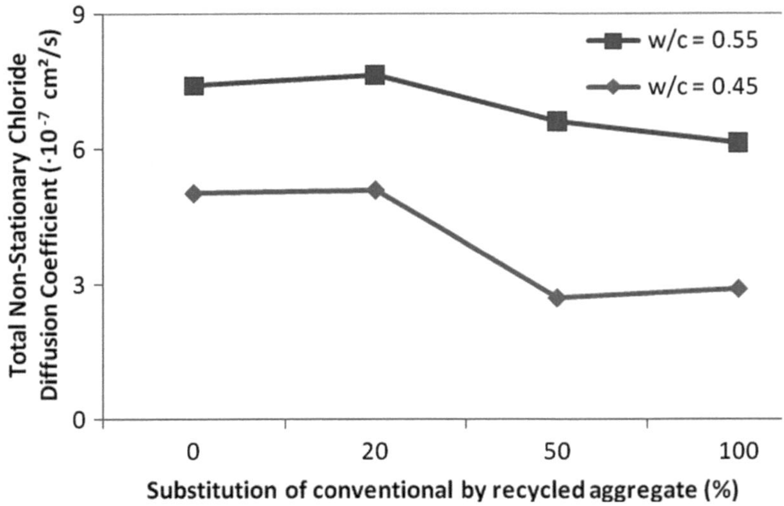

Fig. 4.5 Chloride diffusion coefficient measured for different mixes

On the other hand, the chloride penetration profile measured according with the NTBuild 443:1995 test was used to estimate the total non-stationary chloride diffusion coefficient by means of an interactive procedure. Figure 4.5 shows the results obtained depending on the w/c and on the content of recycled aggregate.

The diffusion coefficient of mixes with the same w/c and low recycled aggregate content (0 and 20%) show very similar diffusion coefficients. The same is observed between mixes with high recycled aggregate content (50 and 100%). Furthermore, it is observed that the substitution of conventional aggregates in a percentage superior to 20% produces a decrease of 18 and 80% on the average diffusion coefficient for mixes with w/c equal to 0.55 and 0.45, respectively.

Such results show a tendency that goes against the expected due to the permeability of concrete. Indeed, if only the permeability affected the diffusion process, the mixes with higher recycled aggregate content should present higher chloride diffusion coefficients since they tend to be more permeable. Therefore, it is clear that another phenomenon acting in parallel to the simple diffusion of chlorides is responsible for the results shown in Fig. 4.5. This phenomenon is the chloride retention by the aluminates of the paste, which should be bigger in the mixes with more recycled aggregates thus compensating the effect of a bigger permeability of the matrix. Therefore, according to the estimation performed, the use of a high recycle aggregate content had a positive repercussion, leading to a concrete less permeable to chlorides.

It was also observed that the change on the w/c from 0.45 to 0.55 produces an increase on the diffusion coefficient. Such increase is approximately of 49% in mixes with low recycled aggregate content (0 and 20%) and of 227% in mixes with high aggregate content (50 and 100%). Consequently, the latter appear to be more sensible to variations in the w/c.

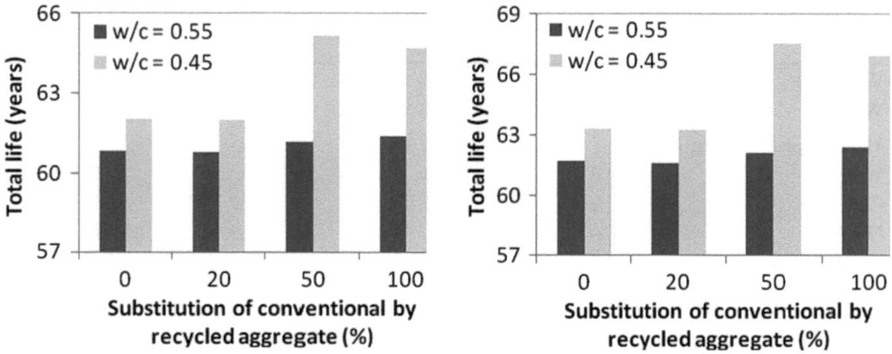

Fig. 4.6 *Left*- Total life predicted for different concrete mixes according to the EHE-08. *Right*- the Helene model

To understand the practical repercussion of the results of chloride diffusion, the durability estimation was performing using the models proposed by the EHE-08 and by Helene. In both of them, the induction and the propagation time was calculated and summed to obtain the total life of a standardized structure. This structure considered in this estimation is constructed with reinforced concrete that present conventional steel bars with 16 mm of diameter with a 35 mm concrete cover. The surficial, the internal and the critical chloride content are 0.25, 0.03 and 0.20% by weight of cement, respectively.

Figure 4.6 shows the total life predicted considering the use of each concrete mix tested previously. It is observed that, for the same w/c, the total life is bigger for mixes with higher content of recycled aggregate. Nevertheless, in this case, the difference in favor of mixes with higher recycled aggregate content is of approximately 0.75% according to the Helene Model and 4.75% according to the EHE-08 model, respectively.

This result demonstrates that the use of recycled aggregates in substitution of more than 50% of the conventional aggregate, may lead to an improvement of the durability of concrete if the mix is properly designed.

Otsuki [23] found that chloride penetration was higher in the conventional aggregate concrete, than in recycled aggregate concrete. Limbachiya et al. [24] studied concretes of 50 MPa and 370 kg/m3, and concluded that even for concretes with 100% of recycled coarse aggregate, the values of D decrease compared to conventional aggregate concrete.

Other studies start from the idea that chloride penetration increases with the recycled aggregate content. They attribute the effect to cracks and fissures produced during the production of recycled aggregate, and that this effect can be compensated using active mineral admixtures.

V. Corinaldesi and G. Moriconi [25] showed that when fly ash was added to recycled aggregate concrete the pore structure was improved, and particularly the macro pores volume was reduced causing benefits in terms of mechanical performance. From the serviceability point of view, the drying shrinkage of recycled

Table 4.10 Concrete mixture proportions

Mixture	NAT-0.6	REC-0.3	REC-WRA-0.3	REC-FA-0.6
W/C	0.60	0.30	0.30	0.60
W/CM	0.60	0.30	0.30	0.30
Mixture proportions, kg/m3				
Water	230	230	165	230
Cement	380	760	550	380
Fly Ash	–	–	–	380
Natural Sand	314	–	–	–
Aggregate	1,338	–	–	–
Fine recycled fraction	–	–	372	–
Coarse recycled fraction	–	1,169	1,060	1,057
Superplasticizer	–	–	5.5	6.8

Table 4.11 Physical properties of the aggregate fractions

Aggregate fractions	Bulk specific gravity	Water absorption	Passing 75 μm
Natural sand	2,620	3	0.9
Crushed aggregate	2,680	2	0.2
Fine recycled fraction	2,150	10	0.5
Coarse recycled fraction	2,320	8	0.3

Table 4.12 Diffusion coefficients of chloride ions into different concretes at 20°C

Mixture	NAT-0.6	REC-0.3	REC-0.3+	REC-FA-0.6
D (cm2 s-1 10–8)	1.90	0.87	0.72	0.46

aggregate concrete did not appear to be a problem since the same risk of crack formation resulted in ordinary concrete. As far as corrosion aspects were concerned, the use of fly ash appeared very effective in protecting galvanized steel reinforcement in porous concrete, as it could occur when recycled aggregates are used, even with cracked concrete. Resistance to chloride ion penetration was investigated concerning the durability of recycled aggregate concretes containing high volumes of fly ash. The Concrete mixture proportions are shown in the next two tables (Tables 4.10 and 4.11).

Collepardi et al. [26] evaluated the chloride penetration into concrete through the $AgNO_3$ and X-ray fluorescense test. Concrete specimens exposed to a 10% NaCl aqueous solution after a wet curing of 1 week and an air curing of 3 weeks at a temperature of 20°C were studied. The diffusion coefficients obtained in different concretes are shown in Table 4.12.

The strong beneficial effect of fly ash on the chloride penetration depth measured from concrete sample REC-FA-0.6 is quite evident. In fact, the diffusion coefficient of chloride ions into REC-0.3 concrete doubled with respect to REC- FA-0.6, the only difference being the replacement of 50% of cement with fly ash.

Lower water/cement ratios also influence positively the concrete resistance against chloride penetration: the diffusion coefficient of chloride ions into NAT-0.6 concrete more than doubled with respect to REC-0.3, even though recycled aggregate was used.

The apparent chloride diffusion coefficients for all RCA-concrete specimens made of mixtures proportioned by the new EMV method were found by Abbas et al. [27] to be of the same order of magnitude as the specimens made of conventional structural-grade concrete (i.e. 10^{-12} m^2/s). The chloride penetration of RCA-concrete was investigated by conducting the acid soluble bulk diffusion test as per ASTM C 1556–04. This test method covers the laboratory determination of the apparent chloride diffusion coefficient for cementitious mixtures by measuring the acid soluble (total) chloride content. In fact, the apparent chloride diffusion coefficients for the RCA-concrete specimens proportioned by the EMV method were lower than those of the specimens made of mixture proportioned by the conventional method. The measured acid soluble initial chloride concentration of the RCA-concrete proportioned by the EMV method was found to be lower than the limits specified by the current standards.

There is a need for research that provides more data on the penetration of chloride in real conditions in situ. The miscellany of recently published papers, with results based in laboratory studies with accelerated tests led to conclusions relatively optimistic about the use of concrete with recycled concrete aggregates in environments with chlorides.

The use of low enough water/cement ratio and quality recycled aggregate free of chlorides and sulfates in the new concrete or the use of mineral admixtures, led to believe that it is possible to use concretes with high contents of recycled aggregate in media containing chlorides.

References

1. Razaqpur, A.G., Fathifazl, G., Isgor, O.B., Abbas, A., Fournier, B., Foo, S.: How to produce high quality concrete mixes with recycled concrete aggregate. In: RILEM PRO 73: Construction Waste Recycling and Civil Engineering Sustainable Development ICWEM2010, pp. 11–35. Bagneux-France (2010)
2. Jiménez, C., Barra, M., Valls, S., Vázquez, E.: The EMV method of mixture proportioning for concretes with RCA. The EMV method of mixture proportioning for concretes with RCA. Authors' communication (2011) authors' communication
3. ACI 211.1-91: ACI 211.1-91-Standard Practice for Selecting Proportions for Normal, Heavyweight, and Mass Concrete. Standard, ACI-American Concrete Institute (2002)
4. Fernández, M.: Hormigón Eighth Edition edn. Colegio de Ingenieros de Caminos, Canales y Puertos, Madrid (2007)
5. Barra, M.: Estudio de la durabilidad del hormigón de árido reciclado en su aplicación como hormigón armado. Dissertation, Universitat Politècnica de Catalunya, Barcelona (1996)
6. Vazquez, E., Varela, E., Güell, A.: Spanish Guide on Recycling Aggregates (Gear Project). GERD, MMA, Madrid (2011)
7. Müller, A., Schnellert, T., Kehr, K.: Gypsum constituents in recycled concrete – origin, effects and removal by jigging, part 2. AT 6/2010. AT Tech. Solut. **51**, 54–69 (2010)
8. Dupré, F.: Kreislaufwirtschaft – Zukunft oder Gegenwart? In: Proceedings of the EQAR-Congress, Recycling of Building Materials in Europe, pp. 16–17. Brussels (2011)
9. Hendriks, C.F., Nijkerk, A., Van Koppen, A.: The Building Cycle. Aeneas Technical Publishers, Delft (2000)

10. Gokce, A., Nagataki, S., Saeki, T., Hisada, M.: Identification of frost-susceptible recycled concrete aggregates for durability of concrete. Constr. Build Mater. **25**(5), 2426–2431 (2011)
11. Mehta, P., Monteiro, P.J.M.: Concrete: Microstructure, Properties, and Materials, 3rd edn. McGraw-Hill Inc., New York (2006)
12. Petkovic, G., Mehus, J., Myren, S.: Recycled Concrete Aggregate – Durability Aspects. Overview of research conducted as a part of the Norwegian Roads Recycled Materials R&D Program (2003)
13. Richardson, A., Coventry, K., Bacon, J.: Freeze/thaw durability of concrete with recycled demolition aggregate compared to virgin aggregate concrete. J. Clean. Prod. **19**(2–3), 272–277 (2010)
14. Kropp, J., Hilsdorf, H.K.: Performance Criteria for Concrete Durability. RILEM report. E & FN Spon, London (1995)
15. Tuutti, K.: Corrosion of Steel in Concrete. CBI forskning/research, vol. 4.82. Swedish Cement and Concrete Research Institute, Stockholm (1982)
16. Tang, L., Nilsson, L.O.: Chloride binding capacity, penetration and pore structure of blended cement pastes with slag and fly ash. In: Proceedings of the International Conference on Blended cements in Construction, Elsevier Applied Science, Sheffield 1991
17. Friedl, L., Volkwein, A., Schiebl, P.: The risk of corrosion of steel in recycled aggregate concrete. Am. Concr. Inst. (ACI) Spec. Pbl. **212**, 1055–1072 (2003)
18. Villagrán-Zaccardi, Y.A.M., Zega, C.J., Di Maio, A.A.: Chloride penetration and binding in recycled concrete. J. Mater. Civ. Eng. **20**, 449 (2008)
19. Huang, Y., Deng, Z.H., Luo, Y.M., Yang, H.F.: Experimental study on shear strength of recycled aggregate concrete (only available in Chinese). Chin. Concr. J. **2**(31), 14–17 (2010)
20. Tang, L., Nilsson, L.O., Basheer, P.A.: Resistance of Concrete to Chloride Ingress. Spon Press, London (2011). ISBN 9780415486149
21. Ulloa, V.A., Barra, M., Serna, P., Pelufo, M.J., Aponte, D., Vazquez, E.: Analysis of chloride diffusion in concrete with recycled concrete aggregates. Authors' Communications (2011)
22. Barra, M., Aponte, D., Pielarrisi, S., Vazquez, E.: Chloride penetration in recycled concrete. Authors' Communications (2011)
23. Otsuki, N.M.: Influence of recycled aggregate on interfacial transition zone, strength, chloride penetration and carbonation of concrete. J. Mater. Civ. Eng **155**, 443–451 (2003)
24. Limbachiya, M., Leelawat, T., Dhir, R.: Use of recycled concrete aggrégate in high strength concrete. Mater. Struct. **33**, 574–580 (2000)
25. Corinaldesi, V., Moriconi, G.: Influence of mineral additions on the performance of 100% recycled aggregate concrete. Constr. Build Mater. **23**, 2869–2876 (2009)
26. Collepardi, M., Marcialis, A., Turriziani, R.: Penetration of chloride ions into cement pastes and concretes. J. Am. Ceram. Soc. **55**(10), 534–535 (1972)
27. Abbas, A., Fathifazl, G., Isgor, O.B., Razaqpur, A.G., Fournier, B., Foo, S.: Durability of recycled aggregate concrete designed with equivalent mortar volume method. special issue of the journal of cement and concrete composites on sustainability of civil engineering structures – durability of concrete. J. Cem. Concr. Compos. **31**(8), 555–563 (2009)

Chapter 5
Use of Fine Fraction

Isabel Martins, Annette Müller, Angel di Maio, John Forth, Jörg Kropp, Sergio Angulo, and Vanderley John

Abstract As an outcome of recycling the mineral part of construction and demolition waste significant quantities of the fine fraction, particle size below 4 mm in Europe, are generated. A common feature of the produced fine recycled aggregates is the high content of cement paste that could impair its quality. With the recent advances on the processing technologies improvement on the characteristics of the fine fraction of C&DW can be envisaged and new perspectives to their application arise.

After a brief review of the new technologies developed for C&DW processing, of the characteristics of fine recycled aggregates and of the requirements they must fulfill, an overview of different solutions for the use of the fine fraction is provided.

I. Martins (✉)
Concrete Division of Materials Department, Laboratório Nacional de Engenharia Civil - LNEC, Av. Brasil 101, 1700-066 Lisbon, Portugal
e-mail: imartins@lnec.pt

A. Müller
Universität Weimar, Chair of Mineral Processing of Building Materials and Reuse, Couydrastr, 7, 99421 Weimar, Germany

A. di Maio
Laboratorio de Entrenamiento Multidisciplinario para la Investigación Tecnológica (LEMIT), Calle 52 e/121 y 122 – (1900) La Plata, Buenos Aires, Argentina

J. Forth
School of Civil Engineering, University of Leeds, Leeds LS29JT, United Kingdom

J. Kropp
Institute for Building Materials, Hochschule Bremen, Neustadtswall 30, 28199 Bremen, Germany

Hochshule Bremen, Institut für Baustofftechnologie, Bremen, Germany

S. Angulo
Instituto de Pesquisas Tecnológicas do Estado de São Paulo, Cidade Universitária, P.O. Box 0141, 01064-970 São Paulo, Brazil

V. John
Laboratório de Microestrutura e Ecoeficiencia de Materiais, Escola Politécnica, University of São Paulo, Av. Prof. Almeida Prado 83, 05508-900 São Paulo, Brazil

E. Vázquez (ed.), *Progress of Recycling in the Built Environment: Final Report of the RILEM Technical Committee 217-PRE*, RILEM State-of-the-Art Reports 8, DOI 10.1007/978-94-007-4908-5_5, © RILEM 2013

The use in the field of precast concrete elements, of rendering and bedding mortars or for producing cement blends constitute some of the highlighted applications.

Keywords Fine recycled aggregate • Concrete • Mortar • Ceramic • Processing

5.1 Introduction

The increased concern on the reuse and beneficial application of construction and demolition waste (CDW) lead to a significant number of research studies mainly related to the use of the coarse fraction leaving behind the fine fraction for which there is little or no market at the moment.

Nowadays, the fine fraction, which accounts for 20–50% of the CDW stream, depending on the characteristics of the processing unit, is usually disposed off but with the increase of building activity this scenario has to be changed in order to avoid depletion of natural aggregates, divert construction and demolition waste from landfill and close the life cycle of building materials.

It is well known that fine recycled aggregates, e.g. from crushed concrete, have a higher content of mortar adhered to its surface than the coarse fraction and this characteristic will influence several properties, like absorption or density, that restrict its use. Envisaging the removal of the binder attached to CDW different processing techniques, more or less committed to the fine fraction, have evolved.

Regarding its application, the use of fine recycled aggregates in the manufacturing of concrete is usually not allowed or very restricted owing to its high absorption capacity that could result in loss of workability, larger shrinkage and creep deformations, low durability and high permeability of the cement paste [1–3]. Depending on the composition of the fine fraction of CDW applications like back-filling material, road material or cement burning material have been proposed [4]. In recent years the use of the fine fraction in mortars and in paving blocks has also been subject of research.

From an environmental point of view fine recycled aggregate are more susceptible to contamination with foreign materials than the coarse fraction. Some measures applied during CDW collection, like selective demolition, are essential to control this feature. Minimization of contamination could also be achieve in recycling plants through implementation of adequate practises, for instance separation of fines coming from primary screening from those of crushing because of the higher content of contaminants, such as gypsum, organic materials and dusts, on the former.

5.2 Processing Fine Recycled Aggregates

The main constraint for producing high quality recycled aggregates from CDW relates to the chosen processing technology. In most countries recycling of construction and demolition waste usually relies on simple technologies, like crushing

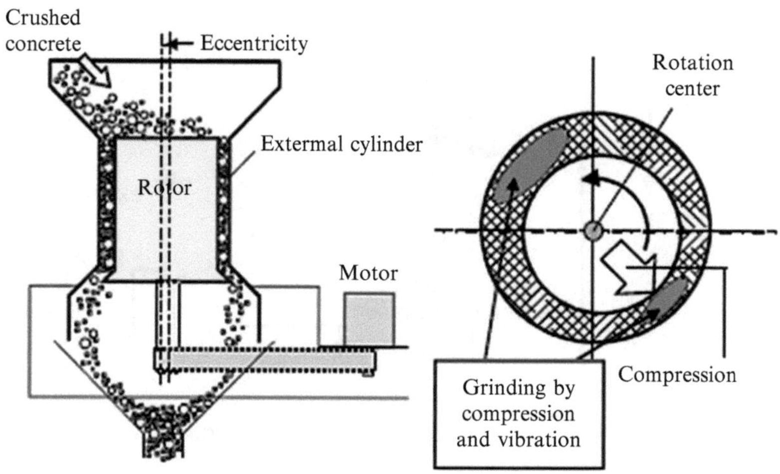

Fig. 5.1 Mechanical scrubbing – eccentric rotor type [5]

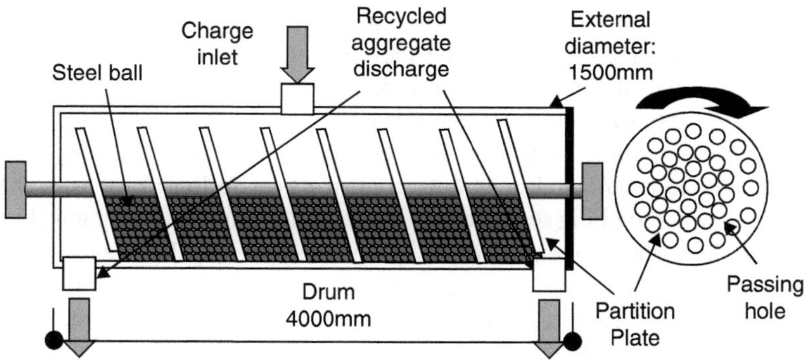

Fig. 5.2 Mechanical scrubbing – screw type [6]

and screening, which are not efficient for taking away the binder stuck on the recycled aggregate surface, leading to low quality aggregates. This problem is much more evident in the fine fraction of construction and demolition waste hindering their application.

Advances in technologies based on mechanical scrubbing lead to the production of recycled concrete aggregates of improved quality. In the mechanical scrubbing – eccentric rotor type, crushed concrete lumps are fed into an eccentric tubular mill unit (Fig. 5.1) and the cement and mortar are removed from the surface of aggregates by eccentric rotation of two cylinders leading to coarse recycled concrete aggregates that comply with the requirements of Japanese standards for natural aggregates. Regarding the properties of concrete manufactured with these coarse aggregates they are similar to those of a concrete using natural aggregates [5]. Fine recycled aggregate produced by this technology does not have the adequate characteristics to be used in structural concrete.

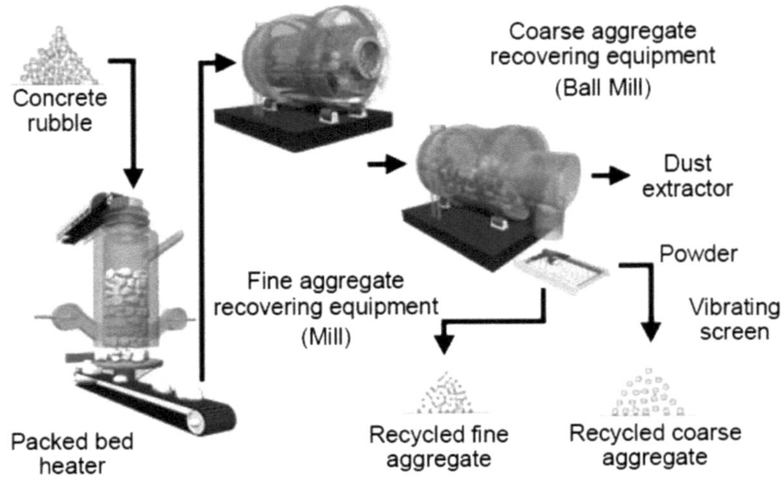

Fig. 5.3 Processing of CDW using heating-scrubbing [8]

Another type of mechanical scrubbing named screw-type is also used for the production of recycled (Fig. 5.2). The partition plates along the drum allow the separation of coarse and fine recycled fractions and attrition will promote reduction of the cement paste stick on aggregate surface.

An heating scrubbing method based rely on the use of hot air, at 300°C, to heat the concrete rubble and dehydrate cement paste, followed by a mechanical treatment, in which rubbing separates the brittle paste from the aggregates (Fig. 5.1) have been developed to enhance the properties of recycled aggregates. The generated coarse and fine recycled concrete aggregate can be used in ready mixed concrete while fine powder can be recycled as raw material for cement, as cement admixture, or as soil stabilizer [7] (Fig. 5.3).

Based on the experiments of Mitsubishi, TNO and KEMA Mulder et al. proposed a sorting process for concrete from CDW that combine a thermal and a mechanical procedure [9]. Heating at temperatures above 700°C will remove the adhered the cement paste and almost all the constituents of the concrete waste will be recovered allowing closing the material life-cycle (Fig. 5.4).

With the aim of improving the quality of fine recycled aggregates Weimann and Müller treated, in a pilot plant, uncontaminated crushed concrete fines from selective demolition by mixing with water and feeding the mix into a stirring unit, to remove the adhered mortar. This processing technology also include a hydrocyclone that allow elimination of particles sized under 100 μm, and a jig that promote the separation of the material into a heavy fraction, with density closer to that of natural fine aggregates, and a light fraction. This separation is based on the difference of materials densities, for example natural aggregates like quartz or dolomite have higher densities (quartz: 2.65 t/m and dolomite: 2.85–2.95 t/m) than concrete (2.2–2.5 t/m) and hardened cement paste (1.5–1.8 t/m) [10]. This treatment changed the grading of the fines and the heavy fraction evidenced a reduction on binder content and an improvement in water absorption allowing a higher replacement level of fine recycled aggregates in concrete [11] (Fig. 5.5).

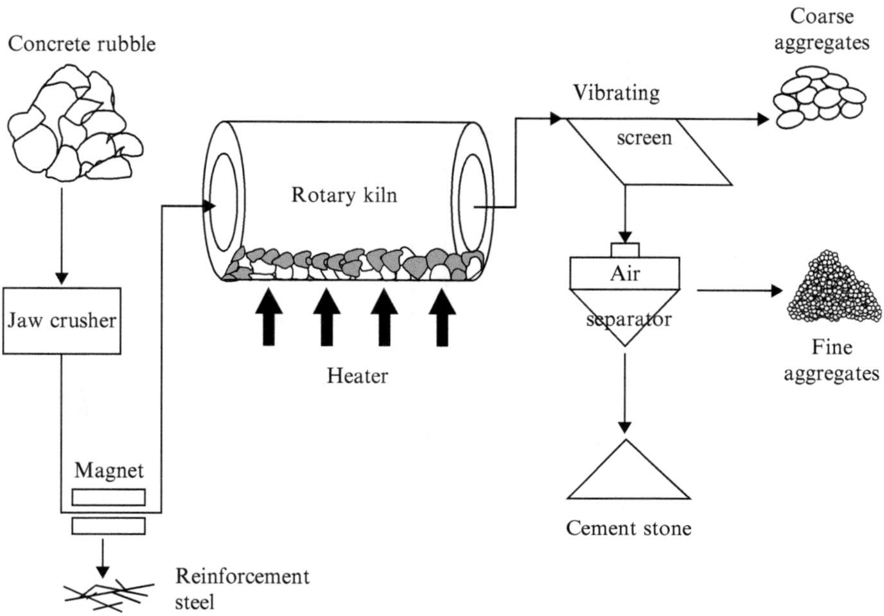

Fig. 5.4 Thermal/mechanical treatment of C&DW [9]

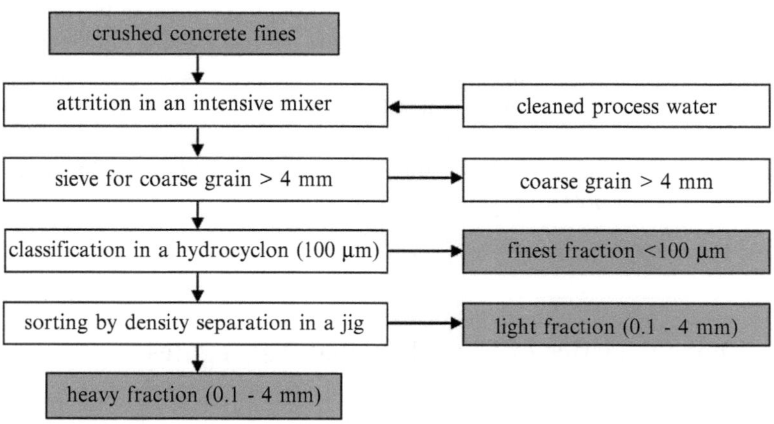

Fig. 5.5 Flow scheme of the wet treatment process for crushed concrete fines [10]

Yamasaki et al. developed a technique for removal of the mortar adhered to fine aggregate from waste concrete comprising direct carbonation with high pressure CO_2 and ball mill crushing (Fig. 5.4). In addition to the recycling of fine aggregate this process can also be seen as a new measure for CO_2 sequestration [12] (Fig. 5.6).

Another approach for removal of the cement paste from the aggregates relies on a proactive methodology that relies on applying new procedures in the design phase

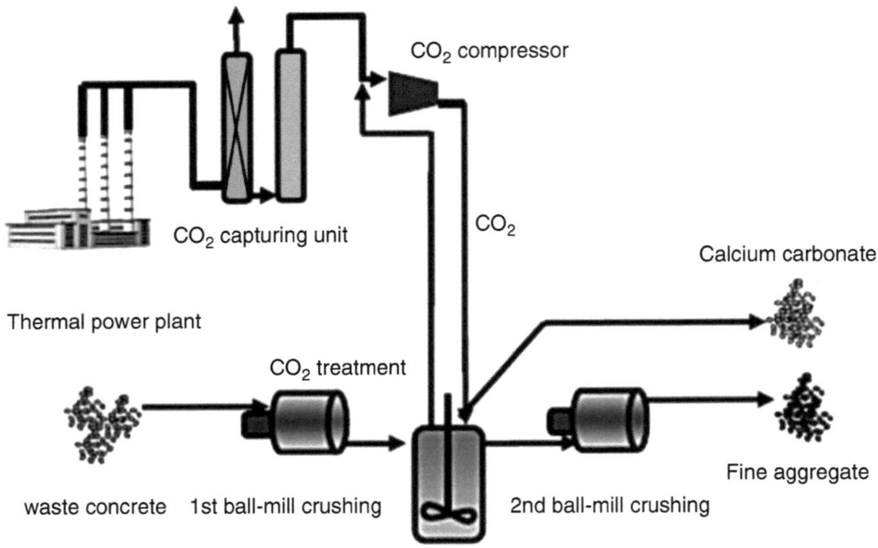

Fig. 5.6 Recycling of fine aggregate using carbonation with high pressure CO_2 [12]

of construction works that will lead to a closed lifecycle of the used materials. Noguchi and Tamura [13] proposed that reducing the adhesive strength of original aggregates to cement paste, which is achieved through chemical or physical coating of the virgin aggregates, will facilitate aggregate recovery and lead to a completely recyclable concrete. Since this procedure for aggregate recycling could conflict with the mechanical behaviour of concrete Noguchi et al. [14] proposed as alternative the modification of coarse aggregate surface using silica fume and a dielectric material that allow increase of concrete strength up to 20% and after microwave heating recovery of 90% of high quality aggregate.

5.3 Characteristics of Fine Recycled Aggregates

The variability of the characteristics of the fine fraction of construction and demolition waste has a great influence on the possible applications of this residue. According to Van der Wegen and Haverkort [15] this fraction collected in different processing phases, after sieving or after crushing, shows diverse characteristics that will influence the performance of the building materials where they are applied. This chapter is devoted to the fine recycling aggregates resulting from CDW crushing, excluding the fine particles from primary screening. A brief review of the main properties for this fraction is presented hereafter.

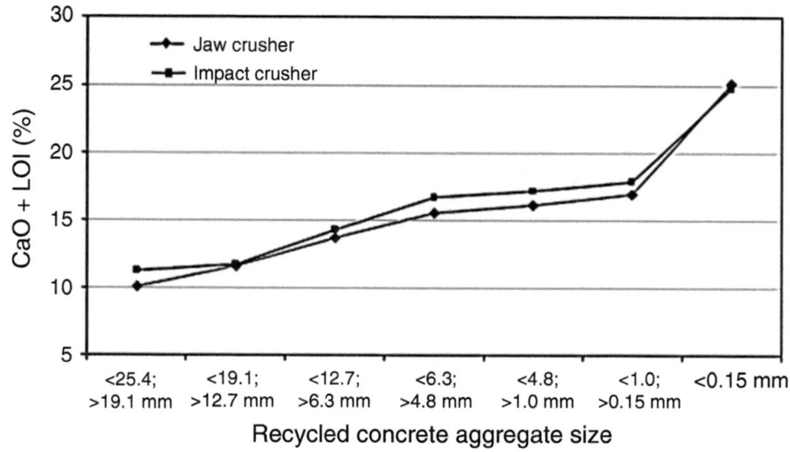

Fig. 5.7 Calcium oxide content plus loss on ignition for different size fractions of recycled concrete aggregate

5.3.1 Content of Cement Paste and Fines

Cement paste content adhered to the surface of recycled aggregates plays a key role on prescribing their field of application. Cement mortar will affect the quality of recycled aggregates by lowering density, increasing water absorption and sulphate content [16].

The content of old cement paste can be as high as 45–65% in the 0–0.3 mm fraction of the fine recycled aggregate [2]. Engelsen et al. [17] reported a cement paste content around 28% in fine recycled concrete aggregates, 0–4 mm, prepared in laboratory. Angulo et al. [18] found an average of 17.7% of cement paste in the fine fraction of mixed CDW and 38% in the power sized below 0.150 mm.

Zega et al. [19] studied the properties of fine recycled aggregates prepared from crushing of concretes produced with different types of coarse natural aggregates. The content of cement paste was lower, 28%, on the fine recycled aggregates with origin on concrete with quartzite aggregates in relation to those of concretes with granite and basalt aggregates source, 39 and 36% respectively, and this fact affect the properties like water absorption and sodium sulphate attack.

Tseng [20] also verified the increase of cement paste content with the decrease of particle size of recycled concrete aggregates, processed with jaw and impact crushers, by relating the calcium oxide content plus the loss on ignition to the binder percentage, in the absence of limestone aggregates (Fig. 5.7).

Reduction of cement paste attached to the surface of fine recycled aggregates could be achieved through processing. Mulder [9] reported a cement paste content of 2% on the sand grains from concrete recycled by a thermal process. Using mining engineer techniques, Ulsen [21] achieved high quality fine concrete aggregates with contents of cement paste lower than 10%.

According to Hansen and Narud [22] the material finer than 75 μm ranges from 0.8 to 3.5% in fine concrete aggregates below 4 mm. Recycled fine aggregates obtained from crushed waste concretes, corresponding to the fraction 0–4.75 mm, presented a content of 4.3% of material finer than 75 μm [23]. This percentage result lower than the limit indicated for natural fine aggregates on the Reglamento Argentino de Estructuras de Hormigón (CIRSOC 201), which is of 5.0%.

5.3.2 Density and Water Absorption

The density of recycled aggregates is lower in relation to fine natural aggregates and the water absorption is higher for recycled versus natural aggregates. These properties reflect not only the composition but also the processing operations during CDW recycling.

Evangelista and Brito [24] reported densities of 1.94 and 2.55 g/cm^3 for fine recycled concrete aggregates and fine natural aggregate respectively. Zega and Di Maio [23] also presented values for density of fine recycled aggregate, 2.48 g/cm^3, which is inferior of that of natural fine aggregate, 2.59 g/cm^3. Densities of 2.05, 2.34 and 2.65 g/cm^3 were found for fine crushed bricks, fine recycled concrete and natural sand [25]. Data regarding density of fine recycled aggregate depend on their source and are in general in the range of 2.0–2.5 g/cm^3 [26].

Water absorption ranging from 8.3 to 12.1% for fine recycled aggregates were presented by Hansen [2]. According to Solyman [26] this range is more extended, it goes from 3.8 to 11.5% in 10 min water absorption test. Fung [27] quoting De Pauw et al. reported that for recycled CDW with grain sizes between 2 and 4 mm water absorption in 30 min amount over 90% of water absorption in 24 h, 10.12 and 10.95% respectively.

Water absorption has significant implication on water/cement ratios for concrete mixes so it is very important to have accurate results. Evaluation of water absorption of recycled fine aggregates is usually difficult to perform. According to Hansen [28] this difficulty is related to the amount of cement paste adhered to the surface of fine CDW. Solyman [26] pointed out some conditions, like the exclusion of particles below 125 μm or achieving saturated particle surface dry, that lead to inaccurate measurements of water absorption and proposed an equation to predict water absorption (WA) from the water demand (MWD) of a mortar with recycled fine aggregates that have the same spread of a standard mortar prepared according to EN 196–3:

$$\mathrm{WA} = 0.059\,\mathrm{MWD} - 12.4 \quad R^2 = 0.90$$

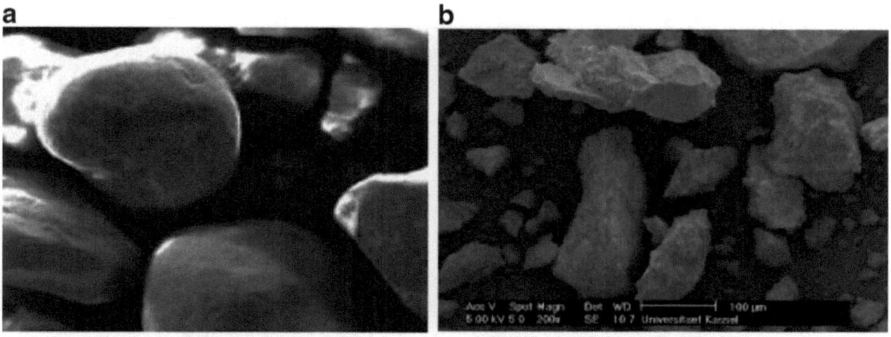

NS (photo width = 0.145 mm) RC 1 (photo width = 0.145 mm)

Fig. 5.8 Texture of (**a**) natural sand (*NS*) and (**b**) recycled sand (*RC*) [26]

5.3.3 Shape

Aggregate shape has an important effect in workability and water requirement of concrete. According to Rashwan and Abourizk [29] the crushing operation of the mineral fraction of CDW results in recycled aggregates having an angular shape that impair the workability when they are used owing to the higher internal friction [29]. For crushed fine recycled aggregates Hansen [2] reported that they are coarser and more angular than it is advantageous for having good concrete mixes.

Using a computer-aided particle analysis system Solyman [26] studied the length-width ratio (L/W) and the spherically (S) values of recycled sands and noticed that the former is similar to that of natural aggregates however spherically values of recycled fine aggregates diverge from those of natural aggregates and have a good correlation with water absorption up to 8%. Concerning surface texture, assessed by scanning electron microscopy (SEM), the smooth surface of natural aggregates no longer exists in recycled fine aggregates, Fig. 5.8. The texture of the recycled sands could be related to the loss of workability of concrete mixes incorporating these aggregates

By use of a vertical shaft impactor, VSI, tertiary crusher, Ulsen [21] increased the roundness of high quality fine concrete aggregates. The particles shape were evaluated by dynamic image analysis on QICPIC real time particle analyzer to appraise the influence of the attrition and abrasion crushing on particles sphericity as well as to compare the effect of the rotor speed as depicted in Fig. 5.9.

5.3.4 Deleterious Substances

Deleterious substances represent major drawbacks for the reuse of the fine fraction of CDW, in particular the finest particles, below 75 μm, which have a higher level of contaminants. The level of contamination is more or less significant depending on

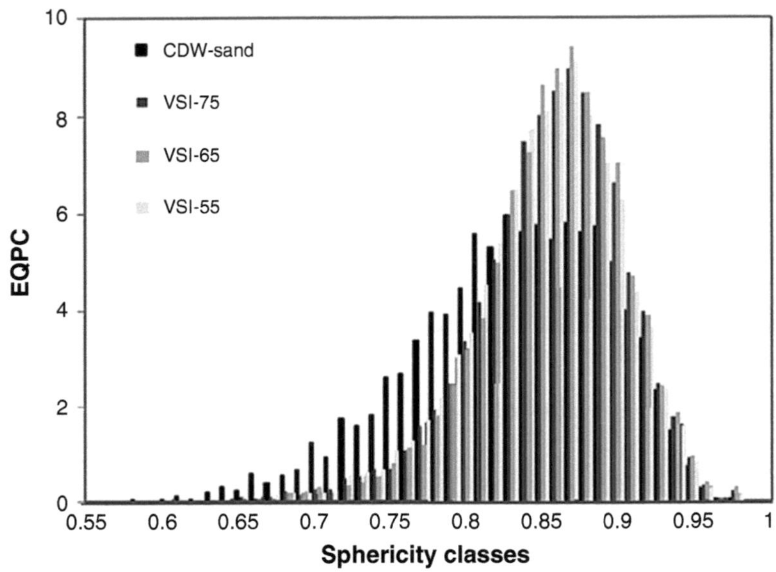

Fig. 5.9 Sphericity of fine concrete recycled aggregate [21]

the delivery requirements of CDW at processing plants as well as on the operations that it will be subjected during processing.

As it is well known contaminants like gypsum or clay materials adversely affect the strength properties and durability of concrete so it is essential to reduce their contents. Owing to the lower density of these contaminants it is possible to minimize its content using separation techniques based on density. According to Montero et al. [30] for mixed construction and demolition waste gypsum is concentrated on densities between 1.59 and 2.28 g/cm^3 so removal of material within this range will improve the contamination level.

5.4 Technical Specifications for Fine Recycled Aggregates

Gonçalves [31] analysed the specifications for recycled aggregates to be used on concrete and summarized some restrictions on their use. In Table 5.1 an updated overview of the limits regarding the use of the fine fraction of construction and demolition waste for that application in several countries is presented. As indicated a few number of countries permit that fine recycled aggregates can be used in concrete, namely Brazil, Denmark, Russia and Switzerland, and only two countries consent 100% substitution of fine natural aggregates for non-structural concrete, with a maximum allowed strength of 15 MPa. As detailed in 0 the research regarding application of fine recycled aggregates in concrete foresee a better implementation Detailed information can be found on the cited specifications.

Table 5.1 Overview of the use of fine recycled aggregates

Specification	Classes	Fine aggregate replacement level	Application clauses	Maximum allowed strength
Belgium [33]	GBSB-I (RMA)	N.d.	Comply with specifications for natural fine aggregates and in non aggressive exposure classes	C16/20
	GGBS-II (RCA)			C30/37
Brazil [34]	RCA RMA	100%	Non-structural concrete. Recycled aggregates must be pre-saturated	15 MPa
Denmark	GP1 with or without testing (RCA)	20%	In non aggressive exposure classes	40 MPa
	GP2(RMA)			20 MPa
Germany [35]	1 (RCA) 2 (RCA)	0%	–	–
	3 (RMA) 4 (RMA)	0%	–	–
Hong-Kong [36]	RCA	0%`	–	–
Netherlands [37]	RCA RMA	20%	Combined with coarse natural aggregates in non aggressive exposure classes.	C40/50 C20/25
Portugal [38]	RCA1 RCA2 RMA	0%	For concrete production	–
Russia	RMA	100% 50%	Not allowed in prestressed concrete	15 MPa 20 MPa
Spain [39]	RCA	0%	–	–
Switzerland [40]	RC-B (RCA)	N.d	Accept the use of fine recycled aggregate	–
	RC-M (RA)	100%	Non-reinforced concrete	N.d
United Kingdom [41]	RCA RA (RMA)	0%	–	–
RILEM [42, 43]	I (RMA) II (RCA) III (M)	N.d	Comply with specifications for natural fine aggregates	C16/20 C50/60 No limit

Exposure classes defined according to EN 206-1
RCA recycle concrete aggregates, *RMA* Recycled mixed aggregate, *M* mix of natural and recycled aggregates, *N.d.* not defined

In recent years the Japanese Standards Association has established standards for high quality recycled aggregate for concrete, H, that are produced using some of the more advanced processing technologies and for concrete with low quality recycled aggregates, L, respectively JIS A5021 and JIS A 5023 [32]. The physical requirements for fine aggregate and the limits of deleterious substances for high quality aggregates are listed in Tables 5.2 and 5.3

Table 5.2 Physical requirements for high quality fine recycle aggregate, H [32]

Items	Fine aggregate
Oven dry density (g/cm^3)	Not less than 2.5
Water absorption (%)	Not more than 3.0
Abrasion (%)	NA
Solid volume percentage for shape determination (%)	Not less than 53
Amount of material passing test sieve 75 μm (%)	Not more than 7.0
Chloride ion content (%)	Not more than 0.4

Table 5.3 Limits of deleterious substances for high quality recycled aggregate, H [32]

Category	Deleterious substances	Limits (mass %)
A	Tile, brick, ceramics, asphalt	2.0
B	Glass	0.5
C	Plaster	0.1
D	Inorganic substances other than plaster	0.5
E	Plastics	0.5
F	Wood, paper, asphalt	0.1
Total		3.0

5.5 Fine Recycled Concrete Aggregates and Fine Recycled Mixed Aggregates

The application of CDW aggregates in the construction and building sector is largely related with the technology used in its processing. A general approach point to the use of recycled aggregate in concrete, in mortars and in roads but taking in account their specific properties some limits for incorporation have been indicated.

In some studies the lack of information about the type of fine recycled aggregates and their condition – dried, pre-wetted, pre-saturated – at the moment of mixing could influence the drawn conclusions when intercomparisons were performed,

5.5.1 Concrete

Studies regarding the combined used of fine recycled concrete aggregates and coarse recycled concrete aggregates in the manufacture of concrete evidence that the fine fraction, in particular the one below 2 mm, has a negative effect on mechanical properties and also on durability of concrete and so its use have been restricted or even prohibited for this purpose [2, 44, 45],

For concrete produced with coarse, fine or a mixture of coarse and fine recycled concrete aggregates. Kenai et al. [46] reported that for constant slump an increase of mixing water is needed, resulting on a decrease of mechanical performance and superior shrinkage in relation to a reference concrete. For 100% replacement of natural sand by recycled aggregate the compressive strength decrease 30% and shrinkage increase 70%. In order to overcome durability problems a maximum limit of 50% of fine recycled aggregates was indicated.

According to Corinaldesi and Moriconi [47] the lower mechanical performance of concrete incorporating mixed recycled aggregates, coarse and fine fraction, can be surpassed by the use of mineral additions. In this research, fly ash or silica fume replaced 15 or 30% by weight of cement and natural aggregates are totally replaced by recycled aggregates, with fine fraction reaching 26% of the whole aggregate. The mixing water is added in two steps, "two stage mixing procedure", allowing the development an initial layer of cement slurry on the surface of recycled aggregates that strengthen the interfacial transition zone [48]. The mechanical performance of concrete with recycled aggregate is similar to the reference concrete, with natural aggregates and without mineral additions. By decreasing the w/c ratio and maintaining the workability with a superplasticizer it is possible to achieve compressive strengths similar or better than those of the conventional concrete.

The difficulties associated with the combined use of fine and coarse fraction of CDW are partially overcome when only the fine fraction is replaced by recycled concrete aggregates. Dhir et al. [49] found that substitution of 20% of sand by fine recycled concrete aggregates on concrete production does not influence compressive strength. Regarding carbonation depth, freeze-thaw resistance and expansion due to sulphate exposure the achieved results do not compromise the intended use.

Masood et al. [50] produced concrete with partial replacement, 10, 20 and 30%, of the fine aggregate fraction of mixed CDW (lime concrete and lime mortar). Concretes with 30% substitution of the sand have a decrease around 26% on the 28 day compressive strength and 20% on modulus of elasticity. Flexural strength is also lower in relation to a conventional concrete but, according to the authors, the results allow the use of fine recycled aggregates in concrete in a defined range for replacing natural aggregates. Partial replacement of cement by demolished waste, up to 30%, evidence larger decreases on compressive strengths and modulus of elasticity, mainly for the higher level.

Evangelista and Brito [51] evaluated the replacement of fine recycled concrete aggregates in the production of concrete and verify that it is viable to replace up to 30% of the fine aggregates without significant differences on shrinkage and on water absorption by capillarity in relation to a reference concrete produced with natural aggregates with the same grading curve. Concerning compressive strength there were only small differences even for total replacement of the fine fraction. The lack of a significant effect of concrete crushed fines on compressive strength was previously stated by Hansen [2]. Data regarding splitting tensile strength and modulus of elasticity show a reduction of 5.2 and 3.7% respectively for a replacement level of 30%, but with entire substitution of fine aggregates the decrease reaches 30.5 and 18.5% [52]. Regarding durability further studies, including assessment of carbonation resistance and non-steady-state chloride migration coefficient, highlight the feasibility of producing concrete with 30% fine recycled concrete aggregate [24].

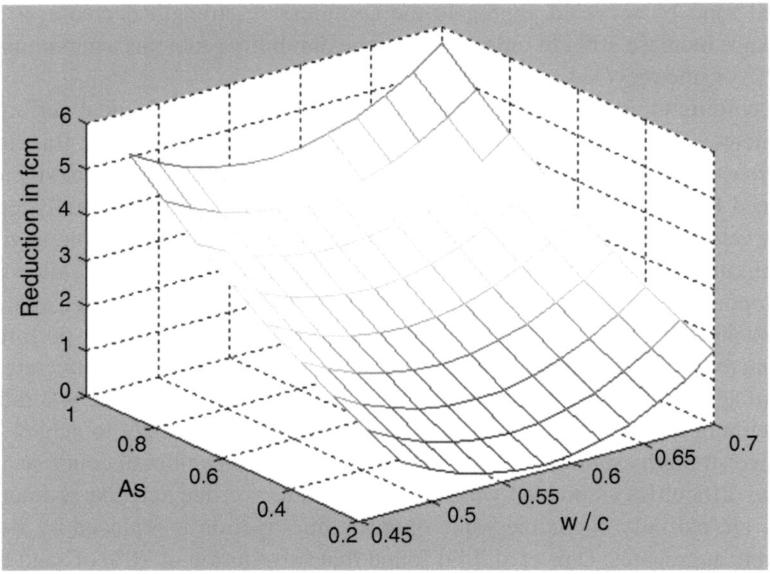

Fig. 5.10 Reduction on compressive strength versus sand content and w/c ratio [26]

Similar results on compressive strength were obtained when crushed concrete was used as fine recycled aggregates in concrete production: a decrease in compressive strength reaching 30% occurs when an integral substitution of fine aggregates is performed [25]. An increase on shrinkage and expansion was observed upon substitution of natural aggregates by crushed concrete fines.

According to Solyman [26] when recycled sand, mainly composed by crushed concrete and brick, replace natural fine aggregates on concrete production the reduction on compressive strength can be predicted, for w/c ratios between 0.48 and 0.66, by the equation:

$$R = 97.5(w/c)^2 - 112.5(w/c) + 5.3A_s - 2.6{A_b}^2 + 6.7A_b + 31.3$$

where: w/c – water to cement ratio (0.48–0.66)
A_s – recycled concrete volume/total sand volume
A_b – recycled brick volume/total sand volume

Figure 5.10 depicts the reduction on compressive strength when only crushed concrete, A_s, was used, with water to cement ratio ranging from 0.48 to 0.66.

This research also highlighted the possibility of producing concretes up to C35/45 when recycled concrete fines were used to replace natural sand, with a considerable decrease on consistency using pre-wetted recycled aggregates. Regarding the splitting-tensile strength, fsp, and the dynamic modulus of elasticity, E_{dyn}, the following relations with compressive strength, fcm, apply:

$$fsp = 0.3\ fcm^{2/3} \text{and}$$

$$E_{dyn} = 5400\ fcm^{0.551}$$

Concrete durability in the presence of recycled fine aggregates, addressed through the carbonation depth after 1 year and the frost de-icing salt resistance, was not noticeable when compared to reference concrete.

Concretes made with fine recycled concrete aggregate, in percentage higher than 25%, replacing the natural fine aggregate, showed a significant decrease of the slump, despite the increase in the dosage of plasticizer additive [23]. The compressive strength of recycled concretes made with 25, 50 and 75%, in volume, of fine recycled aggregates, obtained from crushed waste concretes with unknown characteristics, determined at the age of 7, 28 and 70 days, was similar or higher than that of concrete made totally with natural aggregates and the static modulus of elasticity of recycled concretes was similar to that of reference concrete.

To assess the effect of fine recycled aggregates on reinforced concrete Reis et al. performed pull-out tests on concrete, with 10 and 16 mm bars, incorporating 25 and 50% of fine recycled concrete aggregates and. The bond strength decrease up to 30% in relation to a reference concrete, being this effect more noticeable when using higher diameter bars [53].

Durability studies concerning concrete with partial replacement of fine aggregates, up to 30%, by crushed concrete fines was performed by Zega and Di Maio [54]. The effective water to cement ratio of recycled concretes is lower in relation to the control concrete and this condition justify the similar compressive strengths, carbonation depths and water penetration under pressure achieved. Compressive strength at 28 day was 43.6 MPa for control concrete and 41.4 MPa for recycled concrete with 30% of fine recycled aggregate. Analogous drying shrinkage results, at 180 days, were attained for control concrete and recycled concrete with 20% of fine crushed concrete and a slightly lower value is observed with 30% of fine recycled aggregates. From the quantification of capillary absorption by sorptivity it was concluded that recycled concretes comply with the Argentinean Regulation requisites, i.e., below 4 $g/m^2/s^{1/2}$.

The use of the fine fraction of concrete waste can also be beneficial in the production of self-compacted concrete (SCC). Corinaldesi et al. [55] found that the use as filler of the finest fraction of mixed CDW, below 90 μm, in self-compacting concrete (SCC) improve flowability as well as viscosity and segregation resistance as evidenced by slump and L-box tests. Ultra-sonic pulse velocity measurements corroborate these findings. The use of this mineral addition in combination with coarse recycled aggregates could also lead also to the development of a sustainable SCC.

Kou and Poon [56] produced SCC with 100% replacement of coarse recycled concrete aggregates and increasing level of replacement of the fine fraction and found that the mechanical properties present a maximum for substitutions between 25 and 50%. Regarding resistance to chloride penetration it increases with increase of fine recycled aggregates, probably owing to a filler effect caused by the fraction below 300 μm. Drying shrinkage also increase with the content of fine recycled aggregate but it can be controlled by the water to binder ratio.

Kim et al. [57] used recycled fines from waste concrete, with a fineness modulus of 2.8, to replace 25, 50, 75 and 100% of fine aggregate on SCC. The workability decrease with increased substitution of the aggregate and the mechanical properties, compressive and flexural strength, indicate the viability of using up to 50% of recycled fine aggregates.

Corinaldesi and Moriconi [58] evaluated the use of powder, below 150 μm, from concrete and masonry waste on SCC as an alternative to the use of a viscosity modifying agent. Rheological studies on cement pastes with rubble powder showed an increase on yield strength and plastic viscosity and a reduction on thixotropy, in relation to a reference cement paste, which evidence the relevance of this addition for application in SCC. Preparation of distinct SCC, regarding the type of addition and aggregates used, confirm the effectiveness of using the rubble powder mainly when coarse recycled aggregates replace natural aggregates. SCC mixes were proportioned to attain a volume, in the order of 190 l/m^3, of very fine particles. On the fresh state superior flowability and flow segregation resistance were achieved while on the hardened state there is no significant change on compressive strength. For SCC with total replacement of fine natural aggregates by fine recycled aggregates combined with the use of rubble powder poor 28 day compressive strengths were achieved.

5.5.2 Mortar

The use of the fine fraction of CDW in the production of mortars has been explored by several researchers. In order to improve the performance of rendering mortars made up with construction and demolition waste Miranda and Selmo [59, 60] used a mix design based in two parameters, "aggregates and plasticizing materials to cement ratio" and "total materials finer than 75 μm". The results of this research show that mortars with 25% of finer particles or more are prone to cracking and that low levels of these particles lead to good workability.

Improvement of the quality of crushed concrete fines by a wet treatment, developed within the European project RECDEMO, was analysed through studies on mortars incorporating 100% of processed concrete fines with the same particle size, 0.1–4 mm, and different densities [10]. For the heavy fraction a decrease of the cement paste stuck on the particle surface takes place and therefore the water absorption decrease, in opposition to what was observed on the light fraction. The decrease on binder content of the heavy fraction lead to an enhancement of the dynamic modulus of elasticity at 28 days for mortars prepared with this fraction as depicted on Fig. 5.11.

Further tests on concrete mixes using 20 and 50% of the same heavy fraction demonstrate improvements on compressive strength, dynamic modulus of elasticity and shrinkage, Fig. 5.12, evidencing the advantage of the wet treatment.

Studies regarding durability of mortars with fine recycled aggregates were evaluated by testing the sulphate resistance, in sodium and magnesium sulphate

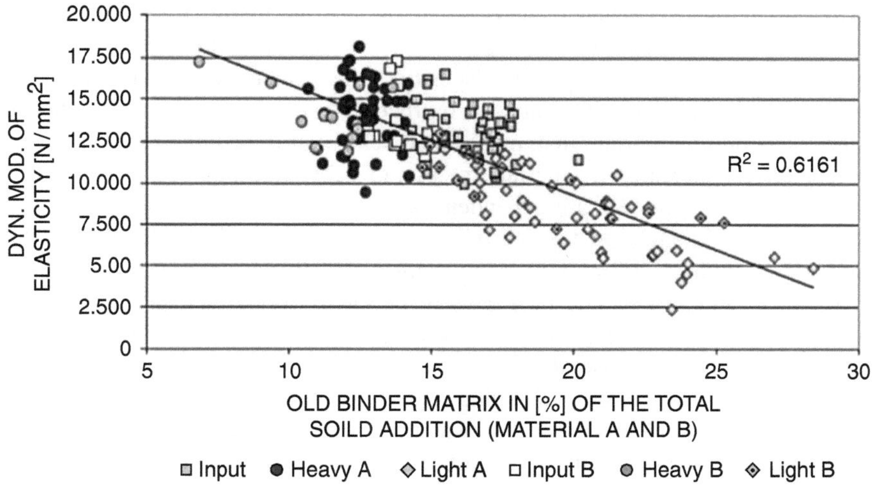

Fig. 5.11 Relation between the dynamic modulus of elasticity and the old cement paste content [10]

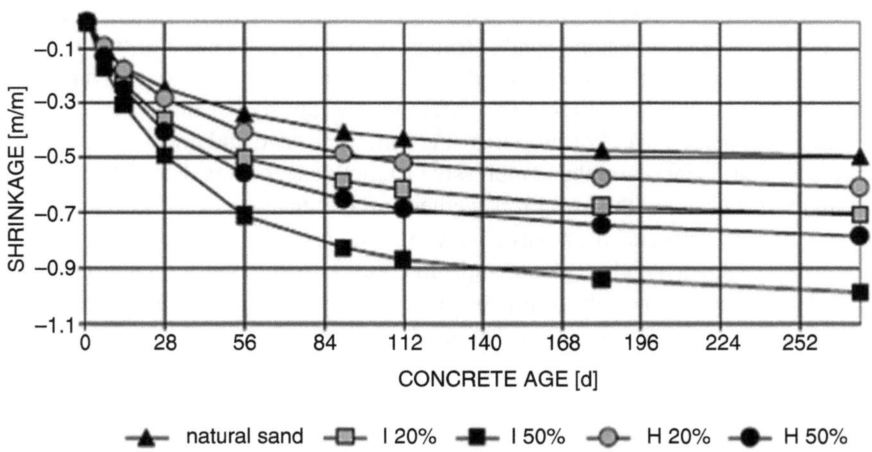

Fig. 5.12 Shrinkage of concrete samples, with 20 and 50% of untreated (I) and wet treated (H) concrete fines in comparison to concrete without concrete fines [10]

solutions, of mortars produced with fine recycled concrete aggregates up to 100%. Better sulphate resistance was achieved on mortars with 50% replacement of fines in relation to a conventional mortar; the worst situation occurs for a 100% replacement level in which the observed deterioration is associated with gypsum and thaumasite formation. The level of mortar deterioration increase for fine recycled aggregates with higher water absorption [61, 62].

Corinaldesi and Moriconi studied the use of fine recycled aggregates in bedding mortars in relation to a mortar with natural aggregate and observed that there was a decrease on the compressive and flexural strength of mortars containing fine recycled aggregate. However, the bond strength of the interface recycled mortar/brick was improved probably owing to the anti-thixotropic effect achieved by the decrease on yield stress and plastic viscosity of the cement paste in the presence of fine recycled aggregates, as evaluated by rheological tests. Because the enhanced adhesion is more important than mechanical behaviour in bedding mortars this can be a suitable application for the fine fraction of CDW [63, 64].

Mix design of mortars are dependent on the type of fine aggregate used [65]. Leite et al. [66] analysed the feasibility of incorporating 50 and 100% of fine recycled aggregates on mortars using compensations rates for water absorption between 50 and 100%. The best balance between workability and compressive strength was achieved using a mortar mix design with a water compensation ratio of 80% but restrain on the use of fine recycled aggregates on mortars could occur for compensation ratios below 70% owing to loss of consistency.

Vegas et al. [67] observed that water absorption of fine recycled concrete aggregates and also sulphur content are critical for their use on masonry mortars and replacement of natural sand by these recycled aggregates, up to 25%, is not unfavourable regarding mechanical performance, workability and shrinkage.

Dapena et al. [68] incorporated fine recycled concrete, from structural concrete with compressive strengths in the range 15–35 MPa, in cement mortars to replace siliceous sand and limestone sand. Mortars with siliceous sand and fine recycled aggregates, up to 20%, could be adequately compacted and their compressive strengths are reduced in relation to a reference mortar, by 0.7 times the aggregate replacement percentage. For mortars using limestone sand and fine recycled aggregate a similar decrease on the compressive strength also occurs. Regarding flexural strength of mortars with partial replacement of aggregates by fine recycled aggregate a more accentuated drop was observed in the presence of siliceous sand when compared to those with limestone sand. On concrete specimens with up to 100% of recycled aggregates the substitution by 10% fine recycled aggregate has no relevant effect on compressive strength and on modulus of elasticity.

5.5.3 Blocks

The grading of aggregates required for the production of concrete blocks makes this application suitable for fine recycled aggregates. Soutsos et al. [69] studied the replacement of natural aggregates by recycled concrete aggregates in precast concrete blocks and observed that the substitution of 30% of the fraction 0 to 4 mm by recycled aggregates has a detrimental effect on compressive strength but increasing the cement content will allow to keep the level strength (Fig. 5.13). Recent studies indicate that 20% replacement of natural fine aggregates by concrete or masonry recycled aggregates have no significant influence on concrete block production [70].

Poon et al. [71] used coarse and fine recycled aggregates to replace natural aggregates on concrete bricks and paving blocks and concluded that substitution up

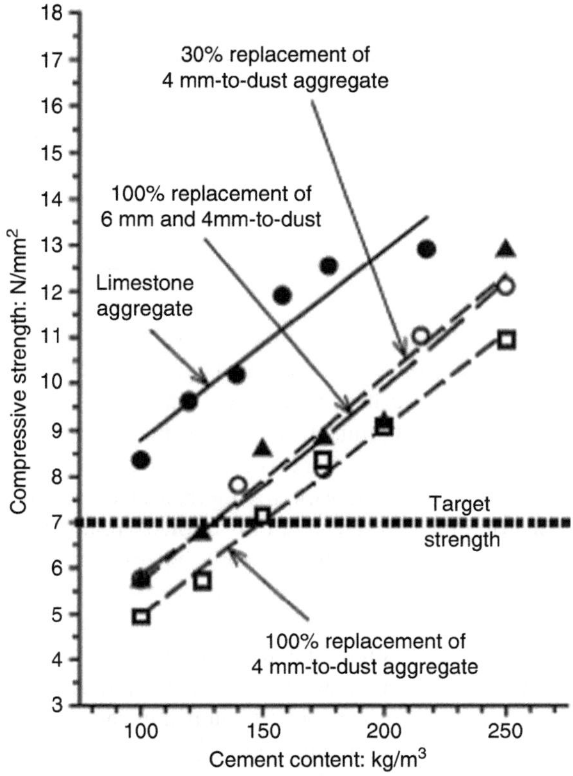

Fig. 5.13 Compressive strength against cement content for concrete blocks with replacement of the fine fraction [69]

to 50% of the coarse and fine fraction has little influence on compressive strength but for higher replacement level this strength will be reduced. Regarding the flexural strength it increases with the increase on the recycled aggregates content.

Poon and Chan [72] also studied different compositions with recycled concrete aggregate and crushed clay bricks to be used as paving blocks. The paving blocks produced only with aggregates sized below 5 mm show an increase on tensile splitting strength and on water absorption and a reduction on density and on resistance to abrasion in relation to paving blocks produced using the fine and coarse fraction of recycled concrete aggregates, with the same replacement level by crushed clay bricks. Despite these changes the paving blocks made only with fine fraction comply with the requirements of Hong-Kong for paving blocks to be used in traffic areas.

A study regarding the feasibility of using fine aggregates from construction waste on the production of paving blocks, for residential applications, was performed by Chan and Poon [73]. Comparison of the results of compressive strength tests with the strength requirements of the Hong Kong and Australian/New Zealand standards, for the above application, allowed establishing recommendations on the maximum amount of waste to be used. As expressed in Table 5.4 only the use of recycled concrete aggregates is not restricted.

Table 5.4 Maximum allowable construction waste contents in paving blocks [74]

Materials	Maximum % by weight of total aggregate
Recycled concrete aggregate	100.00
Clay brick	15.00
Ceramic tile	20.00
Recycled concrete masonry aggregate	20.00
Lightweight aggregate	0.50
Timber	0.25
Bamboo	0.25

Comparison of paving blocks performance produced with fine recycled concrete aggregates and with fine recycled aggregates contaminated with crushed clay brick, crushed tile, crushed waste glass, and wood chips aggregate allow to advise an increase on contaminant to 10% for this application [75].

Dosho [76] found that it is feasible the use of fine recycled aggregates on precast concrete products after improving their characteristics through adequate processing, wet grinding with an eccentric rotor. A relative absorption index that typify the aggregate quality, estimated from the absorption and absolute volume of recycled aggregate, must be used in concrete design with these aggregates. The precast concrete products with fine recycled aggregates using steam curing meet the requirements of strength and durability.

Recycled aggregates from concrete precast elements can be incorporated in prefabricated units for non-structural applications. Substitution of crushed aggregates 0/4 by recycled aggregates up to 50%, with a similar size distribution, resulted in concrete with comparable properties [77].

5.5.4 *Roads*

The large quantities of aggregates used on road construction along with the increase of CDW led to the sustainable practice of incorporating these residues as substitutes of natural aggregates. Reports on recycling construction and demolition wastes in road layers usually cover the use of the fine and the coarse fraction.

Huang et al. studied the fine fraction of construction and demolition waste, coming from a mechanical sorting plant of C&DW, through sieve analysis, LA abrasion test, freezing test, organic content test, and fineness test and concluded that it is appropriate as a material for roadbeds if contamination is removed in advance or it could be used to cover daily waste streams disposed in landfills [78].

Cho and Yeo analyzed the use of coarse and fine recycled aggregates in concrete pavement and verified that the important decrease in flexural strength prevented its use in this application; however the strength requirements for lean concrete base are fulfilled using coarse and fine aggregates if the level of impurities can be controlled below 25% [79].

The use of different recycled sands, up to 21% of total aggregate mass, in hot bituminous mixtures evidenced the feasibility of their use in base and binder

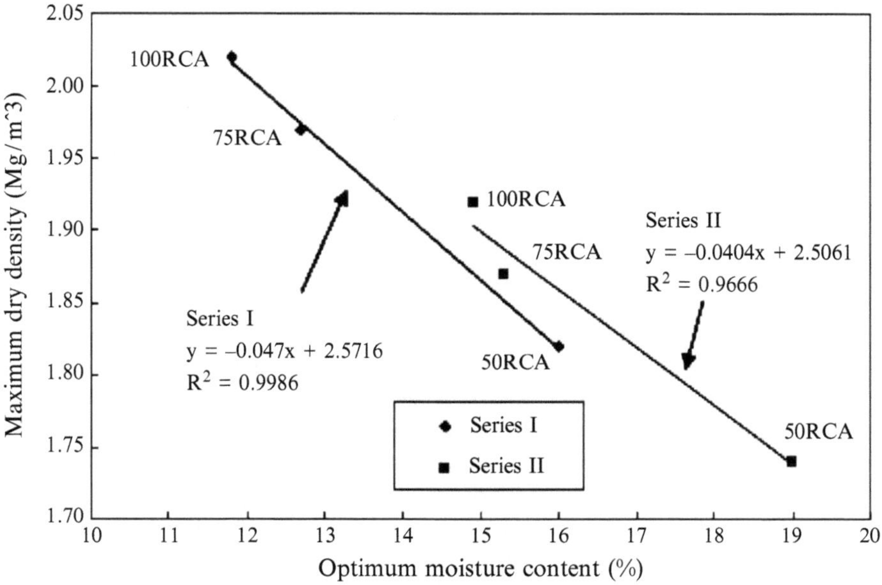

Fig. 5.14 Trend in maximum dry density and optimum moisture content: Serie I with fine recycled concrete aggregate; Serie II with fine crushed clay bricks [80]

courses [26]. Reduction on density, increase on mix stability and decrease of voids on mineral aggregate were observed. The optimum binder content is directly proportional to the water absorption of recycled aggregates.

Poon and Chan [80] researched the use of recycled concrete and crushed clay bricks as coarse fraction mixed with fine recycled concrete aggregate or fine crushed clay bricks as unbound sub-base materials for use in road construction. Blends with fine recycled concrete were more susceptible to moisture changes and exhibit higher maximum dry densities, lower optimum moisture content and higher CBR values toward those with fine crushed clay bricks, Fig. 5.14. The recycled sub-bases fulfil the minimum strength requirements of Hong-Kong.

It has been observed that the combined use of coarse and fine recycled concrete aggregates as unbound sub-base materials is sometimes coupled with an increase of strength over the time. This self-cementing capability, noticed by Arm [81] after triaxial tests in laboratory and FWD tests in the field with crushed concrete in unbound layers, was analysed by Poon et al. [73]. These authors studied the size fractions <0.15, 0.15–0.30, 0.3–0.6, 0.6–1.18 mm and <5 mm of fine recycled concrete aggregates and ascribed the self-cementing characteristic primarily to the fractions <0.15 and 0.3–0.6 mm. The withdrawn conclusion was based on evidence, from XRD experiments, that the particles sized below 0.15 mm have a higher C_2S content, and from that the fraction 0.3–0.6 mm in contact with water have a higher pH indicating a higher content of amorphous hydration products possibly supplying lime for supplementary reaction. The dominating factors associated to self-cementing properties are the age, grade and mix proportions of the original concrete as well as compaction and grading of the sub-base materials.

The recycled fine powder from recycling concrete aggregates was used by Chen et al. [82] as filler in asphalt mixtures to replace limestone powder. Testing results evidenced a better water sensitivity and resistance to fatigue in relation to the control mix. The asphalt mixture with recycled powder complies with China specification for pavement construction even with a lower temperature performance.

5.5.5 Other Applications

In a laboratory study Plaza C. et al. [83] verified that the recycled concrete, with particle size less than 2.5 cm, can be used as a cover layer in CDW landfill to reduce the emission of hydrogen sulphide, H_2S, generated from biological reduction of sulphate from gypsum drywalls. The authors suggest that the sorption of H_2S into the fine fraction followed by conversion to sulphide minerals could be the mechanism underlying the decrease of emissions, higher than 99%.

Shui et al. [84] assessed the use of fine recycled concrete aggregates after thermal treatment and grinding below 75 μm as cementitious constituent of mortars. After dehydration, on heating up to 500°C, the fine recycled aggregates and the hardened cement paste were used to produce mortars and the rehydrated phases were investigated. The new phases, calcium silicate hydrate (C-S-H gel), ettringite and portlandite (CH), present a looser microstructure in relation to those of the original hardened cement paste that is partially responsible for the low values of mortar compressive strength. Enhanced mechanical response could be achieved by fly ash and Portland cement addition.

An alternative use of the fine fraction of CDW is based on hydrothermal curing. Hlawatsch et al. [85] developed their research on construction products obtained by autoclave hardening fine aggregates under elevated temperature on an atmosphere of saturated water vapour. The main material under investigation were crushed concrete fines but other fine recycled aggregates gained from crushed clay bricks, hard fired clay bricks and calcium silicate bricks were also studied. Addition of lime or silica was used as a process to increase the fines reactivity.

On Fig. 5.15 the compressive strength resistance for some of the products after autoclave hardening at 200°C during 12 h is presented. The strength values reached by autoclaved hardened pure concrete fines are attractive for producing construction products such as masonry units. For a finer grain size distribution, i.e. 40% of grains lower than 0.125 mm at a maximum grain size of 4 mm, the observed compressive strength increase up to 18.2 MPa. On carbonated concrete crushed fines the hardening process does not take place owing to the unavailability of calcium oxide; however upon addition of white lime the hydrothermal treatment was successful. Assessment of the pore structure by mercury intrusion porosimetry evidenced that augmenting the duration of the hardening period using concrete fines increase the total porosity and coarser pores with a decrease on compressive strength.

The chemical composition of concrete fines includes siliceous fines and cement hydration products. Under hydrothermal conditions these compounds become

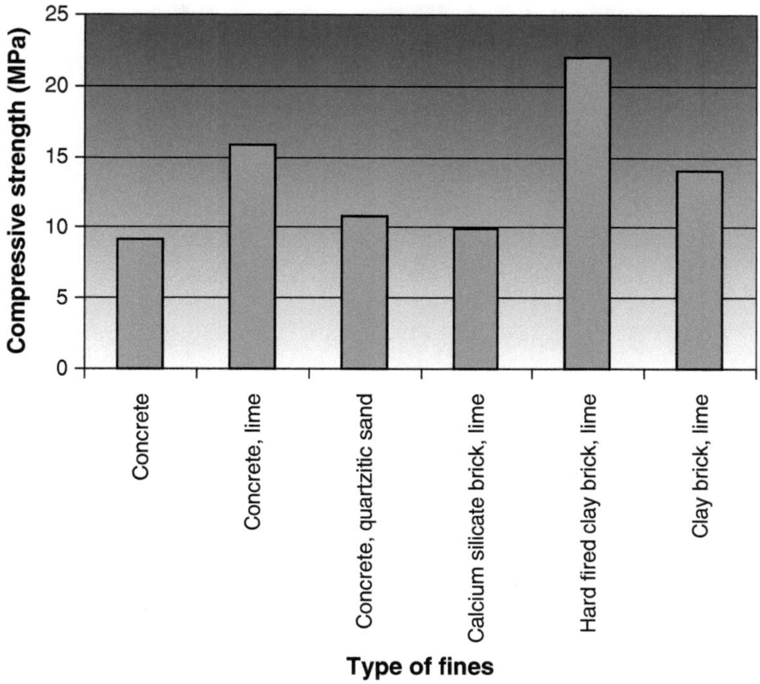

Fig. 5.15 Strength characteristics of fine fractions of CDW with or without additions after autoclave hardening at 200°C for 12 h [85]

reactive and microscopical observation on thin sections evidenced that the surface zones of quartzitic grains exhibited reaction rims where the quartz entered into reactions with CaO and water to form calcium silicate hydrates, as illustrated in Fig. 5.16. The new calcium silicate hydrate phases (CSH) were formed with different CaO/SiO_2 ratios, which depend on the treatment duration, explaining the formation of the crystalline phases like tobermorite and gyrolite, observed in X-ray diffraction analysis.

The autoclave hardening of crushed concrete fines was also used by Al-Otaibi et al. [86] for production of silica-lime bricks with promising results. Reference bricks, with 100% of crushed concrete fines, or mixtures of this residue with different supplementary cementitious materials and water, allowing enough consistency, were moulded and cured at 190–200°C under an adequate pressure program. Bricks using 20 and 40% of ground granulated blastfurnace slag (GGBS) or fly ash are within the water absorption limit of the standard specification for calcium silicate brick ASTM C73 [87]. Regarding compressive strength it was principally enhanced when GGBS up to 40% replace crushed concrete fines.

Achtemichuk et al. [88] found that it is feasible to produce controlled low strength materials (CSLM) using fine recycled concrete aggregates combined with slag or high-calcium fly ash instead of Portland cement. The increase of strength is related to hydraulic and pozzolanic reactions of the industrial by-products, being the last ones promoted by the alkalis present on the cement paste adhered to recycled aggregates surface. Some applications like permanent structural fill and road bases not

Fig. 5.16 Inner and outer reaction rims around a quartzitic grain with different CaO/SiO_2 ratios [85]

requiring short hardening time and bedding for conduits with small spacing were envisaged for CLSM. If the coarse fraction on recycled concrete aggregates is also used the field of application become enlarged.

5.6 Recycled Ceramics

According to Oikonomou [89] ceramic waste represents approximately 30% of demolition wastes. For Juan et al. [90] the ceramic materials – bricks, wall tiles, sanitary ware, etc. – constitute the major fraction of construction and demolition waste, around 54%. Even with this discrepancy the fact is that ceramic wastes correspond to an important part of CDW.

Research on the application of fine ceramic wastes relates to waste generated by the construction sector and by the ceramic industry so that the literature presented hereinafter cover wastes from these different sources.

The high proportion of clay minerals on ceramic wastes makes them a useful source of addition for the production of blended cements. Firing, with adequate temperatures, during fabrication of ceramic products could activate some clay minerals that become pozzolanic materials [91]. The use of glazed ceramic tiles for the production of blended cement was analyzed by Ay and Ünal [92]. After grounding the tiles they were mixed with cement in percentages between 25 and 40%. The ground ceramic waste conforms to Turkish standard regarding pozzolanic properties. Results from bending and compressive strength, at 7 and 28 days, indicate the viability of producing cement with grounded waste tiles up to 35% addition.

Turanli et al. [93] researched the influence of 10, 20 and 30% of crushed clay brick, used as partial cement substitute, on alkali-silica reaction (ASR) of mortars using a reactive fine aggregate. Reference mortar bars, tested according to ASTM C1260 [94], evidenced a severe cracking pattern in comparison to mortars containing crushed bricks. The expansion curves prove that the decline of ASR increase with the increase of cement replacement, achieving reductions around 70%, after 30 days, for the maximum incorporation of crushed clay bricks.

Lin et al. [95] used waste brick, with particle size lower than 75 μm, as cement replacement in cement pastes and verified that pozzolanic reactions begin to develop after 28 days curing with a decrease on portlandite content and a densification of the cementitious matrix.

Red roof tiles waste, non glazed, natural glazed and black glazed, were powdered below 44 μm and were analyzed to evaluate their feasibility of being used as additions in blended cements, on percentages between 20 and 40% [96]. Results from the Fratini pozzolanic test were positive and they were corroborated by reduction on the Ca-OH peak, related to portlandite content, on infrared spectra acquired by FTIR spectroscopy. Compressive strength results point to 20–30% replacement of cement by these wastes without compromising the mechanical resistance.

Waste bricks, from brick manufacture, were used to replace 5–20% of clinker [97] on cement to be used in mortars. The increase of brick waste reduced the cement grinding time up to 10%, which represents a decrease on energy consumption. Cement pastes show a decrease on setting time with increase of waste content that is related to the higher water absorption of the residue. Compressive and flexural tests on mortars indicate that similar resistances were achieved at 90 d for a substitution of clinker by 10% of ceramic bricks suggesting that this is the adequate level of replacement.

Replacement of ceramic wastes, fired at different temperatures, as fine recycled aggregate on mortar, substituting 20% of sand, resulted in slight differences on flexural and compressive resistances [98]. The total porosity does not undergo great changes and in the most unfavourable situation, ceramic fired at higher temperature, the difference on porosity does not reflects on the apparent density neither on the average pore size of mortar. Data from a ceramic factory of concrete tiles in which ceramic wastes replaced 5 and 10% of the siliceous sand show a maximum loss of 5% on transversal flexion strength for the high level of sand replacement. Freeze and thaw and impermeability tests results comply with the requirements of the European standard. EN 490 – Concrete roofing tiles and fittings for roof covering and wall cladding – Product specifications.

Khatib [25] researched the use of crushed bricks as fine recycled aggregates for concrete. Using crushed bricks no compressive strength reduction was observed for a 50% level of replacement and for a 100% level of replacement a reduction of only 10% occurs. Higher shrinkage and expansion were recorded for concrete with fine crushed brick.

Poon and Chan [99] studied the use of fine crushed bricks and tiles in concrete and verified that for 20% replacement of fine aggregates there is a decrease in mechanical strength, in relation to a reference concrete, that could be partially overcome using a double mixing procedure, i.e., mixing recycled aggregates, addition of half of the water, mixing step, addition of cement, mixing step, addition of the rest of the water and another mixing step. Non structural concrete applications, such as road curbs and road barriers are suggested for the application of concrete with fine crushed bricks and tiles aggregates.

Debieb and Kenai [100] produced concrete mixes using different replacement levels of coarse and of fine recycled brick aggregates and, from the decrease on the performance of concrete, set up limits for the use of the coarse fraction and of the fine fraction, respectively 25 and 50%, and restrict the application of concrete to pavement blocks and other manufactured elements.

Precast concrete paving blocks, produced at laboratory using vibration and compaction to simulate industrial processing, were used to investigate the feasibility of total replacement of natural aggregates by simultaneous use of recycled ceramic wastes sized 0–4 and 4–8 mm, with or without superplasticizer [101]. Water absorption, abrasion resistance and tensile splitting strength tests allowed to verify the conformity of the fabricated blocks with the requirements for paving blocks of EN 1338 [33]. The water absorption was recognized as a critical characteristic but the use of chemical admixture allowed surpassing this issue.

Durability studies of mortars with crushed clay bricks partially replacing the aggregates, evidenced a decrease on flowability with the increase on bricks content, no significant effect regarding compressive strength, up to 20% replacement, a decrease on shrinkage, in a narrow range, with augmenting crushed clay bricks from 10 to 20% and a better freeze-thaw resistance for higher replacement level [102]. Alkali silica reaction was positive with these recycled aggregates when the mortar bar test was performed [87, 94] and the worst result was assigned to a crushed brick content of 30%, nevertheless there is no similar report on the literature on this subject so it should be further investigated.

The use of recycled ceramic aggregates, bricks and tiles, on plastering mortar was studied by Silva et al. [103]. The authors reported that 50% substitution of sand by brick waste resulted in mortars performing as well as a reference mortar, exception made to dimensional stability. For the modified mortar with 20% brick waste the main advantages rely on flexural and compressive strength and water permeability due to capillary action in relation to the reference mortar.

Pacheco-Torgal and Jalali [104] studied the use of different ceramic wastes in concrete as cement and as aggregate substitutes. In the first case concretes with 20% of cement replaced by ceramic powder, sized less than 75 μm, evidenced a slight decrease on compressive strength, no relevant effect on oxygen permeability and an improvement on water permeability and on resistance to chloride penetration in relation to a reference concrete. Concrete mixes with ceramic wastes taking the place of sand present a gain on compressive strength along with a superior durability performance.

Puertas et al. [105] found that the use of white and red ceramic waste as raw materials for Portland cement clinker increase reactivity and burnability in relation to conventional raw mixes when particle size is lower than 90 μm, in particular for red ceramic waste.

5.7 Conclusions

Recycling of the fine fraction from construction and demolition waste is essential since the increased use of the recycled coarse fraction result in massive quantities of the former that is still disposed of in landfill.

Similarly to other materials the fine recycled aggregates characteristics, mainly the cement paste content, influence their uses in the production of materials to be applied in civil engineering works but do not inhibit it. Within a recycling perspective the environmental issues cannot be set aside. This subject is of special concern regarding the fine fraction from construction and demolition waste owing to the potential presence of contaminants in the particles sized below 75 μm.

The research concerning the fields of application of the fine fraction of CDW highlights the feasibility of replacement of a considerable mass of fine natural aggregates by fine recycled concrete or fine mixed aggregates on different segments, with special emphasis in concrete, mortars, blocks and roads.

For concrete production most of the studies deal with fine recycled concrete aggregates, in percentages up to 30%, without significant loss of mechanical properties and similar durability characteristics in comparison to a reference concrete. Improvement of concrete characteristics could be attained in the presence of fine recycled aggregates by adjusting the mixing procedure. Self-compacted concrete prepared with recycled concrete or mixed aggregates is also a valid option for fine recycled aggregates, with replacement levels around 50%, as verified through the rheological behaviour.

Mortars produced with fine recycled aggregates up to 25% could have adequate properties depending on the type of aggregates. Improved durability and enhanced adhesion are some valuable characteristics for rendering and bedding mortars.

The production of blocks allows that fine recycled aggregates be successfully applied, alone or combined with coarse recycled aggregates, being viable to replace the whole natural aggregates when recycled concrete aggregates were used.

In roads vast quantities of fine recycled aggregates can be incorporated in base and sub-base layers, with the advantage that fine crushed concrete could present self-cementing properties resulting on strength increase over time. Application on hot bituminous mixtures used as base and binder course represents another alternative.

Some special applications for the fine fraction of CDW such production of masonry units by hydrothermal curing or production of controlled low strength materials is under development.

For fine ceramic wastes the viability of their use as a pozzolanic material for production of cement blends could be regarded as an opportunity to attain a decrease on the clinker factor, with the consequent reduction on CO_2 emissions. In general, cement substitution up to 30% result on structure densification and possible prevention of alkalis-silica reaction without compromising compressive strength. The use of fine ceramic wastes in partial replacement of natural aggregates, in percentages in the range 20–50%, on mortars, concrete and precast concrete products revealed to be practicable.

Regardless of the opportunities identified for the use of the fine fraction of CDW and of the environmental advantages that could be achieved the lower prices of fine natural aggregates, the lack of dissemination of reliable scientific information and the absence of particular specifications are some factors that still compromise the use of fine recycled aggregates and must be overcome.

References

1. Etxeberria, M., Marí, A., Vázquez, E.: Recycled aggregate concrete as structural material. Mater. Struct. **40**(5), 529–541 (2007)
2. Hansen, T.: Recycled aggregate and recycled aggregate concrete. Second state-of-the-art. Report developments. Mater. Struct. **19**(3), 1845–1985 (1986)
3. Poon, C.S., Azhar, S., Kou, S.C.: Recycled aggregates for concrete applications. In: MaSTEC2003 – Materials Science and Technology in Engineering Conference – Now, New and Next, Hong Kong, 15–17 Jan 2003
4. Yanagibashi, K.: A new concrete recycling process for coarse aggregate to be used in concrete structure. In: Kashino, N., Ohama, Y. (eds.) RILEM International Symposium on Environment-Conscious Materials and Systems for Sustainable Development, pp. 137–143 (2004)
5. Yanagibashi, K.: A new concrete recycling process for coarse aggregate to be used in concrete structure. In: Kashino, N. (ed.) RILEM International Symposium on Environment-Conscious Materials and Systems for Sustainable Development, pp. 137–143 (2004)
6. Sakai, K.: Concrete and environment. In: Industry, M. (ed.) Regional Meeting of Infrastructures, Merida (2008)
7. Shima, H., Tateyashiki, H., Matsuhashi, R., Yoshiba, Y.: An advanced concrete recycling technology and its applicability assessment through input–output analysis. J. Adv. Concr. Technol. **3**(1), 53–67 (2005)
8. Noguchii, T., Kitagaki, R., Tsujino, M.: Minimizing environmental impact and maximizing performance in concrete recycling. Struct. Concr. **12**(1), 36–46 (2011)
9. Mulder, E., Dejong, T., Feenstra, L.: Closed cycle construction: an integrated process for the separation and reuse of C&D waste. Waste Manag. **27**(10), 1408–1415 (2007)
10. Weimann, K., Müller, A.: Effects of wet processed crushed concrete fines as secondary aggregates in building materials. In: Dhir, R., Hewlett, P.C., Csetenyi, L., Newlands, M. (eds.) Proceedings of the International Conference Role for Concrete in Global Development, pp. 499–510. Dundee, UK (2008)
11. Weimann, K., Müller, A.: Properties of building materials gained from wet processed crushed concrete fines. In: Pbl.S.A.R., R. (ed.) Proceedings of the International RILEM Conference on the Use of Recycled Materials in Building Structures, Barcelona (2004)

12. Yamasaki, A., Iizuka, A., Kakizawa, M., Katsuyama, Y., Nakagawa, M., Fujii, M., Yanagisawa, Y.: Development of a carbon sequestration process by the carbonation reaction of waste streams containing calcium or magnesium. In: 5th Annual Conference on Carbon Capture & Sequestration, Alexandria, USA (2006)
13. Noguchi, T., Tamura, M.: Concrete design towards complete recycling. Struct. Concr. **2**(3), 155–167 (2001)
14. Noguchi, T., Kitagaki, R., Nagai, H., Tsujino, M.: Completely recyclable concrete of aggregate-recovery type by using microwave heating technology. In: RILEM, (ed.) International RILEM Conference on Progress of Recycling in the Built Environment, S. Paulo (2009)
15. Van der Wegen, G., Haverkort, R.: Recycled construction and demolition wastes as a fine aggregate for concrete. In: Proceedings of International Symposium – Sustainable Construction: Use of Recycled Concrete Aggregate, London (1998)
16. Sánchez de Juan, M., Alaejos, P.: Study on the influence of attached mortar content on the properties of recycled concrete aggregate. Constr. Build Mater. **23**(2), 872–877 (2009)
17. Engelsen, C., Van der Sloot, H.A., Wibetoe, G., Petkovic, G., Stoltenberg-Hansson, E., Lund, W.: Release of major elements from recycled concrete aggregates and geochemical modelling. Cem. Concr. Res. **39**(5), 446–459 (2009)
18. Angulo, S., Ulsen, C., John, V., Kahn, H., Cincotto, A.: Chemical-mineralogical characterization of C&D waste recycled aggregates from São Paulo, Brazil. Waste Manag. **29**(2), 721–730 (2009)
19. Zega, C., Sosa, M., Di Maio, A.: Propiedades de los agregados finos reciclados procedentes de hormigones elaborados con diferentes tipos de agregados gruesos naturales. In: I Congreso Hormigón Premezclado de las Américas 2010. XII Congreso Iberoamericano del Hormigón Premezclado y IV Congreso Internacional de Tecnología del Hormigón, Mar del Plata, Argentina (2010)
20. Tseng, E.: Concrete pavement recycling for use as aggregates in new concrete for pavements – Mario Covas metropolitan road ring case. Dissertation, Escola Politécnica da Universidade de São Paulo, São Paulo (2010)
21. Ulsen, C.: Characterization and separability of fine aggregates from construction and demolition waste. Dissertation, Universidade de São Paulo, São Paulo (2011)
22. Hansen, T., Narud, H.: Strength of recycled concrete made from crushed concrete coarse aggregates. Concr. Int. Des. Constr. **5**(1), 79–83 (1983)
23. Zega, C., Di Maio, A.: Comportamiento de hormigones elaborados con agregado fino reciclado. In: Memorias 16º Reunión Técnica de la Asociación Argentina de Tecnología del Hormigón 2006, pp. 47–54. Mendoza, Argentina (2006)
24. Evangelista, L., Brito, J.: Durability performance of concrete made with fine recycled concrete aggregates. Cem. Concr. Compos. **32**(1), 9–14 (2010)
25. Khatib, J.: Properties of concrete incorporating fine recycled aggregate. Cem. Concr. Res. **35**(4), 763–769 (2005)
26. Solyman, M.: Classification of recycled sands and their applications as fine aggregates for concrete and bituminous mixtures. Ph.D. dissertation, University of Kassel (2005)
27. Fung, W.: The use of recycled concrete in construction. Dissertation, University of Hong-Kong (2005)
28. Hansen, C.T.: Recycling of Demolished Concrete and Masonry (RILEM Report 7 DRC). E & FN Spon, London (1992)
29. Rashwan, M., Abourizk, S.: The properties of recycled concrete. Concr. Int. **19**(7), 56–60 (1997)
30. Montero, A., Tojo, Y., Matsuo, T., Matsuto, T., Yamada, M., ASAKURA, H., Ono, Y.: Gypsum and organic matter distribution in a mixed construction and demolition waste sorting process and their possible removal from outputs. J. Hazard. Mater. **175**(1–3), 747–753 (2010)
31. Gonçalves, P.: Concrete with recycled aggregates. Annotated review of existing regulations. Dissertation, Instituto Superior Técnico- Technical University of Lisbon, Lisbon (2007)

32. JSCE: New standards on recycled aggregates and molten-slag in Japan. Available at http://www.jsce.or.jp/committee/concrete/e/ (2007) Accessed 2007
33. COPRO. PTV 406.Version 2.0:2003.: Granulats recyclés: granulats pour béton-granulats pour matériaux traités aux liants hydrauliques et matériaux non traités utilisés pour les travaux de génie civil et pour la construction des chaussées. Organisme Impartial de Contrôle de Produits pour la Construction, Belgique (2003)
34. ABNT. NBR 15116:2004: Agregados reciclados de resíduos sólidos da construção civil – Utilização em pavimentação e preparo de concreto sem função estrutural – Requisitos. Brasil (2004)
35. DIN 4226–100:2002–02: Gesteinskörnungen für Beton und Mörtel. Teil 100: Rezyklierte Gesteinskörnungen. Aggregates for Mortar and Concrete, Part 100: Recycled Aggregates. Germany. Standard, Berlin (2002)
36. EPD. WBTC No. 12/2002.: Specifications Facilitating the Use of Recycled Aggregates. Hong-Kong (2002)
37. CUR: Recycled Concrete and Recycled Masonry Aggregate for Concrete CUR Publication 125. The Netherlands (1986)
38. LNEC-E471: Guia para a utilização de agregados reciclados grossos em betões de ligantes hidráulicos. Laboratório Nacional de Engenharia Civil (2009)
39. EHE: Instrucción de Hormigón Estructural. Spanish Minister of Public Works (2008)
40. SIA.: Cahier Technique 2030: 2010. Béton de recyclage, Suisse (2010)
41. BSI. BS 8500–2: Concrete – Complementary British Standard to BS EN 206–1 – Part 2: Specification for Constituent Materials and Concrete. United Kingdom (2006)
42. RILEM: Concrete specifications for recycled aggregates, RILEM TC 121DRG. Mater. Struct. **27**, 557–559 (1994)
43. RILEM: Specifications for concrete with recycled aggregate. Mater. Struct. **27**(9), 557–559 (1994)
44. Buyle-Bodin, F., Hadjieva-Zaharieva, R.: Influence of industrially produced recycled aggregates on flow properties of concrete. Mater. Struct. **35**, 504–509 (2002)
45. Fraaij, A., Pietersen, H., de Vries, J.: Performance of concrete with recycled aggregates. In: International Conference: Sustainable Concrete Construction, pp. 187–198. Dundee, Scotland (2002)
46. Kenai, S., Debieb, F., Assouz, L.: Mechanical properties and durability of concrete made with coarse and fine recycled aggregate. In: Proceedings of the International Conference: Sustainable Concrete Construction, pp. 383–392. Dundee, Scotland (2002)
47. Corinaldesi, V., Moriconi, G.: Influence of mineral additions on the performance of 100% recycled aggregate concrete. Construct. Build Mater. **23**, 2869–2876 (2009)
48. Tam, V., Gao, X., Tam, C.: Microstructural analysis of recycled aggregate concrete produced from two-stage mixing approach. Cem. Concr. Res. **35**(6), 1195–1203 (2005)
49. Dhir, R., Limbachiya, M., Leelawat, T.: Suitability of recycled concrete aggregate for use in BS 5328 Designated mixes. Proc. ICE Struct. Build **134**(3), 257–274 (1999)
50. Masood, A., Ahmad, T., Arif, M., Madhi, F.: Waste management strategies for concrete. Environ. Eng. Policy **3**(1–2), 15–18 (2002)
51. Evangelista, L., Brito, J.: Criteria for the use of fine recycled concrete aggregates in concrete production. In: Proceedings of the International RILEM Conference on the Use of Recycled Materials in Building Structures, Barcelona (2004)
52. Evangelista, L., Brito, J.: Mechanical behaviour of concrete made with fine recycled concrete aggregates. Cem. Concr. Compos. **29**(5), 397–401 (2007)
53. Reis, C., Leite, M., Lima, P.: Influence of the diameter of the bar and CDW content on the bond behavior of recycled reinforced concrete. In: Progress of Recycling in the Built Environment, S. Paulo (2009)
54. Zega, C., Di Maio, A.: Use of recycled fine aggregate in concretes with durable requirements. Waste Manag. **31**(11), 2336–2340 (2011)

55. Corinaldesi, V., Moriconi, G., Tittarelli, F.: SCC: a way to sustainable construction development. In: SCC'2005-China: 1st International Symposium on Design, Performance and Use of Self-Consolidating Concrete, China. RILEM Publications SARL (2005)
56. Kou, S., Poon, C.: Properties of self-compacting concrete prepared with coarse and fine recycled concrete aggregates. Cem. Concr. Compos. **31**(9), 622–627 (2009)
57. Kim, H.Y., Chun, B.S., Park, T.H., Ryou, J.S.: An investigation of the recycling of waste concrete as a cementitious material. J. Ceram. Process. Res. **12**(2), 202–206 (2011)
58. Corinaldesi, V., Moriconi, G.: The role of industrial by-products in self-compacting concrete. Constr. Build Mater. **25**(8), 3181–3186 (2011)
59. Miranda, L., Selmo, S.: CDW recycled aggregate renderings: part I – analysis of the effect of materials finer than 75 [mu]m on mortar properties. Constr. Build Mater. **20**(9), 615–624 (2006)
60. Miranda, L., Selmo, S.: CDW recycled aggregate renderings: part II – analysis of the effect of materials finer than 75 [mu]m under accelerated aging performance. Constr. Build Mater. **20**(9)), 625–633 (2006)
61. Lee, S.T.: Influence of recycled fine aggregates on the resistance of mortars to magnesium sulfate attack. Waste Manag. **29**(8), 2385–2391 (2009)
62. Lee, S.T., Swamy, R., Kim, S.S., Park, Y.G.: Durability of mortars made with recycled fine aggregates exposed to sulphate solutions. J. Mater. Civ. Eng. **20**(1), 63–70 (2008)
63. Corinaldesi, V.: Mechanical behavior of masonry assemblages manufactured with recycled-aggregate mortars. Cem. Concr. Compos. **31**(7), 505–510 (2009)
64. Corinaldesi, V., Moriconi, G.: Behaviour of cementitious mortars containing different kinds of recycled aggregate. Constr. Build Mater. **23**(1), 289–294 (2009)
65. De Schutter, G., Poppe, A.: Quantification of the water demand of sand in mortar. Constr. Build Mater. **18**(7), 517–521 (2004)
66. Leite, M., Lima, P., Santos, A.: Study of the consistence and mechanical behaviour of recycled mortars. In: International RILEM Conference on Progress of Recycling in the Built Environment, S. Paulo (2009)
67. Vegas, I., Azkarate, I., Juarrero, A., Frias, M.: Diseno y prestaciones de morteros de albanilería elaborados con áridos reciclados procedentes de escombro de hormigón. Mater. Constr. **59**(295), 5–18 (2009)
68. Dapena, E., Alaejos, P., Lobet, A., Pérez, D.: Effect of recycled sand content on characteristics of mortars and concrete. J. Mater. Civ. Eng. **23**(4), 414–422 (2011)
69. Soutsos, M., Millard, S., Bungey, J., Jones, N., Tickell, R., Gradwell, J.: Using recycled demolition waste in concrete building blocks. Proc. ICE Eng. Sustain. **157**(3), 139–148 (2004)
70. Soutsos, M., Tang, K., Millard, S.: Concrete building blocks made with recycled demolition aggregate. Constr. Build Mater. **25**(2), 726–735 (2010)
71. Poon, C., Kou, S., Lam, L.: Use of recycled aggregates in molded concrete bricks and blocks. Constr. Build Mater. **16**(5), 281–289 (2002)
72. Poon, C., Chan, D.: Paving blocks made with recycled concrete aggregate and crushed clay brick. Constr. Build Mater. **20**(8), 569–577 (2006)
73. Poon, C.-S., Qiao, X., Chan, D.: The cause and influence of self-cementing properties of fine recycled concrete aggregates on the properties of unbound sub-base. Waste Manag. **26**(10), 1166–1172 (2006)
74. Chan, D., Poon, C.: Using recycled construction waste as aggregates for paving blocks. Proc. ICE Waste Resour Manag. **159**(2), 83–91 (2006)
75. Poon, C.S., Chan, D.: Effects of contaminants on the properties of concrete paving blocks prepared with recycled concrete aggregates. Constr. Build Mater. **21**(1), 164–175 (2007)
76. Dosho, Y.: Development of a sustainable concrete waste recycling system. J. Adv. Concr. Technol. **5**(1), 27–42 (February 2007)

77. Pelufo, M., Serna, P., Jacquin, C., Ulloa, V., López, F.: Execution of concrete waffle slab recycled aggregate to be used in prefabricated elements in Spain. In: International RILEM Conference on Progress of Recycling in the Built Environment, S. Paulo (2009)
78. Huang, W.L., Lin, D.H., Chang, N.B., Lin, K.S.: Recycling of construction and demolition waste via a mechanical sorting process. Resour. Conserv. Recycl. **37**(1), 23–37 (2002)
79. Cho, Y.H., Yeo, S.H.Y.: Application of recycled waste aggregate to lean concrete subbase in highway pavement. Can. J. Civ. Eng. **31**(6), 1101–1108 (2004)
80. Poon, C., Chan, D.: Feasible use of recycled concrete aggregates and crushed clay brick as unbound road sub-base. Constr. Build Mater. **20**(8), 578–585 (2006)
81. Arm, M.: Self-cementing properties of crushed demolished concrete in unbound layers: results from triaxial tests and field tests. Waste Manag. **21**(3), 235–239 (2001)
82. Chen, M., Lin, J., Wu, S.: Potential of recycled fine aggregates powder as filler in asphalt mixture. Constr. Build Mater. **25**(10)), 3909–3914 (2011)
83. Plaza, C., Xu, Q., Townsend, T., Bitton, G., Booth, M.: Evaluation of alternative landfill cover soils for attenuating hydrogen sulfide from construction and demolition (C&D) debris landfills. J. Environ. Manag. **84**(3), 314–322 (2007)
84. Shui, Z., Xuan, D., Wan, H., Cao, B.: Rehydration reactivity of recycled mortar from concrete waste experienced to thermal treatment. Constr. Build Mater. **22**(8), 1723–1729 (2008)
85. Hlawatsch, F., Schlutter, F., Berger, M., Kropp, J.: Construction products prepared from autoclave hardened fines from C&DW. In: International RILEM Conference on Progress of Recycling in the Built Environment, S. Paulo, Brazil (2009)
86. Al-Otaibi, S., El-Hawary, M., Abdul-Jaleel, A.: Recycling crushed concrete fines to produce lime-silica bricks. Proc. ICE Waste Resour. Manag. **163**(3), 123–127 (2010)
87. ASTM. C73: Standard Specification for Calcium Silicate Brick (Sand-Lime Brick). ASTM International (2010)
88. Achtemichuk, S., Hubbard, J., Sluce, R., Shehata, M.H.: The utilization of recycled concrete aggregate to produce controlled low-strength materials without using Portland cement. Cem. Concr. Compos. **31**(8), 564–569 (2009)
89. Oikonomou, N.: Recycled concrete aggregates. Cem. Concr. Compos. **27**(2), 315–318 (2005)
90. Juan, A., Medina, C., Morán, J.M., Guerra, M., Aguado, P.J., Sanchez De Rojas, M., Frías, M., Rodriguez, O.: Re-use of ceramic wastes in construction. In: Intech. Ceramic Materials. Available at http://www.intechopen.com/books/ceramic-materials/re-use-of-ceramic-wastes-in-construction(2010) Accessed 2010
91. Medina, C., Sanchez De Rojas, M.I., Frías, M., Juan, A.: Using ceramic materials in ecoefficient concrete and precast concrete products. In: Intech. Advances in Ceramics – Electric and Magnetic Ceramics, Bioceramics, Ceramics and Environment. Available at http://cdn.intechopen.com/pdfs/18289/InTech-Using_ceramic_materials_in_ecoefficient_concrete_and_precast_concrete_products.pdf (2011) Accessed 2011
92. Ay, N., Unal, M.: The use of waste ceramic tile in cement production. Cem. Concr. Res. **30**(3), 497–499 (2000)
93. Turanli, L., Bektas, F., Monteiro, P.J.M.: Use of ground clay brick as a pozzolanic material to reduce the alkali-silica reaction. Cem. Concr. Res. **33**(10), 1539–1542 (2003)
94. ASTM. C1260: Standard Test Method for Potential Alkali Reactivity of Aggregates (Mortar–Bar Method). ASTM International (2006)
95. Lin, K.L., Wu, H.H., Shie, J.L., Hwang, C.L., An, C.: Recycling waste brick from construction and demolition of buildings as pozzolanic materials. Waste Manag. Res. **28**(7), 653–659 (2010)
96. Lavat, A., Trezza, M.A., Poggi, M.N.: Characterization of ceramic roof tile wastes as pozzolanic admixture. Waste Manag. **29**(5), 1666–1674 (2009)
97. Naceri, A., Hamina, M.C.: Use of waste brick as a partial replacement of cement in mortar. Waste Manag. **29**(8), 2378–2384 (2009)
98. Frías, M., Marín, F.P., Rivera Lozano, J., Sánchez De Rojas, M.: Viabilidad de utilización de materiales de desecho procedentes de productos cerámicos en prefabricados de hormigón

(Viability of utilization of waste materials from ceramic products in precast concretes). Mater. Constr. **51**(263–264), 149 (2001)

99. Poon, C.S., Chan, D.: The use of recycled aggregate in concrete in Hong Kong. Resour. Conserv. Recycl. **50**(3), 293–305 (2007)
100. Debieb, F., Kenai, S.: The use of coarse and fine crushed bricks as aggregate in concrete. Constr. Build Mater. **22**(5), 886–893 (2008)
101. Martins, I., Gonçalves, A.: Incorporation of ceramic wastes in precast concrete paving blocks (in Portuguese). In: 2º Congresso Nacional de Prefabricação em Betão, Lisboa, Portugal (2008)
102. Bektas, F., Wang, K., Ceylan, H.: Effects of crushed clay brick aggregate on mortar durability. Constr. Build Mater. **23**(5), 1909–1914 (2009)
103. Silva, J., De Brito, J., Veiga, R.: Recycled red-clay ceramic construction and demolition waste for mortars production. J. Mater. Civ. Eng. **22**(3), 236–244 (2010)
104. Pacheco-Torgal, F., Jalali, S.: Reusing ceramic wastes in concrete. Constr. Build Mater. **24**(5), 832–838 (2010)
105. Puertas, F., García-Díaz, I., Barba, A., Gazulla, M., Palacios, M., Gómez, M., Martínez-Ramírez, S.: Ceramic wastes as alternative raw materials for Portland cement clinker production. Cem. Concr. Compos. **30**(9), 798–805 (2008)

Chapter 6
Quality Assurance of Recycled Aggregates

Pilar Alaejos, Marta Sánchez de Juan, Jorge Rueda, Roberto Drummond, and Ignacio Valero

Abstract Recycled aggregates differ from the natural ones in several properties, making their quality lower in general, but suitable enough to be used in different applications inside the construction field in many cases.

On the other hand, recycled aggregates are quite more heterogeneous compared to natural ones, varying their characteristics along the production process. Therefore, quality assurance is extremely important in order to provide users enough confidence in the material, having properties satisfying requirements, and also with defined tolerances on variation throughout a delivery.

Main factors affecting the quality of recycled aggregates are described in this chapter: influence of the production process, types of treatments, Impurities content, quality of the original debris. The composition and the nature of its components as well as the presence of impurities are the principal parameters affecting the final quality of the recycled aggregates. However, some specific methods and operations may be taken into account during the production process in order to obtain suitable recycled aggregate for use in higher-value applications. The control of the incoming deliveries to the recycling plant is also a basic measure to avoid an excessive heterogeneity in recycled aggregates.

In several countries, the standards have been amended in order to cover specific aspects of recycled aggregates, and even some certification systems have been also developed. Different international standards on quality assurance of recycled aggregates are reviewed in this chapter.

P. Alaejos (✉)
Lab. Central de Estructuras y Materiales, CEDEX, Madrid, Spain
e-mail: Pilar.Alaejos@cedex.es

M.S. de Juan • J. Rueda
CEDEX, Madrid, Spain

R. Drummond
Kleemann gmbh, Madrid, Spain

I. Valero
Gestora de Runes, Barcelona, Spain

E. Vázquez (ed.), *Progress of Recycling in the Built Environment: Final Report of the RILEM Technical Committee 217-PRE*, RILEM State-of-the-Art Reports 8,
DOI 10.1007/978-94-007-4908-5_6, © RILEM 2013

Keywords Aggregate • Recycled • Quality • Standard • Certification

6.1 The Quality of Recycled Aggregates

Recycled aggregates differ from the natural aggregates in several properties (density, absorption, LA coefficient, etc.), making their quality lower in general, but good enough in many cases to be used in different applications inside the construction field.

The composition and the nature of its components are the principal parameters affecting the final quality of the recycled aggregates. This quality can be controlled mainly by means of density and absorption characteristics, as different International Standards establish (Table 6.1).

Table 6.1 shows different grade qualities of recycled aggregates depending on the country: densities may range from 1,500 to 2,500 kg/m^3 and absorption from 20 to 3%. Best qualities are related to aggregates consisting mainly of stone particles, while the low qualities include those mixed aggregates with high contents of porous ceramic particles.

Other properties are also important regarding the final behaviour of recycled aggregates, as the presence of impurities (plastic, gypsum plaster, glass, etc.) which

Table 6.1 Density and absorption specifications of different International Standards

Standard		Density (kg/m^3)	Absorption (%)
Rilem	Type I	≥1,500	≤20
	Type II	≥2,000	≤10
	Type III	≥2,400	≤3
Japan	Fine aggregate (low quality)	≥2,000	≤13
	Fine aggregate (medium quality)	≥2,500	≤10
	Fine aggregate (high quality)	≥2,500	≤3
	Coarse aggregate (low quality)	≥2,200	≤7
	Coarse aggregate (medium quality)	≥2,500	≤5
	Coarse aggregate (high quality)	≥2,500	≤3
Belgium	GBSB-I	≥1,600	≤18
	GBSB-II	≥2,100	≤9
Australia	–	≥2,100	≤6
Hong Kong	(Tipo II)	≥2,000	≤10
Germany	Type 1	≥2,000	≤10
	Type 2	≥2,000	≤15
	Type 3	≥1,800	≤20
	Type 4	≥1,500	–
Brasil	RCA	–	≤7
	MRA	–	≤12
Portugal	ARB1	≥2,200	≤7
	ARB2	≥2,000	≤7
	ARC	≥2,000	–
Spain	RCA	–	≤7

Fig. 6.1 Heterogeneous composition of recycled aggregates [1]

can affect both mechanical and durability properties of concrete. The content of these contaminants (see Sect. 3.1.3) must be controlled to reach a suitable quality level.

6.2 Variation in the Properties of Recycled Aggregates

Recycled aggregates are quite more heterogeneous compared to natural aggregates, varying their characteristics along the production. This variation is present even when the aggregate comes from pure crushed concrete, mainly due to the different quality of the concrete debris stocked at the recycling plant, but also to the incidental presence of contaminants.

For masonry wastes, this heterogeneity also affects to the composition of the aggregate and can be appreciated even at first sight (Fig. 6.1).

Tables 6.2 and 6.3 show variations measured in different properties of both concrete and mixed recycled aggregates, taken from the references indicated.

For aggregates coming from masonry debris, the composition is an important characteristic to be controlled and that can be related to a great dispersion. Thus, the tests show that the coefficient of variation can reach up to 17.6% (Fig. 6.2).

Some standards establish limits to the variation of specific properties of recycled aggregates. DIN 4226-100 Standard requires for density a tolerance of ±150 kg/m^3 and BRE 433 is more strict as requires a variation ±1%.

In Belgium the BENOR volunteer mark requires the following tolerances for density and absorption both for natural and recycled aggregates (Table 6.4).

Table 6.2 Variation of concrete recycled aggregates properties

Property	Mean value	Standard deviation	Coefficient of variation (%)	Sampling period	Reference
Absorption (%)	5.44 [4.7– 6.5]	0.52	9.6	6 months (11 samples)	[2]
	4.87 [4.4–5.28]	0.30	6.2	3 months (10 samples)	[3]
	5.79 [1.75–10.07]				[4]
	[4.8–9.6]		20.8	12 months (15 samples)	[5]
Granulometric modulus (%)	6.48 [6.34–6.57]	0.09	1.4	3 months (10 samples)	[3]
	6.85 [6.64–7.21]				[4]
	[6.7–7.2)		1.7	12 months (15 samples)	[5]
Saturated surface-dry density (kg/m^3)	2.394 [2.319–2.494]	0.06	2.6	6 months	[2]
	2.462 [2.440–2.450]	0.12	0.8	3 months (10 samples)	[3]
	[2.30–2.45]		1.77	12 months (15 samples)	[5]
Bulk density (kg/m^3)	1.400 [1.330–1.540]	0.08	5.7	3 months (10 samples)	[3]
Crushing value (%)	23.3 [21.2–25.7]	1.39	6.0	6 months	[2]
Attached mortar (%)	12.8	1.2	9.4	3 months (10 samples)	[3]
	[27.2–54.3]		19.6	12 months (15 samples)	[5]
Lightweight particles (%)	2.79 [0.55–5.02]	1.75	62.7	3 months (10 samples)	[3]
	[0.0–5.85]		151	12 months (15 samples)	[5]
Impurity content (plaster, brick, glass, asphalt, wood, etc.) (%)	0.97 [0.4–1.7]	0.43	42.7	6 months	[2]

Table 6.3 Variation of recycled aggregates composed of ceramic waste and concrete

Property	Mean value	Deviation	Coefficient of variation (%)	Sampling period	Reference
Absorption (%)	7.49 [4.5–9.7]	1.43	19.1	3 months (10 samples)	[3]
	5.5 [0.6–11.8]	3.07		13 samples	[6]
Granulometric modulus (%)	6.44 [6.24–6.78]	0.18	2.5	3 months (10 samples)	[3]

(continued)

Table 6.3 (continued)

Property	Mean value	Deviation	Coefficient of variation (%)	Sampling period	Reference
Satured surface-dry density (kg/m^3)	2.362 [2.226–2.462]	0.06	2.5	3 months (10 samples)	[3]
Bulk density (kg/m^3)	1.300 [1.180–1.420]	0.07	5.4	3 months (10 samples)	[3]
Attached mortar (%)	19.7	2.3	11.7	3 months (10 samples)	[3]
Lightweight particles (%)	7.55 [2.24–15.88]	4.17	55.2	3 months (10 samples)	[3]

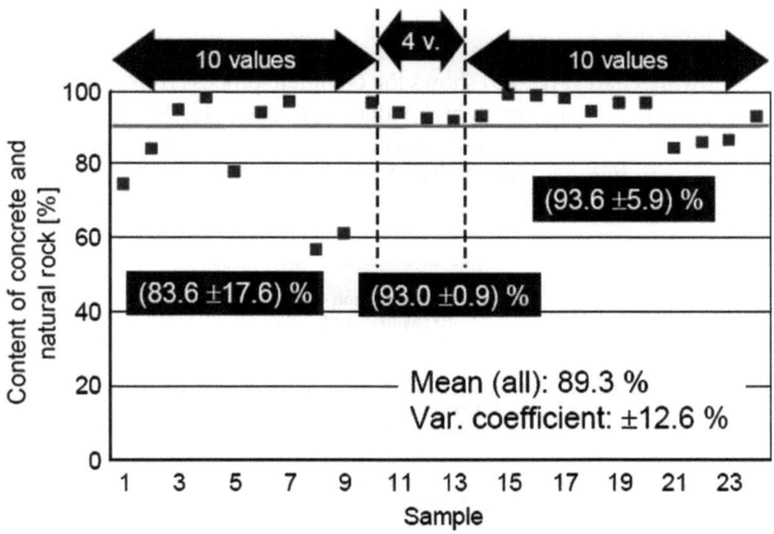

Fig. 6.2 Coefficient of variation of composition of mixed recycled aggregates from three different recycling plants [1]

Table 6.4 Tolerances for density and absorption of recycled aggregates

	Gravillons selon NBN EN 1097-6, §8		Sables alluvionnaires ou marins selon NBN EN 1097-6, §9	
	Masse volumique réelle	Tolérance max. admise valeur individuelle	Masse volumique rélle	Tolérance max. admise valeur individuelle
Masse volumique rélle ρ_{rd}	Valeur moyenne = valeur déclarée	±70 kg/m^3	Valeur moyenne = valeur déclarée	±100 kg/m^3
Absorption d'eau WA_{24}	Valeur moyenne = valeur déclarée	• Si $\leq$ 1.0%: tolérance ±0.3% • Si > 1.0%: tolérance ±0.7%	Valeur moyenne = valeur déclarée	±1.0%

6.3 Factors Affecting the Quality of the Recycled Aggregate

6.3.1 Influence of the Production Process

To obtain recycled aggregate properties suitable for use in higher value applications, operations should be put in place to [7]:

- Restrict the type of CDEW that is used.
- Separate materials before crushing.
- Screen out oversize materials.
- Crush twice: using primary and secondary crushers.
- Separate out impurities.
- Wash aggregates.

The following chart (Fig. 6.3) shows the complete process from a case study of a Recycling Plant Process to obtain high value aggregates (40, 20, 10 mm and sand).

Apart from the basic treatment (crushing, screening, use of magnets), some specific aspects to improve the quality of the aggregate are:

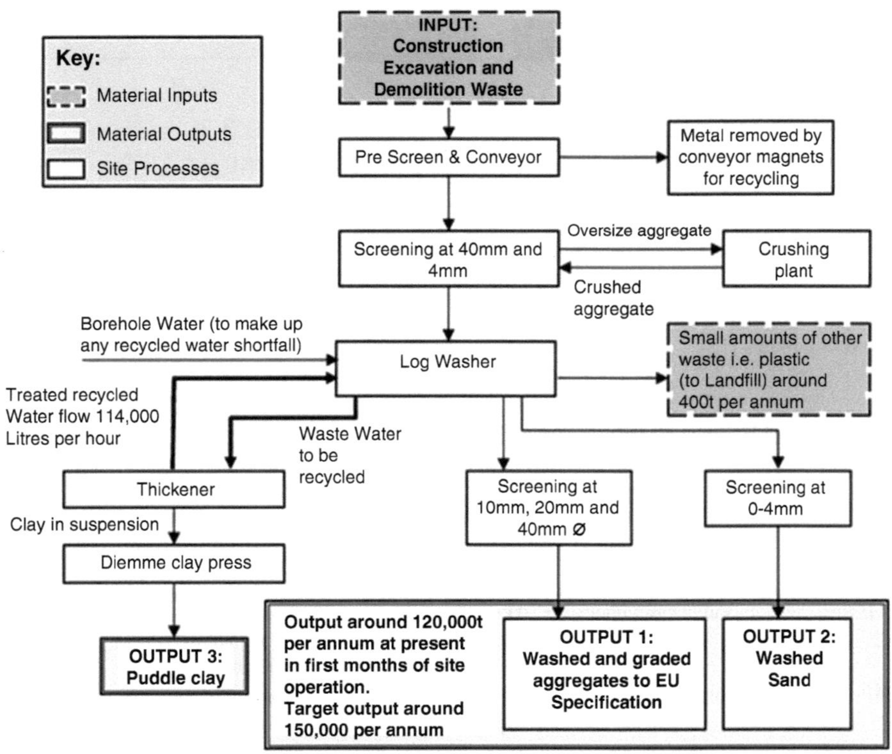

Fig. 6.3 Recycling plant process to obtain high value aggregates

- The use of a log washer to eliminate clay
- Plastic removed by air knifes.
- After washing the aggregate, the use of a thickener where a flocculent is added. The resulting water is then recycled by means of a Diemme clay press, obtaining a puddle clay which is a saleable end product used for landfill linings.
- Following the washing process, the material is sent to a static screening unit consisting of a 12 × 5 screen with a 3-way split. The 3-way split allows products of 10/10, 20/20 and 40/40 mm to be collected. In addition, sand of 0/4 mm is collected.

6.3.1.1 Aspects Related to the Incoming Wastes

An initial basic measure to be taken in order to avoid an excessive heterogeneity in recycled aggregates is the control of the incoming deliveries to the recycling plant, clearly separating the different wastes depending on its nature: pure concrete, mix with ceramic, asphalt, etc.

In practice, the recycling plants usually distinguish the truck load depending on its weight at the entrance. Three possible classes can be used [7]:

- Class I: Wastes with a bulk density over 1.45 t/m^3
- Class II: Wastes with a bulk density over 1.10 t/m^3
- Class III: Wastes with a bulk density under 1.10 t/m^3

Different Waste Taxes apply for each case as the low density loads usually involve wastes more contaminated, which require a more complex processing to eliminate impurities and an additional cost to carry the rejected material to dump.

At the same time of weighing, the truck must be inspected at the entrance, both the visual inspection and the weight helps for a correct and uniform stocking (Fig. 6.4). Finally, also a visual inspection is recommended at the discharging point, to check the nature of the whole load.

Many of the leading recycled aggregate producers in Europe have recognised that in order to obtain RA with properties suitable for use in higher-value applications it is necessary to restrict the type of wastes put through the crushing plant. An example of good practice in Europe is to separate all materials arriving at a recycling centre into three distinct colour-coded groups [8]:

- Black: Asphalt (Fig. 6.5a)
- White: Concrete (Fig. 6.5b)
- Red: Mixed waste, but generally comprised of brick (Fig. 6.5c)

The activities of contractors bringing construction and demolition wastes to the recycling centre could be monitored (sometimes using CCTV), and any contractors found to be violating the rules be banned. The success of these operations is marked by the lack of asphalt and concrete in the “red” pile. Clearly, the success of such an operation is dependent on demolition and excavating contractors to separate materials at source.

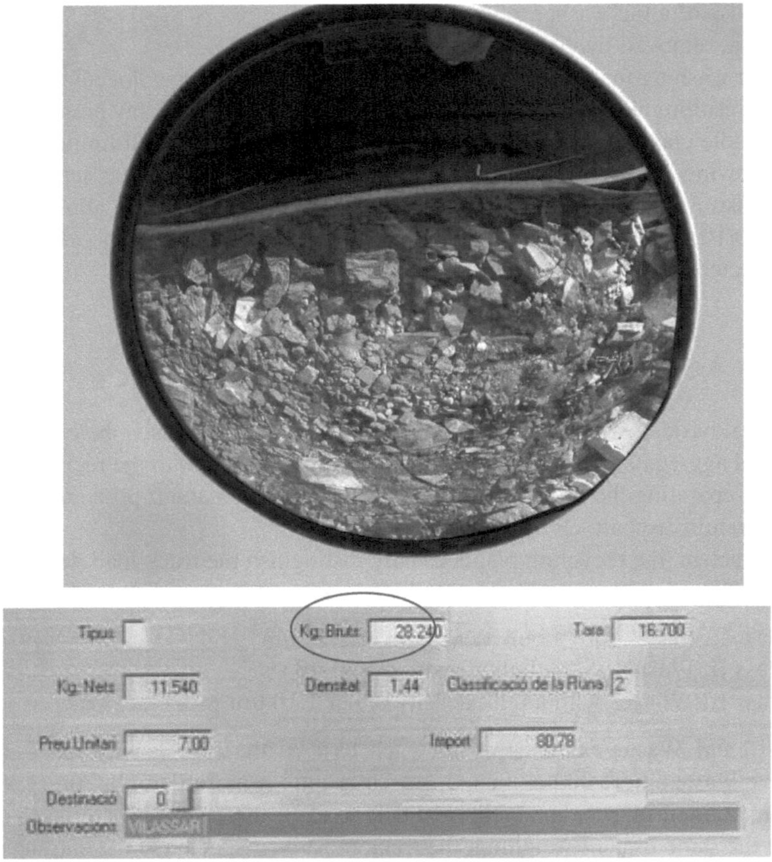

Fig. 6.4 Image of the truck load on the mirror at the entrance cabin of the recycling plant and weight data of the computer with an estimation of the load bulk density[7]

Fig. 6.5 Materials separated into distinct colour-coded groups

The recycled concrete aggregate (RCA) is created from the "white waste". The recycled aggregate (RA) is produced from the "red waste" and is largely brick-based. RCA may sometimes be added to brick-based RA to improve the performance of the product.

Clearly, reliance on external contractors imposes an element of risk in ensuring the consistent and overall quality of recycled aggregates. For this reason some large demolition contractors produce recycled aggregates using only CDEW sourced from their own work. The resulting RCA and RA are usually of very high quality [8].

Whilst best-practice leads to good separation of CDEW and good quality end-products, many smaller recycling companies are unable to put sufficient pressure on contractors to segregate materials at source, nor the capability to police the contractors on-site. Therefore RA produced in many operations, whilst nominally brick based often has an asphalt content too high for use in cement bound applications.

Moreover, in order to have confidence in the final RCA products with respect to dangerous substances and thus to avoid extensive testing of the final product, it is also important to have as much knowledge as possible of the construction and demolition material to be recycled. Chapter 4 gives the minimum data that should be collected for every incoming load to the Recycling Plant, according to European Standards. However, if a high grade quality of recycled aggregate wants to be achieved, it is also advisable to obtain complete information from the waste source [3, 4]:

- Demolition site details, at least type of structure: building, bridge, pavement, marine structure, etc.
- Reason of the demolition
- Original concrete characteristics:
    - Type: mass, reinforced, prestressed
    - Compressive strength
    - Possible use of aluminous cement

This information may be useful with different purposes:

- To identify original concretes that can produce low quality aggregates (possible existence of chemical concrete pathologies, concrete affected by fire, etc.).
- To separate recycled aggregates that should not be used for high grade applications.
- To separate incoming wastes that can be supplied for certain categories of end use with little or no processing [9].
- To optimize the potential use of the recycled aggregates (for example high quality recycled aggregates obtained from original prestressed concrete could be specifically used for the production of a new structural concrete).

6.3.1.2 Influence of the Crushing Process

To achieve the most desirable grading curves and shape for recycled aggregates, a series of successive crushers (or crushes) and screening should be used; with oversized material returned to the respective crusher [8].

Table 6.5 Characteristics of different crushers to produce recycled aggregates

Property	Jaw crusher	Cone crusher	Impact crusher
Efficiency	High	Low	Medium
Production cost	Low	Medium	High
Wearing	Low	Low	Medium
Aggregate quality	Low	Medium	High
Fines content	Low	High	High
Energy consumption	Low	Medium	High

Both the jaw crushers (usually called primary crushers as they can handle larger fragments) and impact crushers (secondary crushers) may be suitable to process CDEW, depending on different factors: presence of impurities, nature of the waste (stone, concrete, bricks, etc.), size of the incoming blocks, kind of final product [10].

The main characteristics of impact crushers are as follows [10]:

- They give RA a more rounded shape which is beneficial for engineering performance [8].
- They provide aggregates with very good grading.
- The modern equipments allows the same incoming size block than the jaw crushers obtaining up to 75% artificial graded aggregate in one step.
- On the contrary, the impact crushers require more maintenance as they suffer more wearing than the jaw crushers.

The jaw crushers main characteristics are [10]:

- Longer service life, as they suffer less wearing.
- They admit a size block up to 1,500 mm but do not reduce the maximum size under 60 mm.
- The efficiency is better as they just reduce the particle size.
- They give a RA with low fines content (5–8%, 0–40 mm), making necessary to use an impact crusher to produce an artificial graded aggregate for roads.

Table 6.5 shows a summary of the main characteristics of the different crushers for the production of recycled aggregates [10, 11]:

Taking into account the previous characteristics, the selection of the equipment for a recycling plant (jaw crusher, impact crusher or a combination) will depend on [10]:

- The original waste: if it is hard and abrasive or the input block size is higher than 600 mm, the best option is a combination of primary (jaw) and secondary (impact) crushing.
- The final product: if an artificial graded aggregate is desired an impact crushing will be necessary.

Some frequent combinations are shown in Table 6.6 [10]. An example of good practice for a high quality recycled aggregate is to use primary crushers (usually jaw crushers due to its high efficiency and low cost) to reduce the blocks of wastes

Table 6.6 Frequent combinations of equipments to produce recycled aggregates

Primary crushing	Secondary crushing
Jaw crusher	Impact crusher
Impact crusher	Impact or cone crusher

to 50 mm, and secondary crushers (either impact or cone crushers) to reduce the particle size to 32 mm or less [8, 10].

Magnets, air knives (directed blasts of air) and hand picking are usually used between the two crushes to remove impurities. Pre-screens and eddy currents are also occasionally used, whilst use of additional magnets after the secondary crusher will aid removal of wire reinforcement [8].

Screens are used at various stages within the crushing process although usually the main separation by size takes place after the second crusher. It is increasingly common however to screen off the fines before the primary crusher as a preponderance of fines can reduce crushing performance. Care should be taken when crushing recycled aggregate containing a large proportion of brick since this produces more fines than crushing of concrete or primary aggregates [8].

Soil, silt and clay can be particular problems and care is usually taken to avoid these materials entering the stream at source. Where they are present some contractors use a scalping screen ahead of the primary crusher to remove soil and clay balls. Other practices rely on the soil and clay being removed during washing processes. To avoid the resulting soil, silt and clay being directed to landfill, a number of producers use innovative methods to create products for this material. One example is the creation of filter cakes for use in drainage [8].

6.3.1.3 Systems to Eliminate Impurities

Impurities Content

Many of the leading recycled aggregate producers in Europe have recognised that in order to obtain RA with properties suitable for use in higher-value applications it is necessary to restrict the type of demolition waste put through the crushing plant, and create RA with a low content of impurities. Most importantly, many recycled aggregate producers have recognised that to use RA in higher-value applications there is a need to limit the asphalt content (and indeed the proposed EN 12620 classification of constituents for recycled aggregate for use in concrete will limit asphalt content to a maximum of 10% by mass) [8].

In the following tables a comparison in different countries of the maximum content of impurities allowed (and the test standard applied) is shown, both for RCA (Tables 6.7 and 6.8) and RA (Table 6.9):

In Germany, a very detailed group of requirements is applied to constituents of the recycled aggregates, as can be seen in the following Table 6.10.

The limits for impurities content are very strict all over the standards, making the producers have included different systems in the process to eliminate the non desirable substances, as will be described as follows.

Table 6.7 International requirements for recycled concrete aggregate

	Rilem (Type II)	Japan fine aggregate[a]	Japan coarse aggregate[a]	Australia	DIGEST 433 (RCA II)
Density of material ≤1,800 kg/m^3	≤10%	≤10%[b]	≤10%[b]	–	–
	ASTM C123	Experimental method		–	–
Density of material ≤1,000 kg/m^3	≤0.5%	≤2%[c]	≤2%[c]	–	≤0.5%
	ASTM C123	Experimental method		–	BS 812-101
Impurity content (metals, glass, soft materials, asphalt)	≤1%	–		≤1%	≤1%
	Visual	–		Visual	BS 812-101
Lightweight particles	–	–		–	–
	–	–		–	–
Fines content (<0.063 mm)	≤2%	≤8%	≤1%	–	–
	UNE-EN 933-1	JIS 5308		–	–
Sand content (<4 mm)	≤5%	–		–	–
	UNE-EN 933-1	–		–	–
Sulphate content (SO_3)	≤1%	–		–	≤1%
	BS 812-118	–		–	BS EN 1744-1
		JIS 5308		–	–

[a] Limit established in the Japanese guide revision, the limit in the first version is 13%
[b] Density material <1950 kg/m^3 in the Japanese guide, <1600 kg/m^3 in Belgium
[c] Density material <1200 kg/m^3 in the Japanese guide

Basic Treatments

All the stationary recycling plants include hand picking cabins (Fig. 6.6a), where the biggest impurities are thrown into specific containers: metal, glass, plastic, wood and papers (Fig. 6.6b).

Also, the use of magnets to remove reinforcement from the concrete is a basic usual treatment in recycling plants (Fig. 6.7a).

The elimination of lightweight particles is usually carried out by knife airs, being the limit particle size of removal a function of the pressure and direction of the air jet (Fig. 6.7b).

Advanced Methods to Improve the Quality

Grinding of the recycled aggregate to reduce the paste content: In Japan, there have been attempts to further improve the quality of recycled aggregate via, for example the methods briefly stated below [12].

Table 6.8 International requirements for recycled concrete aggregate

	Hong Kong (Type II)	Brasil (RCA)	Portugal (ARB1)	Spain[d]	United Kingdom (RCA)
Density of material	–	–	–	–	–
≤1,800 kg/m³	–	–	–	–	–
Density of material	≤0.5%	–	–	–	≤0.5%
≤1,000 kg/m³	BS 812-101	–	–	–	BS 812-101
Impurity content	≤1%	≤3%	≤0.2%	≤1%	≤1%
(metals, glass, soft materials, asphalt)	BS 812-101	–	NP EN 933-11	UNE-EN 933-1	BS 812-101
Lightweight	–	–	≤1%	≤1%	–
particles	–	–	NP-EN 1744-1	UNE-EN 1744-1	–
Fines content	≤4%	≤10%			(<75 μm)
(<0.063 mm)	≤4%	–	≤5%		
BS EN 933-1	NBR NM 46	NP-EN 933-1	–	BS EN 933-1	
Sand content	≤5%	–	–	≤5%	–
(<4 mm)	BS EN 933-1	–	–	UNE-EN 933-1	–
Sulphate content	≤1%	≤1%	≤0.8%	–	≤1%
(SO_3)	BS 812 -118	NBR 9917	NP-EN 1744-1	–	BS EN 1744-1
		NBR 9917	–	–	BS 1881-124

[d] The 7% limit of absorption in Spain applies in cases which natural aggregates are replaced up 20% by mass by recycled concrete aggregates. In higher replacement precentages, applies the 5% limit of absorption.

1. Heating and rubbing: where by the recycled aggregates are heated to 300 °C to remove weaker mortar and cement particles from the aggregate.
2. Eccentric – shaft rotor method: in which the recycled aggregates are passed between two cylinders eccentrically rotating at high speed which separates coarse aggregates from mortar via a grinding effect.
3. Straightforward mechanical grinding: in which aggregates are placed in a rotating drum containing iron balls.

In the production of recycled aggregate concretes, 55–73% of the concrete waste can be used as recycled coarse aggregate; the remaining could also be used as recycled fine aggregate for precast concrete products. As this fine fraction contains a large amount of original mortar that detrimentally affects its properties, a wet grinding treatment process could be applied to improve the quality, in particular to reduce the absorption of the sand. The process is shown in Fig. 6.8. The aggregate is ground by the rotation of a rotor positioned inside a cylindrical shell. Recycled fine aggregate and fine powder of 5 mm or less is manufactured by passing through

Table 6.9 International requirements for recycled aggregate

	United Kingdom (RA)	Holland[e]	Brazil (MRA)	Portugal (ARB2)
Density of material ≤1,800 kg/m^3	–	–	–	–
	–	–	–	–
Density of material ≤1,000 kg/m^3	≤1%	–	–	–
	BS EN 1744-1	–	–	–
Impurity content (metals, glass, soft materials, asphalt)	≤1%	–	≤3%	≤0.5%
	BS EN 933-11	–	–	NP EN 933-11
Lightweight particles	–	≤0.1%	–	≤1%
	–	NEN 5933	–	NP-EN 1744-1
Fines content (<0.063 mm)	≤3% (<75 μm)	≤3% ≤4%	≤10%	
BS EN 933-1	NEN 5917	NBR NM 46	NP-EN 933-1	
Sand content (<4 mm)	–	–	–	–
	–	–	–	–
Sulphate content (SO_3)	≤1%	≤1%	≤1%	≤0.8%
	BS EN 1744-1	NEN 5936	NBR 9917	NP-EN 1744-1
	BS 1881-124	NEN 5921		–

[e] In Holland the natural aggregares can be replaced by mixed recycled aggregates,, in percentages lower than 20%

Table 6.10 Composition of recycled aggregates

	Maximum or minimum content, as a percentage by mass, for aggregate type			
Constituents	1	2	3	4
Concrete and aggregate as in DIN 4226-1	≥90	≥70	≤20	≥80
Clinker, solid bricks	≤10	≤30	≥80	
Calcareous sandstone			≤5	
Other mineral materials[a]	≤2	≤3	≤5	≤20
Asphalt	≤1	≤1	≤1	
Foreign matter[b]	≤0.2	≤0.5	≤0.5	≤1

[a] E.g. porous bricks, lightweight concrete, aerated concrete, porous concrete, plaster, mortar, porous slag, pumice
[b] E.g. glass, ceramics, non-ferrous slags, gypsum, rubber, plastics, metals, wood, plant remains, paper and similar materials

a screen. Impurities such as fine powder aggregate and wood chips are removed by a wet-type high-speed centrifuge, or cyclone. By means of this treatment, the absorption can be reduced from 11.2 to 7.45%, and the density increased from 1.98 to 2.21 (g/cm^3) [13].

Fig. 6.6 (a) Hand picking cabin; (b) Containers for specific materials [10]

Fig. 6.7 (a) Magnet [10] (b) Knife airs for lightweight particles [10]

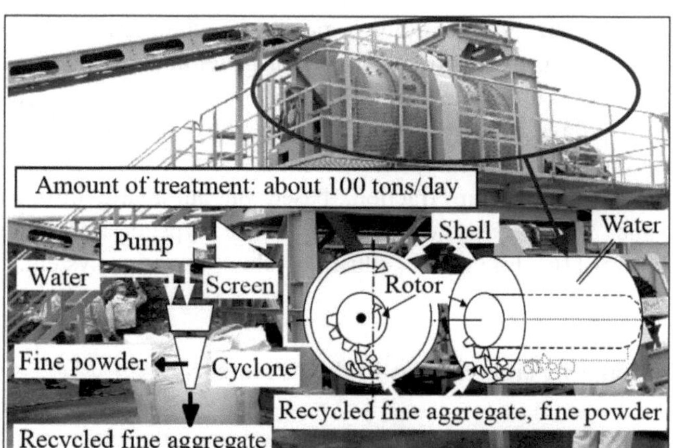

Fig. 6.8 Outline of wet grinding method

Fig. 6.9 System for dry density separation

Dry density separation of mixed recycled aggregate (air jigs) [14]**:** This system employs a pulsating air jet that causes a stratification of a bed of mixed material producing differential displacement of light and heavy particles. The lightest material separated mainly consists on wood, plastics, porous masonry and gypsum (Fig. 6.9). This system has a minimum feed size of 1 mm, and maximum feed size of 50 mm, with a maximum capacity of about 50 tons per meter width per hour.

Systems to remove impurities: Together with recently developed new sensors and advanced industrial image processing, different systems have been developed using X-ray, optical and electromagnetic sensors.

- **Colour sorting.** Scan & Sort The Sense Combi 1200 (Fig. 6.10) developed in Germany consists of an optical system that incorporates a high resolution camera and a special conductivity sensor permitting the identification of materials according to the size, shape, color or position of the particles in the tape. This information is transmitted to a system that generates impulses instructing the nozzles to blow out single particles or allow them to pass. This system is particularly suitable to remove wood, glass and gypsum [15].
- **X-ray sorting.** More recently, a new technology has arisen aiming to a greater separation of the gypsum fraction, based on automatic infra-red detection systems and a subsequent elimination of the gypsum and other impurities using a blower [16]. Figure 6.11 illustrates this separation system.

6.3.1.4 Stationary Plants Versus Mobile Plants [8]

Recycling of construction and demolition waste can take place either at the site from which the material is sourced using mobile crushers, or the material may be

Fig. 6.10 Scan & Sort Combi Sense 1200 system

Fig. 6.11 X-ray sorting

transported to a fixed recycling centre (sometimes referred to as urban quarries) where large stockpiles may be accumulated.

Mobile crushers (Fig. 6.12) whilst often more economical in that they avoid transporting construction and demolition wastes away from site, are rarely sophisticated enough to remove all impurities. For this reason, recycled aggregates produced from mobile crushers are usually used as site fill or capping layers, and used at, or close to, the location they were crushed.

Mobile impact crushers are usually the most economic option regarding licences, set up, production, etc.; so they are increasingly used for in situ processing. The modern equipments allow the removal of ferric materials, obtaining high production

Fig. 6.12 Impact mobile crusher [10]

rates up to 400 t/tour, depending on the original waste and final product. They are also used for furnace slag recycling.

In contrast, recycled aggregates from central recycling facilities undergo a number of processes to ensure higher quality, always depending on the cleanliness of the waste. This may include: magnets, picking stations, trash screen, screens, log washers, water pumps and sludge tanks (Fig. 6.13).

Furthermore central recycling facilities take great care to control the type of C&D waste that is allowed to be stockpiled.

6.3.2 Influence of the Original Concrete

Recycled aggregates obtained crushing a high strength original concrete (and subsequently with a low water/cement ratio and porosity) usually have a lower absorption, as can be observed in Fig. 6.14.

6.3.3 Influence of the Maximum Size on the Quality of the Aggregate

The maximum size of the recycled aggregate has a great influence on its final properties. In the particular case of the coarse recycled aggregate, the less coarse the fraction is, the lowest density (Figs. 6.15 and 6.16), highest attached mortar content

Fig. 6.13 Stationary CDW plant consisting of two lines (a) Contaminated wastes: feeding hopper + two floors pre-screen + air knives + picking station (b) Clean wastes: feeding hopper + two floors pre-screen + picking station + impact crusher + magnet + three floors screen + air knives + returning reject belt conveyor

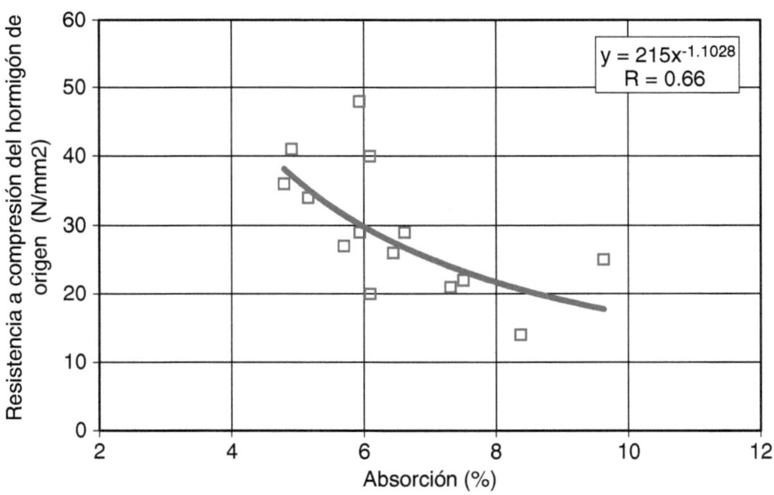

Fig. 6.14 Relationship between the absorption of recycled aggregate and the compression strength of the raw concrete [17]

(Fig. 6.17), and highest absorption (Fig. 6.18) is obtained. This behaviour is even more intense in the case of the recycled sand, which makes not advisable to use it for structural concrete [17]

For mixed recycled aggregates, the amount of ceramic particles has an influence of the shape index (Fig. 6.19) and Los Angeles coefficient (Fig. 6.20). Crushing of bricks produces a high content of flake particles, so the shape index decreases. On the other hand, ceramic bricks are harder than concrete, so they help to decrease the Los Angeles coefficient [18].

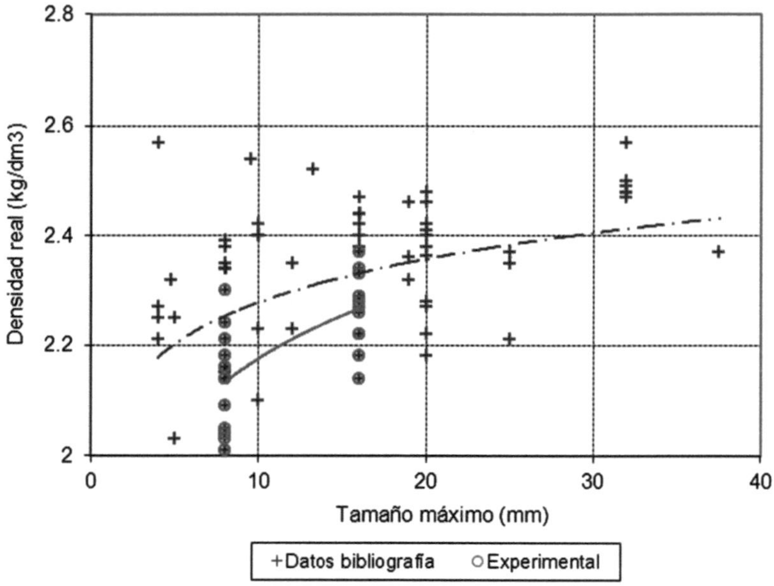

Fig. 6.15 Density versus maximum size of concrete recycled aggregate

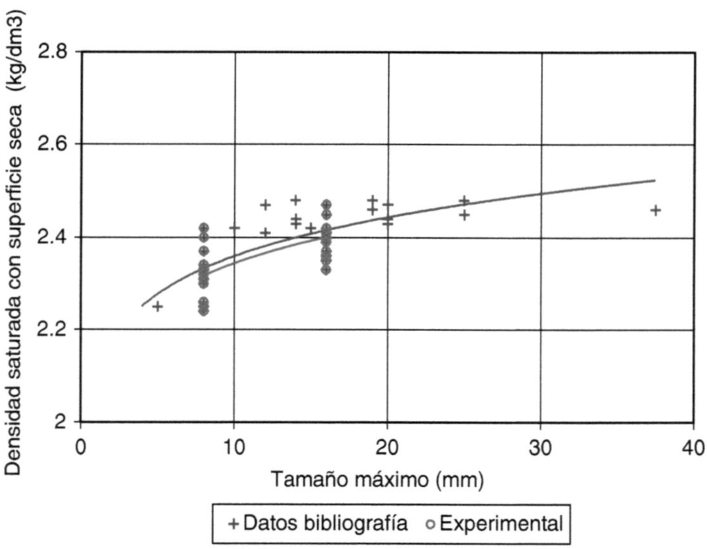

Fig. 6.16 Saturated surface-dry density versus maximum size of concrete recycled aggregate

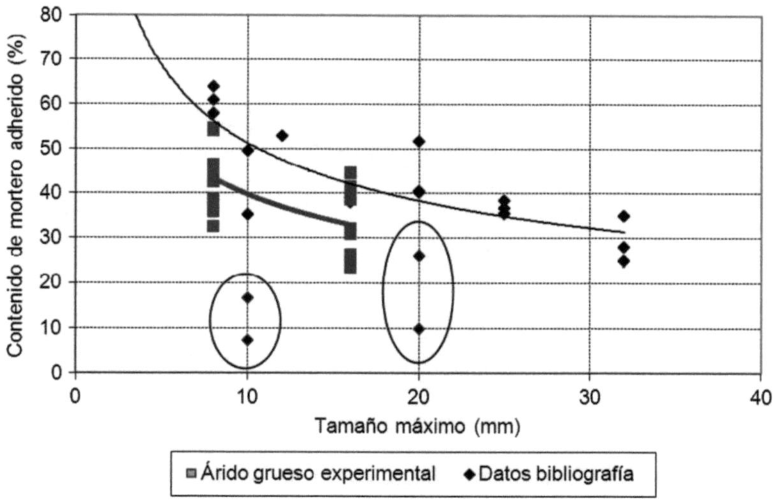

Fig. 6.17 Attached mortar content versus maximum size of concrete recycled aggregate

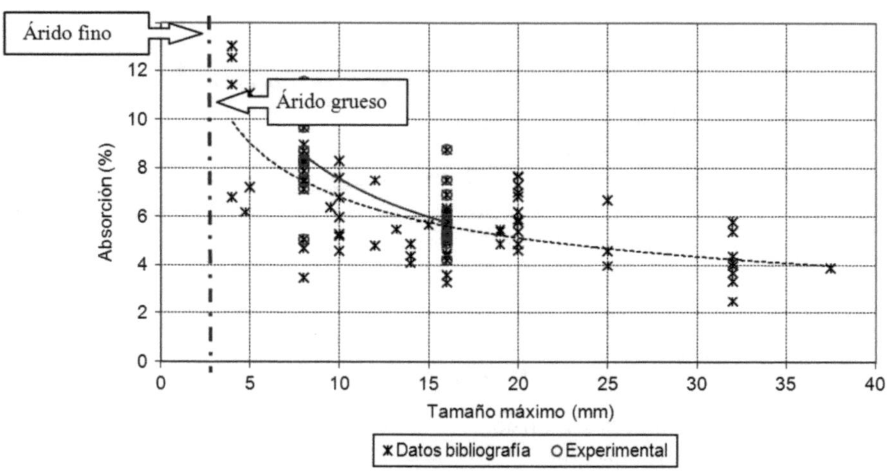

Fig. 6.18 Absorption versus maximum size of concrete recycled aggregate

6.4 Quality Assurance of the Production

The main aim for developing a certification protocol of recycle aggregates is to give users confidence in the material, having properties satisfying requirements, and also with defined tolerances on variation throughout a delivery.

Requirements of the recycled aggregates are set depending on the final use (roads, concrete, etc.). As far as the use of recycled aggregates in the production of concrete concerns, in the European Union they should satisfy, as a minimum, the

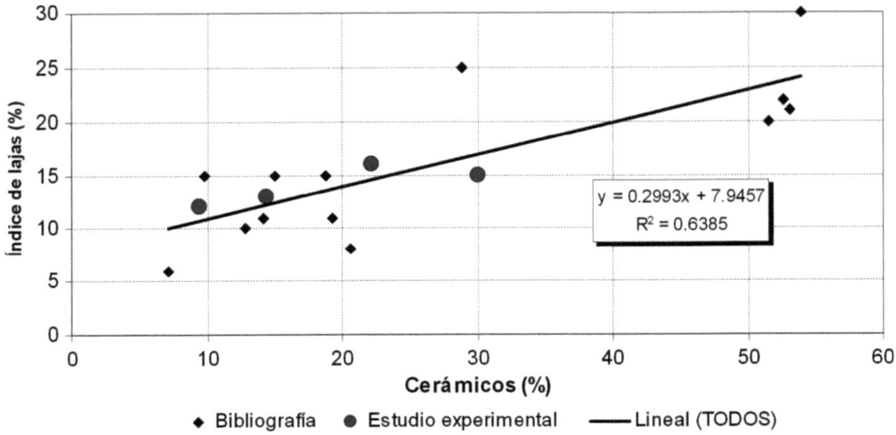

Fig. 6.19 Relationship between shape index and ceramic content (recycled aggregates)

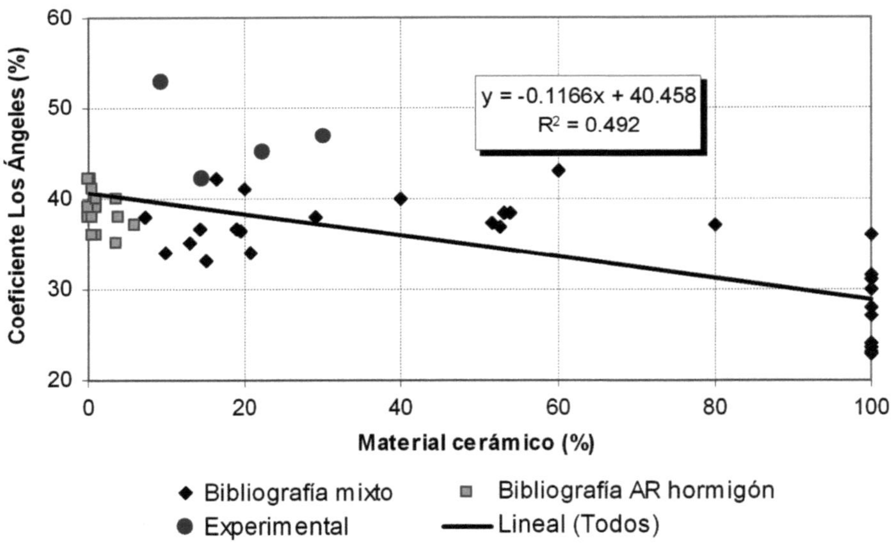

Fig. 6.20 Relationship between Los Angeles coefficient and ceramic content (recycled aggregates)

requirements included in "EN 12620:2002 Aggregates for concrete". However, as the standard itself points out, additional requirements might be needed to cover completely recycled aggregates, which should be established in accordance with local regulations valid at the place of use.

In some countries, as Netherlands and United Kingdom, the European standard has been amended in order to cover specific aspects of recycled aggregates (NEN 12620:2002+A1:2006 and BS EN 12620:2002+A1:2008). Other countries, however, have published independent standards for these aggregates, such is the case of Germany where a specific standard has been developed: DIN 4226-100:2002

"Aggregates for mortar and concrete. Part100: Recycled aggregates". In this section those specific aspects included in all these standards will be detailed.

In the European Union, all the aggregates for concrete must have an EC Declaration of conformity, which entitles the producer to affix the CE marking. The system of attestation of conformity of aggregates for concrete is 2+ and 4, for uses with and without high safety requirements respectively. System 2+ includes an initial certification of the factory production control, and a subsequent continuous surveillance. System 4 only requires the existence of a factory production control and the execution of type initial tests. All these considerations also apply for recycled aggregates.

EN 12620 specifies requirements for the evaluation of conformity and the implementation of a quality control system in the production, which should also be satisfied by recycled aggregates for concrete. Thus, the recycling plants must have in place a system of factory production control consisting of procedures for internal control of the production to ensure that recycled aggregates placed on the market conform the requirements of the standard. Apart from the regulated aspects that the system of production control must accomplished, there are also other measures during processing that could be taken to improve the final quality of the aggregates. All these aspects will be detailed in the following sections, applied to the case of recycled aggregates.

6.4.1 Incoming Debris

As part of the production control of aggregates for concrete, there must be a documentation of the raw material incoming to the factory. This requirement, when applied to recycling plants, means to document the construction and demolition wastes origin, including the following information (NEN and BS EN 12620):

- nature of the raw material
- source (processing depot) and place of origin
- supplier and transporting agent

Additional data, recommended by WRAP [9] includes also the date of each load, description of the quality of the incoming material and its quantity (by weighing/volume).

DIN 4226 also requires organoleptic checks of the material before and after dumping to establish whether its composition is as given on the delivery note. Other necessary information of the incoming debris that should be documented is:

- type of material (e.g. concrete, bricks, etc.)
- name and address of haulage contractor
- information of any suspected contaminants
- details regarding former use to the material (e.g. office building, sugar factory, etc.)

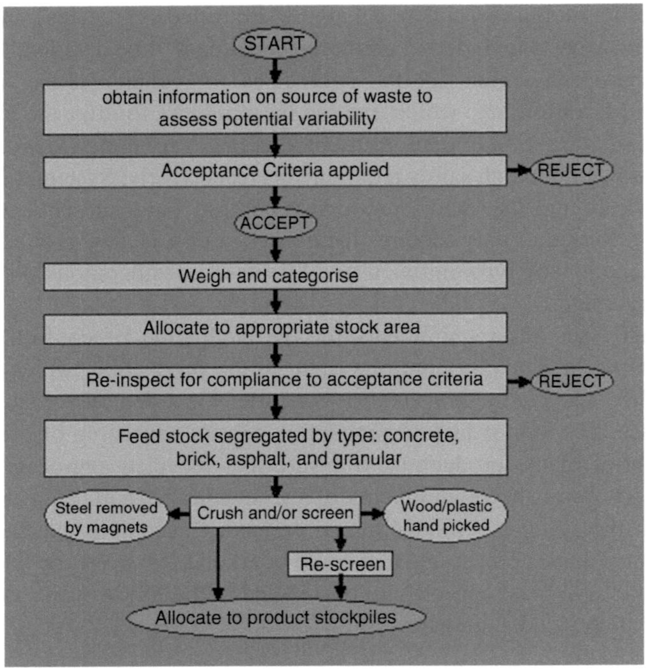

Fig. 6.21 Flow chart for accepting and processing of inert waste

- results obtained from previous tests (e.g. effects of chlorides or alkali-silica reactions)
- supplier's declaration that the material conforms to the data given in the delivery note

The Japanese standard for the use of recycled aggregate and recycled concrete establishes the following recommendations about the stockpile of recycled aggregate:

- To keep separately recycled aggregates from different concrete qualities.
- To keep separately coarse and fine recycled aggregates.
- To keep separately natural and recycled aggregates.
- As the absorption of coarse recycled aggregate is high, it is advisable to use it saturated so the aggregate hopper may have a sprinkler system in order to keep it in moisture conditions.
- Fine aggregate should not be stockpiled for a long time, because it may solidify.

To ensure that only inert waste is accepted, the producer should have a specific procedure on the "acceptance criteria", depending on the local waste management licence. The criteria should describe which types of wastes are accepted and the method of acceptance. An example of flow chart for accepting and processing of inert waste is shown in the following Fig. 6.21, taken from WRAP [9].

Table 6.11 Classification of recycled aggregates (classes H, L and M) in Japan [19]

Class of recycled aggregate	Requirements of recycled aggregate	Suggested concrete applications
Class H	Recycled aggregate which performs advanced processing of a separation, grinding down by friction and classification form the concrete mass generated by demolition of the structures.	It can be used in the main structure part of a concrete structure object on a par with natural river gravel and sand, and the macadam and crushed sand.
Class L	Recycled aggregate which crushed and manufactured concrete waste which arises when a concrete structure object is mainly demolished by the machines for beating and crushing and which has not performed advanced wastewater treatment.	It can be use don concrete without applying energy and costs. Three types of concrete are suggested: a stock item, a salt regulation article, and a technical specification order article.
Class M	Recycled aggregate which is processed by demolition, grinding down by friction and classification.	It can be used for components which cannot be easily influenced by drying shrinkage or freezing and thawing such as a stake, withsatanding-pressure version, a footing beam and steel-tubing in filled concrete.

The control of the incoming material must be done by means of a visual inspection to be carried out on every load, on initial receipt and after tipping, rejecting the consignment whenever the percentage of any contaminant or foreign material is higher than that defined in the acceptance criteria [9].

It is also important that staff responsible for reception inspection be specially trained for this job. In a certification scheme of the recycling plant, this training should be documented [4].

6.4.2 Processing

To optimize the use of recycled aggregate, the Japanese Industrial Standards (JIS) recently issued Recommendations on the use of recycled aggregate concrete, classifying it into three types: H, L and M. The requirements for each class, are shown in Table 6.11 [19].

6.4.3 Recycled Aggregates Stockpiling

Stocking and storage locations identification of the recycled aggregates has to be done in a controlled manner. Also, procedures must be applied to ensure that materials are put into, and taken from the stock in such a way that they do not suffer deterioration. This point is of special importance to recycled aggregates, as they are friable materials, so an unsuitable handling can easily alter the grading or fine contents of the aggregates at the stockpiles.

DIN 4226 specifies that there shall be in the recycling plant procedures to monitor stockpiling and to identify storage sites and their stocks. The products shall be identifiable up to the point of sale as regards source and type.

Some helpful recommendations for correct stockpiling and material handling procedures in order to avoid segregation, contamination or degradation of the aggregates during processing, are shown in the following Figs. 6.22, 6.23, 6.24, 6.25, and 6.26 (USA Guide for aggregate certification).

When recycled aggregates are transported in bulk, it may be necessary to cover or enclose the material so as to prevent contamination. If aggregates are packaged, the system shall not cause the aggregates to be contaminated or segregated (DIN 4226).

6.4.4 Sampling Procedures

In Angulo and Müller [20], it is pointed out that the regulations applied for natural aggregates sampling may be not sufficient, and must be supplemented considering the content of foreign materials. Thus, according to the European Standard EN 932-1, the minimum mass of a sample of aggregates should be calculated only by two parameters: the maximum grain size and the loose bulk density. This means that the composition heterogeneity, typical in recycled aggregates is not considered as a parameter. This study concludes that the mass sample for recycled aggregates should be higher than that recommended by the EN 932-1, especially for the determination of foreign materials such as gypsum and organic materials. For example, for a maximum particle size of 32 mm the standard establishes a representative mass of 50 kg, that gives a coefficient of variation higher than 5% in the composition determination when the concentration of the organic materials is lower than 3% or the concentration of gypsum is lower than 7%. So, the content of such danger foreign components in the recycled aggregates are not determined precisely.

On the other hand, the collection of the sample according to EN 932-1 can be done randomly in a three dimensional lot (conic pile). Reference 20 compares the results of dispersion applying this procedure to those when the collection is made of all section material taken from the conveyor belt, and then a homogenization by horizontal mixing piles. As can be observed in Fig. 6.27, the standard procedure gives higher coefficient of variation.

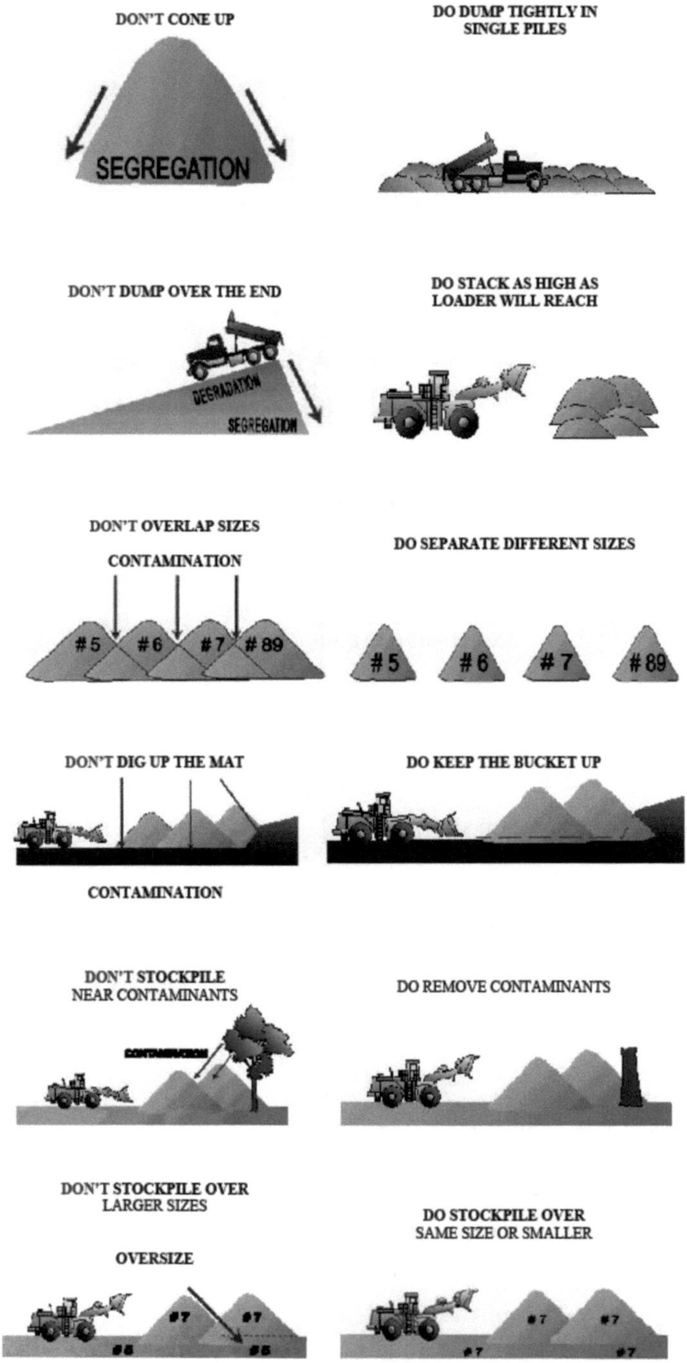

Fig. 6.22 Stockpiling techniques for clean stone

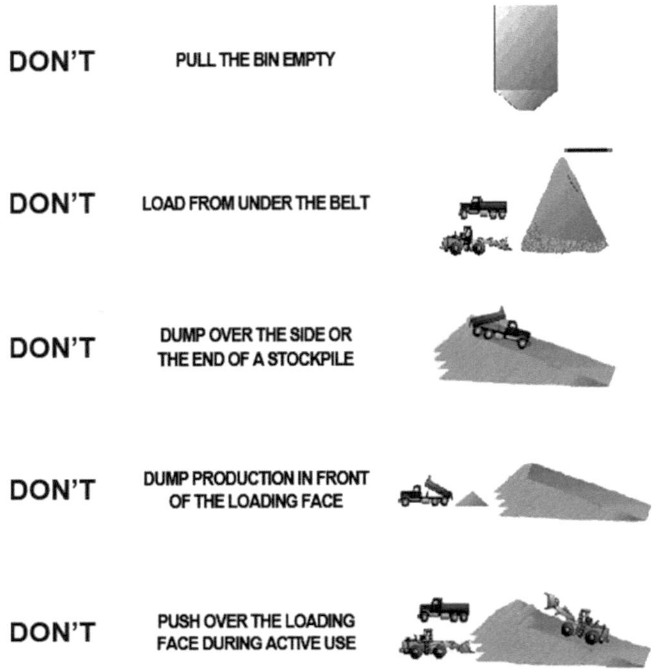

Fig. 6.23 Improper load out methods

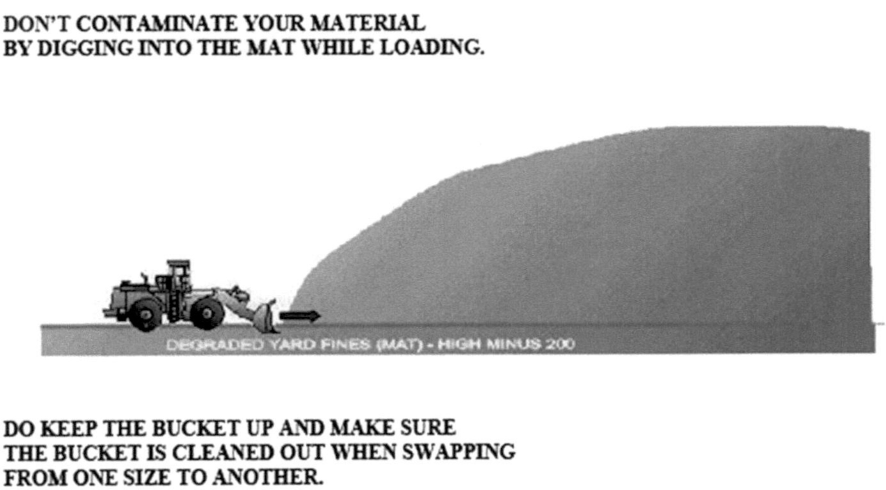

Fig. 6.24 Load out from stockpiles

THE MOST COMMON CONTAMINATIONS THAT OCCUR IN BINS ARE:

A) OVERSIZE THAT HANGS IN THE BINS
B) OVERSIZE THAT HANGS IN THE CHUTES
C) OVERSIZE THAT BOUNCES FROM SCREENS OR CONVEYORS
D) FINDS AND DIRTY WATER THAT STICK TO CONVEYORS AND DRIBBLES OFF ONTO OTHER AGGREGATES

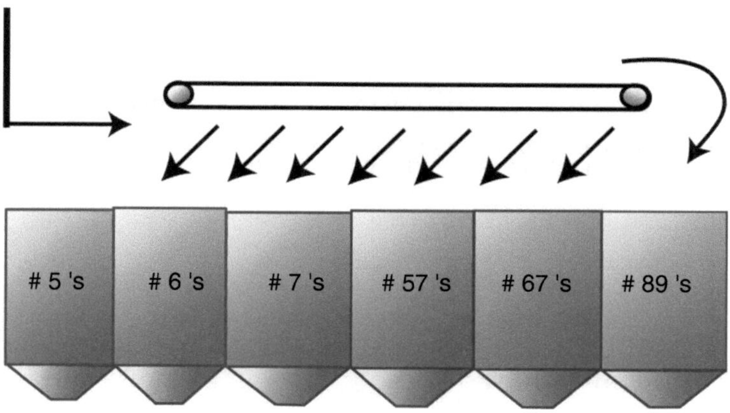

- CONTAMINATION FROM FINES IS TYPICALLY INSIGNIFICANT WHEN MATERIALS ARE STEADILY BEING LOADED OUT.
- CONTAMINATION BECOMES A PROBLEM WHEN A PARTICULAR BIN IS SUBJECTED TO IT FOR TOO LONG–INSPECT DAILY FOR FINES AND OVERSIZE.
- BE SURE TO INFORM THE SUPERVISOR OF WHAT YOU ARE OBVERVING–THERE MAY BE SOMETHING HE/SHE CAN DO MINIMIZE THIS TYPE OF CONTAMINATION.
- THERE IS NOTHING UNIQUE ABOUT THIS; IT HAPPENS ALL THE TIME. THE MAIN POINT HERE IS TO LOOK AT IT.
- BIN OPERATORS SHOULD ALSO CHECK TRUCK BEDS FOR CONTAMINATION PRIOR TO LOADING.

Fig. 6.25 Bin contamination

6.4.5 Tests and Frequency of Sampling

The frequency of sampling and the tests to be carried in the producer control system are shown in Table 6.12, for the general properties of both natural and recycled aggregates, according to EN 12620.

The European Standard establishes three levels of requirement to classify the result of the test:

- Just as a declared value, not under requirement
- It belongs to a specific Category
- The result Pass or Fail a threshold value

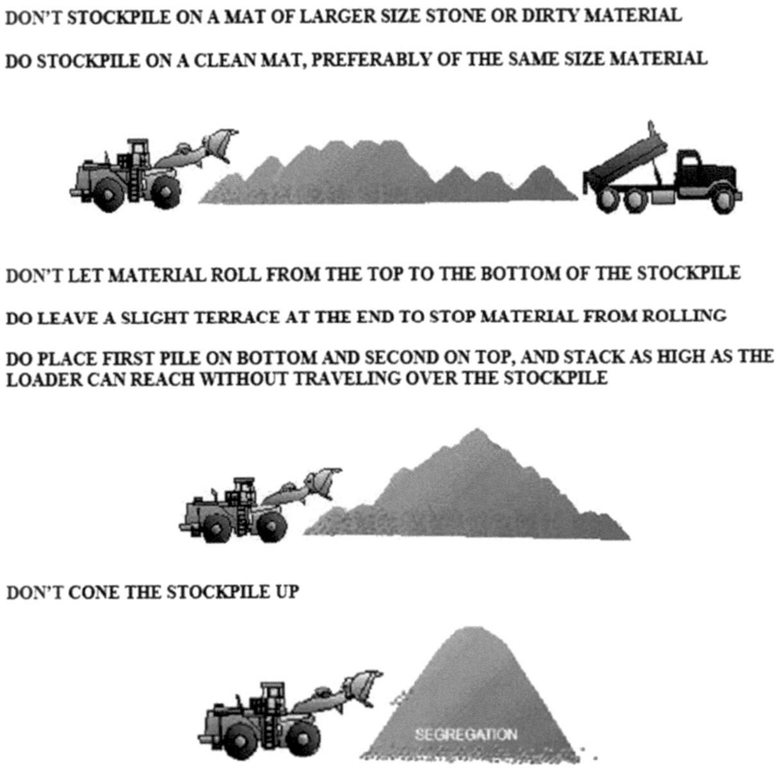

Fig. 6.26 Stockpiling clean stone with a loader

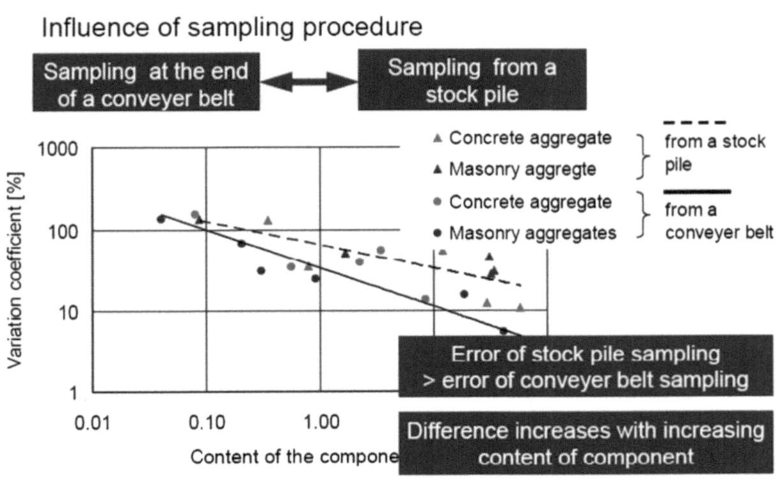

Fig. 6.27 Influence of the sampling procedure on the dispersion of results [1]

Table 6.12 Minimum test frequencies for general properties. Producers control (European standard)

Property	Clause	Notes/references	Test method	Minimum test frequency
1. Grading	4.3.1 4.3.6		EN 933-1 EN 933-10	1 per week
2. Shape of coarse aggregate	4.4	Test frequency applies to crushed aggregates. Test frequency for uncrushed gravel depends on the source and may be reduced	EN 933-3 EN 933-4	1 per month
3. Fines content	4.6		EN 933-1	1 per week
4. Fines quality	4.6	Only when required in accordance with the conditions specified in annex D.	EN 933-8 EN 933-9	1 per week
5. Particle density and water absorption	5.5		EN 1097-6	1 per year
6. Alkali-silica reactivity	5.7.3		[a]	When required and in case of doubt
7. Petrographic description	8.1		EN 932-3	1 per 3 years
8. Dangerous substances[b] In particular: Emission of radioactivity Release of heavy metals Release of polyaromatic carbons	H.3.3 H.4	[b]	[b]	When required and in case of doubt

[a] In accordance with the provisions valid in the place of use
[b] Unless otherwise specified, only when necessary for CE marking purposes (see annex ZA)

Few properties are controlled applying a threshold value, and most of them classify the aggregate in different categories. Depending on the end use of the aggregate, one category or another will be required. In the case of recycled aggregates, due to its nature usually are classified in the lower categories, in special for some properties (LA coefficient, absorption, etc.).

Additional controls, specific for recycled aggregates, are also included in BS EN 12620: influence on setting time of cement, constituents, particle density, absorption and water soluble sulfates, sulphur compounds and chloride content. Minimum test frequency to be carried by the producer is shown in Table 6.13. When the measured value is close to a specific limit, the frequency may need to be increased.

Table 6.13 Minimum test frequencies for aggregates from particular sources[a]. Producers control (European standard)

Property	Clause	Notes/references	Test method	Minimum test frequency
1. Shell content	4.5	Coarse aggregates of marine origin	EN 933-7	1 per year
2. Volume stability – Drying shrinkage	5.7.2		EN 1367-4	1 per 5 years
3. Chloride content	6.2	Aggregates of marine origin	EN 1744-1:1998, Clause 7	1 per week
		Recycled aggregates	EN 1744-5	2 per year
4. Sulfur containing compounds	6.3	Blastfurnace slag and recycled aggregates	EN 1744-1: 1998, Clause 12	2 per year
		Aggregates other than air-cooled blastfurnace slag and recycled aggregates	EN 1744-1:1998, Clause 12	1 per year
5. Organic substances:	6.4.1		EN 1744-1:1998, 15.1	1 per year
Humus content			EN 1744-1:1998, 15.2	1 per year
Fulvo acid (when indicated humus content is high)			EN 1744-1:1998, 15.3	1 per year
			EN 1744-1:1998, 14.2	2 per year
Comparative strength test – stiffening time				
Lightweight organic contaminators				
6. Dicalcium silicate disintegration	6.4.2.1	Blastfurnace slag only	EN 1744-1:1998, 19.1	2 per year
7. Iron disintegration	6.4.2.2	Blastfurnace slag only	EN 1744-1:1998, 19.2	2 per year
8. Influence on initial setting time of cement	6.4.1	Recycled aggregates only	EN 1744-6	2 per year
9. Constituents of coarse recycled aggregate	5.8	Coarse recycled aggregates only	prEN 933-11	1 per month
10. Particle density and water absorption	5.5	Coarse recycled aggregates only	EN 1097-6	1 per month
11. Water-soluble sulfate	6.3	Recycled aggregates only	EN 1744-1	1 per month

[a] For recycled aggregates, the source can be considered as the processing depot

Table 6.14 Minimum test frequencies. Producers control (WRAP)

Property description	BSEM test method	Minimum test frequency
General description	–	Every incoming load by visual Inspection
Aggregate composition including organics	Visual sorting of the plus 8 mm fraction[a]	1 per week[b]
Grading	933-1	1 per week[b]
Fines content	933-1	1 per week[b]
Partide shape[c]	933-3	1 per month[b]

Note: To illustrate suitability for a particular end use the test methods detailed in annex B may prove useful

[a] Test procedure detailed in highways agency specification for highway works clause 710

[b] Time periods relate to production periods not calendar periods

[c] For unbound aggregates PD 6682-6 recommends that 'no requirement' be adopted in the UK for particle shape

Similar minimum frequencies are recommended by WRAP [9] for recycled aggregates, with the exception of constituents, where the test frequency is more strict (1 per week), as can be observed in the following Table 6.14.

DIN 4226 includes a wide factory production control testing, as can be seen in the following Table 6.15. For specific uses, additional controls should be taken as shown in Table 6.15 (general properties) and Table 6.16 (specific uses).

Frequency of tests for bound and unbound use are shown in the following Table 6.17 according to Norway recommendations:

The frequency of testing should be increased or additional tests carried out in the following circumstances (DIN 4226):

- In case of severe variation in composition of the incoming material.
- When the source of the material changes, this being a function of the results of receiving inspection.
- When visual checks indicate any non-conformance indicated
- When measures values are close to a specified limit

On the other hand, the frequency of testing may be reduced (properly documenting the case) under special conditions (DIN 4226):

- highly automated production equipment
- long-term experience as regards constancy of special properties
- a high uniformity of the material supplied
- running an exceptional quality management system

It is considered very important that the producer has a certain minimum of laboratory equipment either on site or nearby, to make use of the results to manage the production [4]. It is mandatory (EN 12620) that the producer be responsible of the calibration and maintenance of the test equipment. Calibration records shall be retained.

As far as external inspections of the producer control system concerns, EN 12620 makes it mandatory for aggregates with high safety requirements (under

Table 6.15 Minimum test frequencies. Producers control (German standard)

Line	Item/Property	Cf. subclause	Notes	Relevant test method(s)	Minimum frequency of testing for aggregated type(s)	
					1, 2, 3	4
1	Processing equipment	A.3	–	Visual check	Daily	Daily
2	Source of material (receiving inspection)	A.4	–	Visual check	Each consignment	Each consignment
3	Grading	4.3 (DIN 4226-1)	–	DIN EN 933-1	Weekly	Weekly
4	Particle shape (coarse aggregate)	4.4 (DIN 4226-1)	–	DIN EN 933-1, DIN EN 933-4	Weekly	–
5	Finess content	4.6 (DIN 4226-1)	–	DIN EN 933-1	Weekly	Weekly
6	Assessment of lines	4.6 (DIN 4226-1)	If conditions set out in Annex D of DIN 4226-1 are met.	DIN EN 933-8, DIN EN 933-9	Weekly	Weekly
7	Composition	4.3	–	Weighing of constituents separated manually	Daily	Weekly
8	Particle density	4.4	–	DIN EN 1097-6	Weekly	Monthly
9	Water absorption	4.4	–	Cf. Annex D	Daily	–
10	Acid-soluble chloride content[a]	4.5	–	Cf. Annex E.	Twice a year	Twice a year
11	Sulfur/sulfate content[a]	6.3 (DIN 4226-1)	–	DIN EN 1744-1	Twice a year	–
12	Organic matter content	6.4.1 (DIN 4226-1)	–	DIN EN 1744-1	To be specified by inspection body, at least once a year,	–
13	Harmful substances[a]	4.8	–	Cf. Annex G.	Twice a year	Twice a year

[a] Not required if performed as part of third party inspection

Table 6.16 Minimum test frequencies for specific uses. Producers control (German standard)

Line	Property	Cf. subclause(s) of DIN 4226-1	Intended use	Relevant test method(s)	Minimum frequency of testing for aggregated type(s) 1, 2, 3	4
1	Resistance to fragmentation	5.2	–	DIN EN 1097-2	Twice a year	–
2	Resistance to wear	5.4	Wearing courses	DIN EN 1097-1	Once every 2 years	–
3	Resistance to polishing	5.5.1	Wearing courses	DIN EN 1097-8	Once every 2 years	–
4	Resistance to abrasion	5.5.2 and 5.5.3	Wearing courses	DIN EN 1097-8, DIN EN 1097-9	Once every 2 years	–
5	Frost resistance[a]	5.8.1	–	DIN EN 1367-1	Twice a year	–
6	Resistance to freezing and thawing[a]	5.8.1	–	DIN EN 1367-1, DIN EN 1367-2	Twice a year	–
7	Calcium carbonate content[a]	6.5	Fine aggregate for concrete wearing surfaces	DIN EN 196-21	Once every 2 years	–
8	Alkali-silica reaction	To be assessed by an expert in accordance with the *DAfstb-Richtlinie über vorbeugende Maßnahmen gegen schädigende Alkaireaktionen im Beton*				–

[a] See Table 6.15

certification System 2+), but no details can be found about how the surveillance of the production control must be carried out. DIN 4226-100, however, establishes minimum frequency of testing (general and specific use) carried out by a third party inspection body, as is shown in the following tables (Tables 6.18 and 6.19).

The inspection body shall check whether the test results conform to the requirements for the relevant category, and whether they are in compliance with those obtained from factory production control tests.

6.4.6 Document and Data Control

It is mandatory according to EN 12620 that the producer keeps documents concerning purchasing, processing, inspection of materials and in general all those affecting the factory production control system. In particular, recording of sampling locations, dates and times and product tested and other relevant information (e.g. weather conditions).

Table 6.17 Minimum test frequencies. Producers control (Norway recommendations)

			Test frequency at continuous production	
Property to be tested	Test method	To be tested by (type of lab.)	Products for bound use (for concrete)	Products for unbound use
Grading	EN 933-1	L[a]	One per week or min. per 2,000 t	
Fines content (materials <0.063 mm calculated on materials <19 mm)	EN 933-1	L[a]	One per week or min. per 2,000 t	
Content of materials <0.020 mm calculated on materials <19 mm		L[a]	–	If required
Material composition	prEN 933-11	L[a]	One per week or min. per 2,000 t	
Organic materials[b]	EN 1744-1	L[a]	One per week or min. per 3,000 t	–
Shape-Flakiness index (of materials >8 mm)	EN 933-3	L/C/E[c]	One per month	
Mechanical properties (Los Angles)	EN 1097-2	L/C/E[c]	–	One per 2 week or min. per 10,000 t
Density	EN 1097-6	L/C/E[c]	One per 2 week or min. per 10,000 t	
Water absorption	EN 1097-6	L/C/E[c]	One per 2 week or min. per 10,000 t	
Chloride content[b]	EN 1744-1	L/C/E[c]	One per 2 week or min. per 10,000 t	If required
Sulfur-containing compounds	EN 1744-1	L/C/E[c]	If required	If required
Chemical analyses[d] (Leaching test)	EN 1744-3	A[e]	–	One per 2 week or min. per 10,000 t

[a] Should be tested at the production site to be able to manage the production on basis of the test results

[b] The test frequency might be reduced by 50% if it is carried out pre-evaluation of the building according to publication no. 26 from the Norwegian Concrete Association

[c] Might be tested either locally, at a central laboratory or external

[d] Alternative method might be accepted

[e] To be tested either at a third party approved or accredited laboratory

Of particular importance are those records related to the incoming wastes and the final products, which shall be kept by the producer.

NEN EN 12620 specifies that the recycled aggregates shall be identified by:

- the source and the manufacturer (if the material has been re-handled in a depot both the source and depot shall be given).
- Type of aggregate (recycled in this case)
- The category according constituents

Table 6.18 Minimum test frequencies for general properties. External control (European standard)

Line	Item/Property	Cf. subclause	Notes	Relevant test method(s)	Minimum frequency of testing for aggregated type(s) 1, 2, 3	4
1	Processing equipment	A.3	–	Visual check	Once a year	Once a year
2	Source of material (receiving inspection)	A.4	–	Visual check	Twice a year	Once a year
3	Grading	4.3 (DIN 4226-1)	–	DIN EN 933-1	Twice a year	Once a year
4	Particle shape (coarse aggregate)	4.4 (DIN 4226-1)	–	DIN EN 933-3, DIN EN 933-4	Twice a year	Once a year
5	Finess content	4.6 (DIN 4226-1)	–	DIN EN 933-1	Once a year	Once a year
6	Assessment of lines	4.6 (DIN 4226-1)	If conditions set out in Annex D of DIN 4226-1 are met.	DIN EN 933-8, DIN EN 933-9	Once a year	Once a year
7	Composition	4.3	–	Assessment of constituents separated manually	Twice a year	Once a year
8	Particle density	4.4	–	DIN EN 1097-6	Twice a year	Once a year
9	Water absorption		–	Cf. Annex D	Twice a year	Once a year
10	Acid-soluble chloride content	4.5	–	Cf. Annex E.	Twice a year	Twice a year
11	Harmful substances content	4.8	–	Cf. Annex G.	Twice a year	Twice a year
12	Organic matter content	6.4.1 (DIN 4226-1)	–	DIN EN 1744-1	Twice a year	Once a year
13	Sulfur/sulfate content	6.3 (DIN 4226-1)	–	DIN EN 1744-1	Twice a year	Twice a year

Table 6.19 Minimum test frequencies for aggregates from particular sources. Producers control (European standard)

Line	Property	Cf. subclause(s) of DIN 4226-1	Intended use	Relevant test method(s)	Minimum frequency of testing for aggregated type(s)	
					1, 2, 3	4
1	Resistance to fragmentation	5.2	–	DIN EN 1097-2	Twice a year	–
2	Resistance to wear	5.4	Wearing courses	DIN EN 1097-1	Once every 2 years	–
3	Resistance to polishing	5.5.1	Wearing courses	DIN EN 1097-8	Once every 2 years	–
4	Resistance to abrasion	5.5.2 and 5.5.3	Wearing courses	DIN EN 1097-8, DIN EN 1097-9	Once every 2 years	–
5	Frost resistance	5.8.1	–	DIN EN 1367-1	Twice a year	–
6	Resistance to freezing and thawing	5.8.1	–	DIN EN 1367-1, DIN EN 1367-2	Twice a year	–
7	Calcium carbonate content	6.5	Fine aggregate for concrete wearing surfaces	DIN EN 196-21	Once every 2 years	–
8	Alkali-silica reaction	To be assessed by an expert in accordance with the *DAfstb-Richtlinie über vorbeugende Maßnahmen gegen schädigende Alkaireaktionen im Beton.*				–

The basic information to be provided by the supplier of the recycled aggregate according to DIN 4226 includes:

(a) Source of supply
(b) Material designation
(c) Aggregate type and size
(d) Average particle density
(e) Average water absorption

The purchaser should inform the supplier, at the time of ordering, of any special requirements for a particular end use and of any additional information needed of any property shown in Tables 4.4 and 4.5.

When requested by the purchaser, the producer should provide [21]:

(a) test results
(b) test procedures
(c) outline details of the factory production control manual

Based on the experience from his production and the test results, the producer shall work out a document of declared values for the requested properties dependent on the end use. This document shall also state within tolerances or maximum/ minimum levels the user can expect his material to be delivered [4].

All cases of non-conformity shall be recorded by the producer, investigated and if necessary corrective action shall be taken (EN 12620), controlling its effectiveness (DIN 4226). If the producer obtains results outside the declared values, he has to inform his customer [4].

Non conforming aggregates shall be reprocessed, assigned to another application for which be suitable or rejected. The manufacturer shall document and investigate the cause of nonconformity checking if the tests have been carried out properly, analyzing the processes, operations, quality records, etc. (DIN 4226).

6.5 Control Tests for Recycled Aggregates

The type of control tests for quality assurance of recycled aggregates, as for the natural aggregates, depends on its final utilization: roads, concrete, etc. According to WRAP [9] Recommendations, the following test methods may be used as a means of either deciding or illustrating suitability for a particular end use (Table 6.20).

Many of the controlled properties of recycled aggregates apply the same tests procedures than the natural aggregates, but some specific tests should also be carried out (DIN 4226-100):

- Determination of 10 min water absorption
- Determination of acid-soluble chloride content
- Evaluation for harmful substances
- Verification of the frost resistance of aggregate by testing concrete
- Effect of recycled aggregates on soil and groundwater

NEN 12620 indicates that new test methods necessary for recycled aggregates are at an advanced stage of preparation. On the other way some specific tests, different from those used for natural aggregates, are applied to recycled aggregates:

- Acid-soluble chloride content EN 1744-5 (instead of water–soluble chloride content of EN 1774-1:1998, clause 7)
- Water soluble sulphate content EN 1744-1:1998 (instead of acid-soluble sulphate content, EN 1744-1:1998, clause 12)
- Inorganic constituents that may adversely affect the rate of setting and hardening of concrete may be inorganic in recycled aggregates, so the procedure given in EN 1744-6 should be used.
- The magnesium sulphate test EN 1367-2 is unsuitable for recycled aggregates with cement-bound fractions. Resistance to freezing-thawing must be carried out under EN 1367-1 or using a test on concrete.

Table 6.20 Test methods for particular end uses

	Test reference	
All end uses	BS EN	BS
Particle density	1097-6	
Resistance to fragmentation:		
Los Angeles	1097-2	–
Bulk density	1097-3	
Use in concrete/hydraulically bound materials		
Water absorption	1097-6	
Magnesium sulfate	1367-2	–
Abrasion resistance:		
AAV	1097-8	
Drying shrinkage	1367-4	
Chlorides	1744-1	
Sulfate and sulfides	1744-1	
Alkali silica reaction*	–	–
Organic contamination	1744-1	–
*All RCA must be classed as highly reactive		
Uses as fill		
Water absorption	1097-6	
CBR	–	1377: Part 4
Plasticity of fines		1377: Part 2
Uses as unbound, pipe bedding		
Particle density	1097-6	
Resistance to fragmentation:		
Los Angeles	1097-2	–
Plasticity of fines	–	1377: Part 2
Frost heave		817: Part 124
Water soluble sulfate	1744-1	
Magnesium sulfate	1367-2	
Use in asphalt		
Particle density	1097-6	
Water absorption	1097-6	
Resistance to fragmentation:		
Los Angeles	1097-2	
Abrasion Resistance (AAV)	1097-8	
Polishing resistance	1097-8	
Resistance to heat	1367-5	

– To prevent alkali-silica reaction when using recycled aggregates, it will be necessary to ascertain that the original concrete does not contain reactive (or reacting) aggregate, and its alkali content will be determined and taken into account. The possibility of unpredictable compositional variability should be considered.

Regarding the control of the composition of recycled aggregates, the test based on visual analysis can introduce errors due a wrong identification of particles:

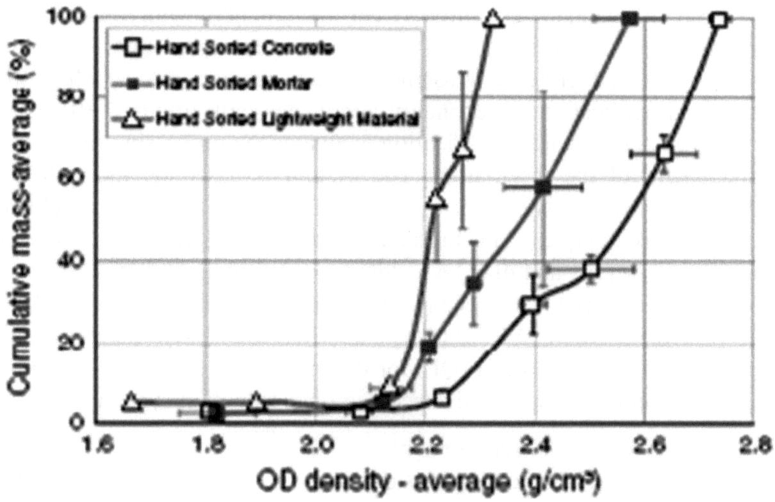

Fig. 6.28 Average cumulative OD density distributions and span of the measurements for densities and cumulative masses

between mortar and concrete; between gypsum lumps and plaster, mortar calcium silica bricks, etc.; errors due to unknown materials. This test requires experience of the laboratory technician who carries out it.

Reference 19 points out that a more accuracy results could be obtained if the control is carried out by means of the determination of the oven dry density of the aggregate instead of a visual analysis of its composition.

Moreover, a density distribution of particles could give the most complete information of the quality of the aggregate. Next Fig. 6.28 shows an example of the average cumulative oven dry density distribution for concrete, mortar and lightweight material particles of one sample of recycled aggregate.

The distribution of densities was done by a sequential sink float density separations, using sodium polytungstate solutions of different densities.

6.6 Voluntary Marks for High Grade Recycled Aggregates

Several countries have developed certification systems aimed to the production of high grade recycled aggregates.

In Belgium, the BENOR-mark certifies the conformity of the aggregates for concrete, including the recycled aggregates themselves and testifies that the product is in accordance with the European Standard for aggregates (EN 12620). In order to obtain this voluntary quality mark, the aggregates must already have been certified under the CE-Mark level 2+. The BENOR mark includes an autocontrol system for the producer, periodically confirmed by external certified bodies. There are more strict frequency control tests and also described procedures to manage non conformity products.

In the region of Flanders, Certipro® (a certification body with representatives of the authorities, research institutes and the sector) manages the Quarea certification system. This is an integrated environmental management system, which aims at production and application of secondary aggregates with minimal effect on man and the environment. Quarea-certification evaluates the processing of the material and management of the installation. The system is based on the principles of ISO14001. Technical measures for the installations are based on the selection of Best Available Techniques. The produced secondary aggregates are considered secondary raw materials and may be handled as a product. As such they lose their status of a waste material. The system is operational since September 2007.

In the United Kingdom, WRAP (Waste and Resources Action Programme) has developed a Quality Protocol [9] (QP) for the production of aggregates from inert waste. The Protocol covers several aspects in the production system:

- Waste management legal requirements
- Acceptance criteria of incoming waste
- Mandatory Factory Production Control, including
  - Manual of procedures
  - Manager responsible of FPC
  - Trained personnel
  - Records of all the quality control activities
  - Identifiability of the product up to the point of sale
  - Procedures to maintain the quality of the product during handling, storage, transport and delivery.
  - Minimum testing requirements – Frequencies

The Michigan Department of Transportation's (MDOT) strategy for the recycling and reuse of concrete aggregate is to use it if it enhances or equals the performance of virgin material in the final product. Statewide use of recycled concrete aggregate (RCA) is permitted in the Standard Specifications of Construction, 2003, Aggregate section 902.03 part B, 902.04, and 902.06. It allows the use of RCA as coarse aggregate for Portland cement concrete for curb and gutter, valley gutter, sidewalk, concrete barriers, driveways, temporary pavement, interchange ramps, and shoulders. RCA is also allowed as coarse aggregate in hot mix asphalt and as dense-graded aggregate for base course, surface course, shoulders, approaches and patching. RCA was widely in the pavement structure during the 1980s.

MDOT and local aggregate suppliers have been using RCA for over 19 years and have overcome some barriers. Recommendations given by MDOT and Detroit Metro region for using RCA in state highways for durability, bonding, and strength are:

- Quality assurance and quality control of materials is needed to assure meeting specifications.
- Normally commercial sources of concrete are not allowed for recycling in the crushing plants. Most recycled material comes from the MDOT's reconstruction projects. This assures a consistent source of original aggregate.

- Certification of recycling aggregate producers and the approval of stockpiles, when the primary source is the concrete from highways.
- Changes in the design on the permeable base allow RCA to be used when the density of material is increased and the design of the drainage system is modified.

Finally, in Singapore an accreditation scheme for Recycled Aggregate Supplier has been implemented, as an industrial led effort initiated by the Waste Management and Recycling Association of Singapore (WMRAS). The scheme applies for recycling plants supplying recycled concrete aggregates meeting requirements of SS EN 12620: Specification for Aggregates for Concrete. The accreditation protocol aims to improve the quality and consistency of the waste processors serving the construction industry. The final objective is to increase the Industry Professional and Users' confidence in specifying the use of recycled products in their projects.

The accreditation criteria are based on the following areas:

1. Company Financial Status
    - Requirement for minimum NTA (Net Tangible Asset) investment of S $250,000
    - Land Lease for a Minimal Period throughout the Accredited duration with Landlord and/or NEA approved usage for Recycling Activities (minimum of 2 years).
2. Facilities & Equipment
    - Production Crushing Facility equipped with input, crushing, transfer, output and selective separation device
    - Proper processing plant
    - Proper weighing facility (with calibration certificate)
    - Proper dispatch system for issuance of delivery records
    - Production Facility with Relevant Authority Approval or Support with regards to Pollution Control, Sewerage Discharge and WSHA requirements.
    - Declaration of Waste Stockpile Limit Set by Landlord (Declaration only)
3. Human Resource
    - Qualified Full-Time Trained Personnel to be employed
    - Manager (At least with Diploma in any discipline)
    - Production Supervisors (Experienced/In House OJT* – On the Job Training Programme)
    - QC/QA Supervisors (Experienced/In House OJT* – On the Job Training Programme)
    - *OJT Programme to be submitted for auditing
4. Quality Assurance Testing
    - Company should have a Quality Assurance Testing Program for regular checking and periodic testing of End Product by Accredited Laboratories

- Level of impurities (not more than 5% for Class 1 Recyclers) & Chemical composition of end product generated from plant to comply with Authority Requirement (Test Report to be submitted for Assessment)
- End Product Quality must Comply with SS EN 12620* and other relevant Codes/
- Guidelines in accordance to Project requirements as imposed by authorities from time to time (Quarterly conduct of Prescribed Test, please refer to Specifications and details in Application Form)
- Product Subject to Periodic Sampling Tests by Authority

5. Quality Management System & Documentation

Company should have a Quality Management Plan for the Management of Incoming Waste, Processing and End Product incorporating disposal and segregation of hazardous and non-recyclable material with Proper Documentation and Tracking of Quantities Company should have a nominated personnel for the preparation and submission of documentation and relevant data to the Authorities on a monthly basis (consistency of submission shall be a criteria for subsequent renewal of certification) (ISO Certification will be implemented in the future)

References

1. Mueller, A.: Determination of the composition of C&D recycled aggregates. In: NEWA, Viena Presentation at NEWA (2008)
2. Sagoe-Crentsil, K., Brown, T.: Guide for Specification of Recycled Concrete Aggregates (RCA) for Concrete Production. Ecorecycle, Victoria (1998)
3. Yanagi, K., Nakagawa, M., Hisaka, M., Kasai, Y.: Effect of impurities in recycled coarse aggregate upon a few properties of the concrete produced with it. In: Proceedings of the Second International RILEM Symposium on Demolition and Reuse of Concrete and Masonry, Japan, pp. 613–623 (1988)
4. Karlsen, J., Petkovic, G., Lahus, O.: A Norwegian Certification Scheme for Recycled Aggregate (RCA). Sustainable Buildings (2002)
5. Sánchez de Juan, M., Alaejos Gutiérrez, P.: Estudio sobre las propiedades del árido reciclado: utilización en hormigón estructural CEDEX (2006)
6. Mueller, A., Winkler, A.: Characteristics of processed concrete rubble. In: Use of Recycled Concrete Aggregate, Sustainable Construction, pp. 109–120. Dhir, Henderson y Limbachiya (1998)
7. Gestora de Runes de Cataluña: Data of CDW management. Author communication, Gestora de Runes de Cataluña, Barcelona (2010)
8. Waste and Resources Action Programme WRAP: Performance related approach to use of recycled aggregates. Report of the Aggregate Research Programme, Waste and Resources Action Programme WRAP, Banbury, Oxon, UK (2007)
9. Waste and Resources Action Programme WRAP: The quality protocol for the production of aggregates from inert waste. Final Report, Waste and Resources Action Programme WRAP, Banbury, Oxon, UK (2005)
10. KLEEMANN: Our press releases. In: Kleemann - Wirtgen Group. Available at: http://www.kleemann.info/en/news-and-media/press-releases/pressearchiv/pressreleases_2.html (2012). Accessed 2012

11. Buchner, S., Scholten, L.J.: Demolition and Construction Debris: Questionnaire about an EC Priority Waste Stream, The Hague, The Netherlands (1992)
12. Parekh, D., Modhera, C.: Assessment of recycled aggregate concrete. J. Eng. Res. Stud. **2,** 1–9 (January–March 2011)
13. Dosho, Y.: Development of a sustainable concrete waste recycling system. J. Adv. Concr. Technol. **5**(1), 27–42 (February 2007)
14. De Jong, T.P.R., Fabrizi, L., Kuilman, W.: Dry Density Separation of Mixed Construction and Demolition Waste. Delft (2006)
15. De Jong, T.P.R., Fabrizi, L., Kuilman, W.: Dry density separation of mixed construction and demolition waste. Recycl. Int. 24–27 (December 2006)
16. Vegas, I., Lisbona, A., Sánchez de Juan, M., Carvajal, M.D.: Characterization of mixed recycled aggregates for use in non-structural concrete: influence of treatment process. In: 2nd International RILEM Conference: Progress of Recycling in the Build Environment (2009)
17. Sánchez de Juan, M.: Tesis doctoral UPM. Escuela Técnica Superior de Ingenieros de Caminos. Canales y Puertos, Madrid (2004)
18. Sánchez de Juan, M., Alaejos, P., Geraldes, V., Vegas, I., Girbes, I., Marti, P., Agrela, F., Ayuso, J.: Uso de áridos reciclados mixtos procedentes de Residuos de Construcción y Demolición. Investigación prenormativa. Ihobe, Sociedad Pública de Gestión Ambiental. Departamento de Medio Ambiente, Planificación Territorial, Agricultura y Pesca. Gobierno Vasco (2011)
19. Tam, V.: Comparing the implementation of concrete recycling in the Australian and Japanese construction industries. J. Clean. Prod. **17**, 688–702 (2009)
20. Angulo, S.C., Müller, A.: Determination of construction and demolition recycled aggregates composition in considering their heterogeneity. Mater. Struct. **42**(6), 739–748 (2008)
21. Waste and Resources Action Programme WRAP: Coleman and Company's "Urban Quarry". Production of high value aggregates from construction, demolition and excavation waste. In: AggRegain. Available at: http://aggregain.wrap.org.uk/case_studies/coleman_and.html (2012). Accessed 2012

RILEM Publications

The following list is presenting our global offer, sorted by series.

RILEM Proceedings

PRO 1: Durability of High Performance Concrete (1994) 266 pp., ISBN: 2-91214-303-9; e-ISBN: 2-35158-012-5; *Ed. H. Sommer*
PRO 2: Chloride Penetration into Concrete (1995) 496 pp., ISBN: 2-912143-00-4; e-ISBN: 2-912143-45-4; *Eds. L.-O. Nilsson and J.-P. Ollivier*
PRO 3: Evaluation and Strengthening of Existing Masonry Structures (1995) 234 pp., ISBN: 2-912143-02-0; e-ISBN: 2-351580-14-1; *Eds. L. Binda and C. Modena*
PRO 4: Concrete: From Material to Structure (1996) 360 pp., ISBN: 2-912143-04-7; e-ISBN: 2-35158-020-6; *Eds. J.-P. Bournazel and Y. Malier*
PRO 5: The Role of Admixtures in High Performance Concrete (1999) 520 pp., ISBN: 2-912143-05-5; e-ISBN: 2-35158-021-4; *Eds. J. G. Cabrera and R. Rivera-Villarreal*
PRO 6: High Performance Fiber Reinforced Cement Composites (HPFRCC 3) (1999) 686 pp., ISBN: 2-912143-06-3; e-ISBN: 2-35158-022-2; *Eds. H. W. Reinhardt and A. E. Naaman*
PRO 7: 1st International RILEM Symposium on Self-Compacting Concrete (1999) 804 pp., ISBN: 2-912143-09-8; e-ISBN: 2-912143-72-1; *Eds. Å. Skarendahl and Ö. Petersson*
PRO 8: International RILEM Symposium on Timber Engineering (1999) 860 pp., ISBN: 2-912143-10-1; e-ISBN: 2-35158-023-0; *Ed. L. Boström*
PRO 9: 2nd International RILEM Symposium on Adhesion between Polymers and Concrete ISAP'99 (1999) 600 pp., ISBN: 2-912143-11-X; e-ISBN: 2-35158-024-9; *Eds. Y. Ohama and M. Puterman*

E. Vázquez (ed.), *Progress of Recycling in the Built Environment: Final Report of the RILEM Technical Committee 217-PRE*, RILEM State-of-the-Art Reports 8,
DOI 10.1007/978-94-007-4908-5,

PRO 10: 3rd International RILEM Symposium on Durability of Building and Construction Sealants (2000) 360 pp., ISBN: 2-912143-13-6; e-ISBN: 2-351580-25-7; *Eds. A. T. Wolf*

PRO 11: 4th International RILEM Conference on Reflective Cracking in Pavements (2000) 549 pp., ISBN: 2-912143-14-4; e-ISBN: 2-35158-026-5; *Eds. A. O. Abd El Halim, D. A. Taylor andEl H. H. Mohamed*

PRO 12: International RILEM Workshop on Historic Mortars: Characteristics and Tests (1999) 460 pp., ISBN: 2-912143-15-2; e-ISBN: 2-351580-27-3; *Eds. P. Bartos, C. Groot and J. J. Hughes*

PRO 13: 2nd International RILEM Symposium on Hydration and Setting (1997) 438 pp., ISBN: 2-912143-16-0; e-ISBN: 2-35158-028-1; *Ed. A. Nonat*

PRO 14: Integrated Life-Cycle Design of Materials and Structures (ILCDES 2000) (2000) 550 pp., ISBN: 951-758-408-3; e-ISBN: 2-351580-29-X, ISSN: 0356–9403; *Ed. S. Sarja*

PRO 15: Fifth RILEM Symposium on Fibre-Reinforced Concretes (FRC) – BEFIB'2000 (2000) 810 pp., ISBN: 2-912143-18-7; e-ISBN: 2-912143-73-X; *Eds. P. Rossi and G. Chanvillard*

PRO 16: Life Prediction and Management of Concrete Structures (2000) 242 pp., ISBN: 2-912143-19-5; e-ISBN: 2-351580-30-3; *Ed. D. Naus*

PRO 17: Shrinkage of Concrete – Shrinkage 2000 (2000) 586 pp., ISBN: 2-912143-20-9; e-ISBN: 2-351580-31-1; *Eds. V. Baroghel-Bouny and P.-C. Aïtcin*

PRO 18: Measurement and Interpretation of the On-Site Corrosion Rate (1999) 238 pp., ISBN: 2-912143-21-7; e-ISBN: 2-351580-32-X; *Eds. C. Andrade, C. Alonso, J. Fullea, J. Polimon and J. Rodriguez*

PRO 19: Testing and Modelling the Chloride Ingress into Concrete (2000) 516 pp., ISBN: 2-912143-22-5; e-ISBN: 2-351580-33-8; Soft cover, *Eds. C. Andrade and J. Kropp*

PRO 20: 1st International RILEM Workshop on Microbial Impacts on Building Materials (2000) 74 pp., e-ISBN: 2-35158-013-3; *Ed. M. Ribas Silva (CD 02)*

PRO 21: International RILEM Symposium on Connections between Steel and Concrete (2001) 1448 pp., ISBN: 2-912143-25-X; e-ISBN: 2-351580-34-6; *Ed. R. Eligehausen*

PRO 22: International RILEM Symposium on Joints in Timber Structures (2001) 672 pp., ISBN: 2-912143-28-4; e-ISBN: 2-351580-35-4; *Eds. S. Aicher and H.-W. Reinhardt*

PRO 23: International RILEM Conference on Early Age Cracking in Cementitious Systems (2003) 398 pp., ISBN: 2-912143-29-2; e-ISBN: 2-351580-36-2; *Eds. K. Kovler and A. Bentur*

PRO 24: 2nd International RILEM Workshop on Frost Resistance of Concrete (2002) 400 pp., ISBN: 2-912143-30-6; e-ISBN: 2-351580-37-0, Hard back; *Eds. M. J. Setzer, R. Auberg and H.-J. Keck*

PRO 25: International RILEM Workshop on Frost Damage in Concrete (1999) 312 pp., ISBN: 2-912143-31-4; e-ISBN: 2-351580-38-9, Soft cover; *Eds. D. J. Janssen, M. J. Setzer and M. B. Snyder*

PRO 26: International RILEM Workshop on On-Site Control and Evaluation of Masonry Structures (2003) 386 pp., ISBN: 2-912143-34-9; e-ISBN: 2-351580-14-1, Soft cover; *Eds. L. Binda and R. C. de Vekey*

PRO 27: International RILEM Symposium on Building Joint Sealants (1988) 240 pp., e-ISBN: 2-351580-15-X; *Ed. A. T. Wolf, (CD03)*

PRO 28: 6th International RILEM Symposium on Performance Testing and Evaluation of Bituminous Materials, PTEBM'03, Zurich, Switzerland (2003) 652 pp., ISBN: 2-912143-35-7; e-ISBN: 2-912143-77-2, Soft cover; *Ed. M. N. Partl (CD06)*

PRO 29: 2nd International RILEM Workshop on Life Prediction and Ageing Management of Concrete Structures, Paris, France (2003) 402 pp., ISBN: 2-912143-36-5; e-ISBN: 2-912143-78-0, Soft cover; *Ed. D. J. Naus*

PRO 30: 4th International RILEM Workshop on High Performance Fiber Reinforced Cement Composites – HPFRCC 4, University of Michigan, Ann Arbor, USA (2003) 562 pp., ISBN: 2-912143-37-3; e-ISBN: 2-912143-79-9, Hard back; *Eds. A. E. Naaman and H. W. Reinhardt*

PRO 31: International RILEM Workshop on Test and Design Methods for Steel Fibre Reinforced Concrete: Background and Experiences (2003) 230 pp., ISBN: 2-912143-38-1; e-ISBN: 2-351580-16-8, Soft cover; *Eds. B. Schnütgen and L. Vandewalle*

PRO 32: International Conference on Advances in Concrete and Structures, 2 volumes (2003) 1592 pp., ISBN (set): 2-912143-41-1; e-ISBN: 2-351580-17-6, Soft cover; *Eds. Ying-shu Yuan, Surendra P. Shah and Heng-lin Lü*

PRO 33: 3rd International Symposium on Self-Compacting Concrete (2003) 1048 pp., ISBN: 2-912143-42-X; e-ISBN: 2-912143-71-3, Soft cover; *Eds. Ó. Wallevik and I. Níelsson*

PRO 34: International RILEM Conference on Microbial Impact on Building Materials (2003) 108 pp., ISBN: 2-912143-43-8; e-ISBN: 2-351580-18-4; *Ed. M. Ribas Silva*

PRO 35: International RILEM TC 186-ISA on Internal Sulfate Attack and Delayed Ettringite Formation (2002) 316 pp., ISBN: 2-912143-44-6; e-ISBN: 2-912143-80-2, Soft cover; *Eds. K. Scrivener and J. Skalny*

PRO 36: International RILEM Symposium on Concrete Science and Engineering – A Tribute to Arnon Bentur (2004) 264 pp., ISBN: 2-912143-46-2; e-ISBN: 2-912143-58-6, Hard back; *Eds. K. Kovler, J. Marchand, S. Mindess and J. Weiss*

PRO 37: 5th International RILEM Conference on Cracking in Pavements – Mitigation, Risk Assessment and Prevention (2004) 740 pp., ISBN: 2-912143-47-0; e-ISBN: 2-912143-76-4, Hard back; *Eds. C. Petit, I. Al-Qadi and A. Millien*

PRO 38: 3rd International RILEM Workshop on Testing and Modelling the Chloride Ingress into Concrete (2002) 462 pp., ISBN: 2-912143-48-9; e-ISBN: 2-912143-57-8, Soft cover; *Eds. C. Andrade and J. Kropp*

PRO 39: 6th International RILEM Symposium on Fibre-Reinforced Concretes (BEFIB 2004), 2 volumes, (2004) 1536 pp., ISBN: 2-912143-51-9 (set); e-ISBN: 2-912143-74-8, Hard back; *Eds. M. Di Prisco, R. Felicetti and G. A. Plizzari*

PRO 40: International RILEM Conference on the Use of Recycled Materials in Buildings and Structures (2004) 1154 pp., ISBN: 2-912143-52-7 (set); e-ISBN: 2-912143-75-6, Soft cover; *Eds. E. Vázquez, Ch. F. Hendriks and G. M. T. Janssen*
PRO 41: RILEM International Symposium on Environment-Conscious Materials and Systems for Sustainable Development (2005) 450 pp., ISBN: 2-912143-55-1; e-ISBN: 2-912143-64-0, Soft cover; *Eds. N. Kashino and Y. Ohama*
PRO 42: SCC'2005 – China: 1st International Symposium on Design, Performance and Use of Self-Consolidating Concrete (2005) 726 pp., ISBN: 2-912143-61-6; e-ISBN: 2-912143-62-4, Hard back; *Eds. Zhiwu Yu, Caijun Shi, Kamal Henri Khayat and Youjun Xie*
PRO 43: International RILEM Workshop on Bonded Concrete Overlays (2004) 114 pp., e-ISBN: 2-912143-83-7; *Eds. J. L. Granju and J. Silfwerbrand*
PRO 44: 2nd International RILEM Workshop on Microbial Impacts on Building Materials (Brazil 2004) (CD11) 90 pp., e-ISBN: 2-912143-84-5; *Ed. M. Ribas Silva*
PRO 45: 2nd International Symposium on Nanotechnology in Construction, Bilbao, Spain (2005) 414 pp., ISBN: 2-912143-87-X; e-ISBN: 2-912143-88-8, Soft cover; *Eds. Peter J. M. Bartos, Yolanda de Miguel and Antonio Porro*
PRO 46: ConcreteLife'06 – International RILEM-JCI Seminar on Concrete Durability and Service Life Planning: Curing, Crack Control, Performance in Harsh Environments (2006) 526 pp., ISBN: 2-912143-89-6; e-ISBN: 2-912143-90-X, Hard back; *Ed. K. Kovler*
PRO 47: International RILEM Workshop on Performance Based Evaluation and Indicators for Concrete Durability (2007) 385 pp., ISBN: 978-2-912143-95-2; e-ISBN: 978-2-912143-96-9, Soft cover; *Eds. V. Baroghel-Bouny, C. Andrade, R. Torrent and K. Scrivener*
PRO 48: 1st International RILEM Symposium on Advances in Concrete through Science and Engineering (2004) 1616 pp., e-ISBN: 2-912143-92-6; *Eds. J. Weiss, K. Kovler, J. Marchand, and S. Mindess*
PRO 49: International RILEM Workshop on High Performance Fiber Reinforced Cementitious Composites in Structural Applications (2006) 598 pp., ISBN: 2-912143-93-4; e-ISBN: 2-912143-94-2, Soft cover; *Eds. G. Fischer and V.C. Li*
PRO 50: 1st International RILEM Symposium on Textile Reinforced Concrete (2006) 418 pp., ISBN: 2-912143-97-7; e-ISBN: 2-351580-08-7, Soft cover; *Eds. Josef Hegger, Wolfgang Brameshuber and Norbert Will*
PRO 51: 2nd International Symposium on Advances in Concrete through Science and Engineering (2006) 462 pp., ISBN: 2-35158-003-6; e-ISBN: 2-35158-002-8, Hard back; *Eds. J. Marchand, B. Bissonnette, R. Gagné, M. Jolin and F. Paradis*
PRO 52: Volume Changes of Hardening Concrete: Testing and Mitigation (2006) 428 pp., ISBN: 2-35158-004-4; e-ISBN: 2-35158-005-2, Soft cover; *Eds. O. M. Jensen, P. Lura and K. Kovler*
PRO 53: High Performance Fiber Reinforced Cement Composites HPFRCC5 (2007) 542 pp., ISBN: 978-2-35158-046-2; e-ISBN: 978-2-35158-089-9, Hard back; *Eds. H. W. Reinhardt and A. E. Naaman*

PRO 54: 5th International RILEM Symposium on Self-Compacting Concrete, 3 Volumes (2007) 1198 pp., ISBN: 978-2-35158-047-9; e-ISBN: 978-2-35158-088-2, Soft cover; *Eds. G. De Schutter and V. Boel*

PRO 55: International RILEM Symposium Photocatalysis, Environment and Construction Materials (2007) 350 pp., ISBN: 978-2-35158-056-1; e-ISBN: 978-2-35158-057-8, Soft cover; *Eds. P. Baglioni and L. Cassar*

PRO 56: International RILEM Workshop on Integral Service Life Modelling of Concrete Structures (2007) 458 pp., ISBN 978-2-35158-058-5; e-ISBN: 978-2-35158-090-5, Hard back; *Eds. R. M. Ferreira, J. Gulikers and C. Andrade*

PRO 57: RILEM Workshop on Performance of cement-based materials in aggressive aqueous environments (2008) 132 pp., e-ISBN: 978-2-35158-059-2; *Ed. N. De Belie*

PRO 58: International RILEM Symposium on Concrete Modelling CONMOD'08 (2008) 847 pp., ISBN: 978-2-35158-060-8; e-ISBN: 978-2-35158-076-9, Soft cover; *Eds. E. Schlangen and G. De Schutter*

PRO 59: International RILEM Conference on On Site Assessment of Concrete, Masonry and Timber Structures SACoMaTiS 2008, 2 volumes (2008) 1232 pp., ISBN: 978-2-35158-061-5 (set); e-ISBN: 978-2-35158-075-2, Hard back; *Eds. L. Binda, M. di Prisco and R. Felicetti*

PRO 60: Seventh RILEM International Symposium (BEFIB 2008) on Fibre Reinforced Concrete: Design and Applications (2008) 1181 pp, ISBN: 978-2-35158-064-6; e-ISBN: 978-2-35158-086-8, Hard back; *Ed. R. Gettu*

PRO 61: 1st International Conference on Microstructure Related Durability of Cementitious Composites (Nanjing), 2 volumes, (2008) 1524 pp., ISBN: 978-2-35158-065-3; e-ISBN: 978-2-35158-084-4; *Eds. W. Sun, K. van Breugel, C. Miao, G. Ye and H. Chen*

PRO 62: NSF/RILEM Workshop: In-situ Evaluation of Historic Wood and Masonry Structures (2008) 130 pp., e-ISBN: 978-2-35158-068-4; *Eds. B. Kasal, R. Anthony and M. Drdácký*

PRO 63: Concrete in Aggressive Aqueous Environments: Performance, Testing and Modelling, 2 volumes, (2009) 631 pp., ISBN: 978-2-35158-071-4; e-ISBN: 978-2-35158-082-0, Soft cover; *Eds. M. G. Alexander and A. Bertron*

PRO 64: Long Term Performance of Cementitious Barriers and Reinforced Concrete in Nuclear Power Plants and Waste Management – NUCPERF 2009 (2009) 359 pp., ISBN: 978-2-35158-072-1; e-ISBN: 978-2-35158-087-5; *Eds. V. L'Hostis, R. Gens, C. Gallé*

PRO 65: Design Performance and Use of Self-consolidating Concrete, SCC'2009, (2009) 913 pp., ISBN: 978-2-35158-073-8; e-ISBN: 978-2-35158-093-6; *Eds. C. Shi, Z. Yu, K. H. Khayat and P. Yan*

PRO 66: Concrete Durability and Service Life Planning, 2nd International RILEM Workshop, ConcreteLife'09, (2009) 626 pp., ISBN: 978-2-35158-074-5; e-ISBN: 978-2-35158-085-1; *Ed. K. Kovler*

PRO 67: Repairs Mortars for Historic Masonry (2009) 397 pp., e-ISBN: 978-2-35158-083-7; *Ed. C. Groot*

PRO 68: Proceedings of the 3rd International RILEM Symposium on 'Rheology of Cement Suspensions such as Fresh Concrete' (2009) 372 pp., ISBN: 978-2-35158-091-2; e-ISBN: 978-2-35158-092-9; *Eds. O. H. Wallevik, S. Kubens and S. Oesterheld*
PRO 69: 3rd International PhD Student Workshop on 'Modelling the Durability of Reinforced Concrete' (2009) 122 pp., ISBN: 978-2-35158-095-0; e-ISBN: 978-2-35158-094-3; *Eds. R. M. Ferreira, J. Gulikers and C. Andrade*
PRO 71: Advances in Civil Engineering Materials, Proceedings of the 'The 50-year Teaching Anniversary of Prof. Sun Wei', (2010) 307 pp., ISBN: 978-2-35158-098-1; e-ISBN: 978-2-35158-099-8; *Eds. C. Miao, G. Ye, and H. Chen*
PRO 74: International RILEM Conference on 'Use of Superabsorsorbent Polymers and Other New Additives in Concrete' (2010) 374 pp., ISBN: 978-2-35158-104-9; e-ISBN: 978-2-35158-105-6; *Eds. O.M. Jensen, M.T. Hasholt, and S. Laustsen*
PRO 75: International Conference on 'Material Science – 2nd ICTRC – Textile Reinforced Concrete – Theme 1' (2010) 436 pp., ISBN: 978-2-35158-106-3; e-ISBN: 978-2-35158-107-0; *Ed. W. Brameshuber*
PRO 76: International Conference on 'Material Science – HetMat – Modelling of Heterogeneous Materials – Theme 2' (2010) 255 pp., ISBN: 978-2-35158-108-7; e-ISBN: 978-2-35158-109-4; *Ed. W. Brameshuber*
PRO 77: International Conference on 'Material Science – AdIPoC – Additions Improving Properties of Concrete – Theme 3' (2010) 459 pp., ISBN: 978-2-35158-110-0; e-ISBN: 978-2-35158-111-7; *Ed. W. Brameshuber*
PRO 78: 2nd Historic Mortars Conference and RILEM TC 203-RHM Final Workshop – HMC2010 (2010) 1416 pp., e-ISBN: 978-2-35158-112-4; *Eds J. Válek, C. Groot, and J. J. Hughes*
PRO 79: International RILEM Conference on Advances in Construction Materials Through Science and Engineering (2011) 213 pp., e-ISBN: 978-2-35158-117-9; *Eds Christopher Leung and K.T. Wan*
PRO 80: 2nd International RILEM Conference on Concrete Spalling due to Fire Exposure (2011) 453 pp., ISBN: 978-2-35158-118-6, e-ISBN: 978-2-35158-119-3; *Eds E.A.B.. Koenders and F. Dehn*
PRO 81: 2nd International RILEM Conference on Strain Hardening Cementitious Composites (SHCC2-Rio) (2011) 451 pp., ISBN: 978-2-35158-120-9, e-ISBN: 978-2-35158-121-6; *Eds R.D. Toledo Filho, F.A. Silva, E.A.B.. Koenders and E. M.R. Fairbairn*
PRO 82: 2nd International RILEM Conference on Progress of Recycling in the Built Environment (2011) 507 pp., e-ISBN: 978-2-35158-122-3; *Eds V.M. John, E. Vazquez, S.C. Angulo and C. Ulsen*
PRO 85: RILEM-JCI International Workshop on Crack Control of Mass Concrete and Related Issues concerning Early-Age of Concrete Structures – ConCrack 3 (2012) 237 pp., ISBN: 978-2-35158-125-4, e-ISBN: 978-2-35158-126-1; *Eds F. Toutlemonde and J.M. Torrenti*

RILEM Reports

Report 19: Considerations for Use in Managing the Aging of Nuclear Power Plant Concrete Structures (1999) 224 pp., ISBN: 2-912143-07-1; e-ISBN: 2-35158-039-7; *Ed. D. J. Naus*
Report 20: Engineering and Transport Properties of the Interfacial Transition Zone in Cementitious Composites (1999) 396 pp., ISBN: 2-912143-08-X; e-ISBN: 2-35158-040-0; *Eds. M. G. Alexander, G. Arliguie, G. Ballivy, A. Bentur and J. Marchand*
Report 21: Durability of Building Sealants (1999) 450 pp., ISBN: 2-912143-12-8; e-ISBN: 2-35158-041-9; *Ed. A. T. Wolf*
Report 22: Sustainable Raw Materials – Construction and Demolition Waste (2000) 202 pp., ISBN: 2-912143-17-9; e-ISBN: 2-35158-042-7; *Eds. C. F. Hendriks and H. S. Pietersen*
Report 23: Self-Compacting Concrete state-of-the-art report (2001) 166 pp., ISBN: 2-912143-23-3; e-ISBN: 2-912143-59-4, Soft cover; *Eds. Å. Skarendahl and Ö. Petersson*
Report 24: Workability and Rheology of Fresh Concrete: Compendium of Tests (2002) 154 pp., ISBN: 2-912143-32-2; e-ISBN: 2-35158-043-5, Soft cover; *Eds. P. J. M. Bartos, M. Sonebi andA. K. Tamimi*
Report 25: Early Age Cracking in Cementitious Systems (2003) 350 pp., ISBN: 2-912143-33-0; e-ISBN: 2-912143-63-2, Soft cover; *Ed. A. Bentur*
Report 26: Towards Sustainable Roofing (Joint Committee CIB/RILEM) (CD 07), (2001) 28 pp., e-ISBN: 2-912143-65-9; *Eds. Thomas W. Hutchinson and Keith Roberts*
Report 27: Condition Assessment of Roofs (Joint Committee CIB/RILEM) (CD 08), (2003) 12 pp., e-ISBN: 2-912143-66-7
Report 28: Final Report of RILEM TC 167-COM 'Characterisation of Old Mortars with Respect to Their Repair' (2007) 192 pp., ISBN: 978-2-912143-56-3; e-ISBN: 978-2-912143-67-9, Soft cover;*Eds. C. Groot, G. Ashall and J. Hughes*
Report 29: Pavement Performance Prediction and Evaluation (PPPE): Interlaboratory Tests (2005) 194 pp., e-ISBN: 2-912143-68-3; *Eds. M. Partl and H. Piber*
Report 30: Final Report of RILEM TC 198-URM 'Use of Recycled Materials' (2005) 74 pp., ISBN: 2-912143-82-9; e-ISBN: 2-912143-69-1 – Soft cover; *Eds. Ch. F. Hendriks, G. M. T. Janssen and E. Vázquez*
Report 31: Final Report of RILEM TC 185-ATC 'Advanced testing of cement-based materials during setting and hardening' (2005) 362 pp., ISBN: 2-912143-81-0; e-ISBN: 2-912143-70-5 – Soft cover; *Eds. H. W. Reinhardt and C. U. Grosse*
Report 32: Probabilistic Assessment of Existing Structures. A JCSS publication (2001) 176 pp., ISBN 2-912143-24-1; e-ISBN: 2-912143-60-8 – Hard back; *Ed. D. Diamantidis*

Report 33: State-of-the-Art Report of RILEM Technical Committee TC 184-IFE 'Industrial Floors' (2006) 158 pp., ISBN 2-35158-006-0; e-ISBN: 2-35158-007-9, Soft cover; *Ed. P. Seidler*

Report 34: Report of RILEM Technical Committee TC 147-FMB 'Fracture mechanics applications to anchorage and bond' Tension of Reinforced Concrete Prisms – Round Robin Analysis and Tests on Bond (2001) 248 pp., e-ISBN 2-912143-91-8; *Eds. L. Elfgren and K. Noghabai*

Report 35: Final Report of RILEM Technical Committee TC 188-CSC 'Casting of Self Compacting Concrete' (2006) 40 pp., ISBN 2-35158-001-X; e-ISBN: 2-912143-98-5 – Soft cover; *Eds. Å. Skarendahl and P. Billberg*

Report 36: State-of-the-Art Report of RILEM Technical Committee TC 201-TRC 'Textile Reinforced Concrete' (2006) 292 pp., ISBN 2-912143-99-3; e-ISBN: 2-35158-000-1, Soft cover; *Ed. W. Brameshuber*

Report 37: State-of-the-Art Report of RILEM Technical Committee TC 192-ECM 'Environment-conscious construction materials and systems' (2007) 88 pp., ISBN: 978-2-35158-053-0; e-ISBN: 2-35158-079-0, Soft cover; *Eds. N. Kashino, D. Van Gemert and K. Imamoto*

Report 38: State-of-the-Art Report of RILEM Technical Committee TC 205-DSC 'Durability of Self-Compacting Concrete' (2007) 204 pp., ISBN: 978-2-35158-048-6; e-ISBN: 2-35158-077-6, Soft cover; *Eds. G. De Schutter and K. Audenaert*

Report 39: Final Report of RILEM Technical Committee TC 187-SOC 'Experimental determination of the stress-crack opening curve for concrete in tension' (2007) 54 pp., ISBN 978-2-35158-049-3; e-ISBN: 978-2-35158-078-3, Soft cover; *Ed. J. Planas*

Report 40: State-of-the-Art Report of RILEM Technical Committee TC 189-NEC 'Non-Destructive Evaluation of the Penetrability and Thickness of the Concrete Cover' (2007) 246 pp., ISBN 978-2-35158-054-7; e-ISBN: 978-2-35158-080-6, Soft cover; *Eds. R. Torrent and L. Fernández Luco*

Report 41: State-of-the-Art Report of RILEM Technical Committee TC 196-ICC 'Internal Curing of Concrete' (2007) 164 pp., ISBN: 978-2-35158-009-7; e-ISBN: 978-2-35158-082-0, Soft cover; *Eds. K. Kovler and O. M. Jensen*

Report 42: 'Acoustic Emission and Related Non-destructive Evaluation Techniques for Crack Detection and Damage Evaluation in Concrete' – Final Report of RILEM Technical Committee 212-ACD (2010) 12 pp., e-ISBN: 978-2-35158-100-1; *Ed. M. Ohtsu*

RILEM Publications Published by Springer

RILEM Book Series (Proceedings)

VOL. 1: Design, Production and Placement of Self-Consolidating Concrete (2010) 466 pp., ISBN: 978-90-481-9663-0; e-ISBN: 978-90-481-9664-7, Hardcover; *Eds. K. Khayat and D. Feyes*
VOL. 2: High Performance Fiber Reinforced Cement Composites 6 – HPFRCC6 (2011) 584 pp., ISBN: 978-94-007-2435-8; e-ISBN: 978-94-007-2436-5, Hardcover; *Eds. G.J. Parra-Montesinos, H.W. Reinhardt and A.E. Naaman*
VOL. 3: Advances in Modeling Concrete Service Life – Proceedings of 4th International RILEM PhD Workshop held in Madrid, Spain, November 19, 2010 (2012) 170 pp., ISBN: 978-94-007-2702-1; e-ISBN: 978-94-007-2703-8, Hardcover; *Eds. C. Andrade and Joost Gulikers*
VOL. 5: Joint fib-RILEM Workshop on Modelling of Corroding Concrete Structures (2011) 290 pp., ISBN: 978-94-007-0676-7; e-ISBN: 978-94-007-0677-4, Hardcover; *Eds. C. Andrade and G. Mancini*

For the latest publications in the RILEM Bookseries, please visit http://www.springer.com/series/8781

Rilem State-of-the-Art Reports

VOL. 1: State-of-the-Art Report of RILEM Technical Committee TC 207-INR 'Non-Destructive Assessment of Concrete Structures: Reliability and Limits of Single and Combined Techniques' (2012) 390 pp., ISBN: 978-94-007-2735-9; e-ISBN: 978-94-007-2736-6, Hardcover; *Ed. D. Breysse*
VOL. 2: State-of-the-Art Report of RILEM Technical Committee TC 225-SAP 'Application of Super Absorbent Polymers (SAP) in Concrete Construction' (2012) 165 pp., ISBN: 978-94-007-2732-8; e-ISBN: 978-94-007-2733-5, Hardcover; *Eds. V. Mechtcherine and H-W. Reinhardt*

E. Vázquez (ed.), *Progress of Recycling in the Built Environment: Final Report of the RILEM Technical Committee 217-PRE*, RILEM State-of-the-Art Reports 8,
DOI 10.1007/978-94-007-4908-5, © RILEM 2013

VOL. 3: State-of-the-Art Report of RILEM Technical Committee TC 193-RLS 'Bonded Cement-Based Material Overlays for the Repair, the Lining or the Strengthening of Slabs or Pavements' (2011) 198 pp., ISBN: 978-94-007-1238-6; e-ISBN: 978-94-007-1239-3, Hardcover; *Eds. B. Bissonnette, L. Courard, D.W. Fowler and J-L. Granju*

VOL. 4: State-of-the-Art Report prepared by Subcommittee 2 of RILEM Technical Committee TC 208-HFC 'Durability of Strain-Hardening Fibre-Reinforced Cement-Based Composites' (SHCC) (2011) 151 pp., ISBN: 978-94-007-0337-7; e-ISBN: 978-94-007-0338-4, Hardcover; *Eds. G.P.A.G. van Zijl and F.H. Wittmann*

VOL. 5: State-of-the-Art Report of RILEM Technical Committee TC 194-TDP 'Application of Titanium Dioxide Photocatalysis to Construction Materials' (2011) 60 pp., ISBN: 978-94-007-1296-6; e-ISBN: 978-94-007-1297-3, Hardcover; *Eds. Yoshihiko Ohama and Dionys Van Gemert*

VOL. 7: State-of-the-Art Report of RILEM Technical Committee TC 215-AST 'In Situ Assessment of Structural Timber' (2010) 152 pp., ISBN: 978-94-007-0559-3; e-ISBN: 978-94-007-0560-9, Hardcover; *Eds. B. Kasal and T. Tannert*

For the latest publications in the RILEM State-of-the-Art Reports, please visit http://www.springer.com/series/8780

MIX
Papier aus verantwortungsvollen Quellen
Paper from responsible sources
FSC® C105338

If you have any concerns about our products,
you can contact us on
ProductSafety@springernature.com

In case Publisher is established outside the EU,
the EU authorized representative is:
Springer Nature Customer Service Center GmbH
Europaplatz 3, 69115 Heidelberg, Germany

Printed by Libri Plureos GmbH
in Hamburg, Germany